QUALITY IN THE FOOD ANALYSIS LABORATORY

RSC Food Analysis Monographs

Series Editor: P.S. Belton, *The Institute of Food Research, Norwich, UK.*

The aim of this series is to provide guidance and advice to the practising food analyst. It is intended to be a series of day-to-day guides for the laboratory worker, rather than library books for occasional reference. The series will form a comprehensive set of monographs providing the current state of the art on food analysis.

Dietary Fibre Analysis
by David A.T. Southgate, *Formerly of the AFRC Institute of Food Research, Norwich, UK.*

Chromatography and Capillary Electrophoresis in Food Analysis
by Charlotte Bjergegaard, Hilmer Sørensen, Susanne Sørensen, *Royal Veterinary and Agricultural University, Frederiksberg, Denmark*, and Søren Michaelsen, *Novo Nordisk A/S, Denmark*

Quality in the Food Analysis Laboratory
by Roger Wood, *Joint Food Safety and Standards Group, MAFF, Norwich, UK*, Anders Nilsson, *National Food Administration, Uppsala, Sweden*, and Harriet Wallin, *VTT Biotechnology and Food Research Espoo, Finland*

How to obtain future titles on publication

A standing order plan is available for this series. A standing order will bring delivery of each new volume immediately upon publication. For further information, please write to:

Turpin Distribution Services Ltd.
Blackhorse Road
Letchworth
Herts. SG6 1HN

Telephone: Letchworth (01462) 672555

WITHDRAWN

LIVERPOOL JMU LIBRARY
3 1111 00855 4873

Quality in the Food Analysis Laboratory

Roger Wood
Joint Food Safety and Standards Group, MAFF, Norwich, UK

Anders Nilsson
National Food Administration, Uppsala, Sweden

Harriet Wallin
VTT Biotechnology and Food Research, Espoo, Finland

ISBN 0-85404-566-X

A catalogue record for this book is available from the British Library

Published by The Royal Society of Chemistry,
Thomas Graham House, Science Park, Milton Road,
Cambridge CB4 4WF, UK

For further information see our web site at www.rsc.org

Typeset Computape (Pickering) Ltd, Pickering, North Yorkshire, UK
Printed by MPG Books Ltd, Bodmin, Cornwall, UK

Preface

Users of analytical data need to be assured of the quality of their data and whether they are appropriate, *i.e.* whether they meet 'fit-for-purpose' requirements. Thus the requirement for data to be of the required quality is now recognised by all users and providers (customers and contractors) and this is increasingly being prescribed by legislation. This text surveys the procedures that a laboratory must consider in order to meet such requirements.

General considerations of quality in the food analysis laboratory, and in particular why laboratories need to introduce quality assurance, the quality assurance measures that may be introduced by laboratories, particularly with reference to the need of their customers, and the legislative driving forces, are surveyed in Chapter 1. The formal quality models a laboratory may choose and which will aid the achievement of quality assurance are then described in Chapter 2.

Specific aspects of laboratory practice and the rationale behind the concept of method validation and present analytical concerns are described in Chapters 3 to 8, these commencing with the choice of method of analysis and the considerations to be placed on that choice, *i.e.* the status of its validation, through particular aspects of present analytical concern when reporting results, *e.g.* recovery corrections and measurement reliability. The internal measures that a laboratory should take with regard to internal quality control and proficiency testing are then described, with Chapter 8 being a discussion on the significance of the result with respect to a limit value, albeit statutory or contractual.

Particular points of interest and concern for laboratories seeking formal accreditation to an internationally recognised standard are described in Chapters 9 to 13. However, it should be appreciated that detailed information on the introduction of accreditation measures is not described in this text but a general overview of particular points of interest is given together with an international dimension on how these may be solved. Where particular aspects of accreditation procedures and practices are described it is because we believe that they are of particular interest to laboratories.

By taking due note of all of the points considered in this text, we believe that laboratories will be able to introduce measures to ensure that their data are acceptable to their customers and will be able to justify their introduction.

The information and opinions given in this book reflect our opinions and are not necessarily those of the organisations in which we are employed.

Roger Wood
Anders Nilsson
Harriet Wallin

Contents

Contributors **xii**

Chapter 1 General Considerations on Quality in the Food Analysis Laboratory **1**

1 Introduction 1
2 Difference Between Food Analysis and Food Examination 2
3 Mutual Recognition in the European Union: The Single Market and the Free Movement of Products 2
4 Legislative Requirements 4
5 Needs of Different Laboratories/Customers 10
6 Quality Assurance Measures that may be Introduced by Laboratories 11
7 Sampling 15
References 15

Chapter 2 A Comparison of the Different Quality Models Available **16**

1 Introduction 16
2 Laboratory Accreditation and Certification Models 19
3 Other Standards of Interest 27
4 National Accreditation Agencies 30
5 Guidance on Laboratory Quality Assurance 34
References 35

Chapter 3 Methods of Analysis—Their Selection, Acceptability and Validation **36**

Part A: Introduction and the Acceptability of Methods of Analysis, Types of Validation and the Criteria Approach for Methods of Analysis **38**

1 The Introduction of a New or Unfamiliar Method into the Laboratory 38

2 Requirements of Legislation and International Standardising Agents for Methods of Analysis 44
3 Future ISO Guidance on the Validation of Microbiological Methods 50
4 NMKL Guidelines on Validation of Chemical Analytical Methods 51
5 Future Requirements for Methods of Analysis—Criteria of Methods of Analysis 51

Part B: The Requirements for and Procedures to Obtain Methods that have been Fully Validated Through Collaborative Trials **56**

6 What is a Collaborative Trial? 56
7 Final Report 61
8 Other Points of Note 62
9 Other Procedures for the Validation of a Method of Analysis 62
10 Assessment of the Acceptability of the Precision Characteristics of a Method of Analysis: Calculation of HORRAT Values 64

Part C: The Requirements for and Procedures to Obtain Methods that have been Validated 'In-house' **66**

11 Protocols for the In-house Validation of Analytical Methods 66

Appendix I Validation of Chemical Analytical Methods as Adopted and Published by the Nordic Committee on Food Analysis as NMKL Procedure No. 4, 1996 68
Appendix II Method Validation—A Laboratory Guide (EURACHEM) 88
Appendix III Intralaboratory Analytical Method Validation 93
Appendix IV Procedure for the Statistical Analysis of Collaborative Trial Data 98
References 106

Chapter 4 Use of Recovery Corrections when Reporting Chemical Results **108**

1 Introduction 109
2 Procedures for Assessing Recovery 111
3 Should Recovery Information be Used to Correct Measurements? 114
4 Estimation of Recovery 116
5 Uncertainty in Reporting Recovery 118
6 Recommendations and Conclusions 123
References 124

Chapter 5 Measurement Uncertainty/Measurement Reliability 125

1 Introduction 125
2 ISO/EURACHEM Approach to the Estimation of Measurement Uncertainty in Analytical Chemistry 126
3 Alternative Approaches for the Assessment of Measurement Uncertainty 129
4 Accreditation Agencies 129
5 MAFF Project for the Comparison of Results Obtained by Different Procedures for the Estimation of Measurement Uncertainty 130
6 Codex Committee on Methods of Analysis and Sampling Approach to Measurement Uncertainty 132
7 Conclusions 133
References 134
Appendix I Procedure for the Estimation and Expression of Measurement Uncertainty in Chemical Analysis Developed by the Nordic Committee on Food Analysis 135

Chapter 6 Role of Internal Quality Control in Laboratory Quality Assurance Procedures 148

1 Introduction 148
2 Basic Concepts of IQC 149
3 The Scope of the Internal Quality Control Guidelines 151
4 Internal Quality Control and Uncertainty 152
5 Quality Assurance Practices and Internal Quality Control 153
6 Internal Quality Control Procedures 154
7 IQC and Within-run Precision 157
8 Suitable Materials that can be Used for IQC Purposes 158
9 The Use of Shewhart Control Charts 163
10 Recommendations 167
References 169

Chapter 7 Role of Proficiency Testing in the Assessment of Laboratory Quality 172

1 Introduction 172
2 Elements of Proficiency Testing 175
3 Organisation of Proficiency Testing Schemes 178
4 Statistical Procedure for the Analysis of Results 186
5 An Outline Example of how Assigned Values and Target Values may be Specified and Used in Accordance with the Harmonised Protocol 192
6 Examples of Commercial Proficiency Testing Schemes in the Food Sector 195
References 201

Chapter 8 Respecting a Limit Value 203

1 Introduction 203
2 Evaluation of Analytical Results Obtained Using Validated Methods 203
3 Results Obtained Using Routine Methods and Which are Found to be Close to a Specified Limit 205
4 Procedure for Determining Compliance by a Laboratory with an Established Reproducibility Limit 207
5 Procedure for Obtaining a Provisional Reproducibility Limit 207
6 Procedures to be Adopted when the Results of Analysis are Disputed 208
References 209

Chapter 9 Experiences in the Implementation of Quality Assurance and Accreditation into the Food Analysis Laboratory: Laboratory Aspects 210

1 Requirements on Management Towards Staff and the Organisation of the Laboratory 210
2 Requirements on the Staff in the Laboratory 212
3 Requirements on the Laboratory Premises 215

Chapter 10 Experiences in the Implementation of Quality Assurance and Accreditation into the Food Analysis Laboratory: Sampling, Sample Handling and Sample Preparation 220

1 Introduction: Quality Throughout the Analytical Chain 220
2 Sampling as a Part of the Experimental Plan 221
3 Sampling 222
4 Packing and Transportation 225
5 Receipt of Samples in the Laboratory 225
6 Storage of Samples Prior to and after Analysis 226
7 Pre-treatment of Samples 227
8 The Future 229
References 230

Chapter 11 Experiences in the Implementation of Quality Assurance and Accreditation into the Food Analysis Laboratory: Equipment, Calibration And Computers 231

1 Apparatus and Equipment 231
2 Calibration 238
3 Computers in the Laboratory 246
References 253

Chapter 12 Experiences in the Implementation of Quality Assurance and Accreditation into the Food Analysis Laboratory: Administrative Aspects—Reports and the Chain of Documentation, Internal and External Quality Audits and Management Reviews and Continuous Improvement 254

1 Reports and the Chain of Documentation 254
2 Internal and External Quality Audits and Management Reviews 264
3 Importance of Continuous Improvements 276
References 284

Chapter 13 Experiences in the Implementation of Quality Assurance and Accreditation into the Food Analysis Laboratory: Sensory Analysis 285

1 Introduction 285
2 Different Types of Sensory Tests 286
3 The Sensory Analyst 286
4 The Sensory Assessor 288
5 Accommodation and Environment 291
6 Equipment 291
7 Methods 292
8 Records 292
9 Evaluation of Assessors and the Reliability of Results in Sensory Analyses 292
References 293

Chapter 14 Vocabulary, Terminology and Definitions 295

1 Analytical Terminology for Codex Use 295
2 Terminology of Interest Frequently Used by Various International Organisations of Importance 301

Subject Index 309

Contributors

Roger Wood, *Joint Food Safety and Standards Group, Ministry of Agriculture, Fisheries and Food, CSL Food Science Laboratory, Norwich Research Park, Colney, Norwich NR4 7UQ, UK*

Anders Nilsson, *National Food Administration, Sweeden*

Present address: KF Varumärkesutveckling, Box 15200, SE 104 65 Stockholm, Sweden

Harriet Wallin, *VTT Biotechnology and Food Research, PO Box 1500, FIN-02044 VTT, Finland*

Present address: National Food Administration, PO Box 5, FIN-00531 Helsinki, Finland

CHAPTER 1

General Considerations on Quality in the Food Analysis Laboratory

1 Introduction

Customers of the providers of analytical data need to be assured about the quality of the data that is being given to them. They must be assured that the data they receive are of the required quality for their purposes; ideally they should stipulate before any analysis is undertaken the 'quality of analysis' that they require. This requirement has long been recognised by users and providers (customers and contractors) of analytical data as part of a normal 'contract', but it is increasingly being prescribed by legislation. Laboratories must, therefore, ensure that they are providing data that are appropriate, *i.e.* that they meet 'fit-for-purpose' requirements, by implementing a programme of quality assurance measures. In essence a laboratory must ensure that the results it provides satisfy its customers' needs. Ultimately, the introduction of quality assurance within a food laboratory is beneficial for both the laboratory and its customers.

In addition, it is now both internationally and nationally recognised that for a laboratory to produce consistently reliable data it must implement a programme of appropriate quality assurance measures. Without such quality assurance measures being introduced there will be no confidence in the work of any particular analytical laboratory and that laboratory will be seriously disadvantaged. This book aims to survey the quality assurance measures that it may undertake with particular emphasis being placed on the international and legislative requirements that have been developed as well as the requirements for 'mutual recognition'. Experiences in some particular aspects of the quality assurance measures that may be introduced into a laboratory will be described, these aspects having been found to be of concern and so addressed in some other European countries.

2 Difference Between Food Analysis and Food Examination

In the United Kingdom (UK) a distinction is made between chemical analysis and microbiological examination for the purposes of the Food Safety Act 1990[1] and Regulations made under the Act. This is unusual in that most countries do not make this distinction and, for them, 'analysis' embraces both chemical and microbiological analysis. Thus, in the European Union (EU) general analysis legislation is taken to refer to both chemistry and microbiology. It is important that this is appreciated when non-UK analytical documents are considered. However, the distinction is maintained throughout this book.

3 Mutual Recognition in the European Union: The Single Market and the Free Movement of Products

There is a new philosophy behind EU legislation, namely there has been a change from 'recipe' to 'information' and 'safety'. In the foodstuff sector there has been a change from controlling and preparing controlled detailed compositional standards to encouraging the consumer to purchase what he/she wishes provided that the food is clearly and informatively labelled and is safe. This means that in the foodstuffs area the development of vertical directives laying down detailed compositional standards has been reduced and replaced by the development of a series of horizontal directives. This is a part of the 'New Approach' that the EU (then the European Community) adopted in 1985.

This has a consequential effect for the work of laboratories and the production of analytical data. In particular it is central to the 'New Approach' and of the Single Market of the Union that there is recognition and acceptance of results and certificates from one Member State in other Member States (*i.e.* 'mutual recognition'). Although this is now recognised in principle across the Union, there is some mistrust of mutual recognition by many Member States and in order to promote the recognition and acceptance of other Member States' results, approved laboratories will of necessity have to conform to agreed standards of competency. The standards that have now been developed are described in the following section of this book.

Manufacturers wishing to sell their products in the EU experience problems due to the need to meet different national testing and certification requirements. In the run-up to the Single Market, the European Organisation for Testing and Certification was set up specifically to overcome these problems by encouraging the development of mutual recognition agreements. The free circulation of products throughout the EU is a key objective of the Single Market.

To achieve this goal a multiplicity of obstacles to trade will have to be or have been dismantled. In particular, what are referred to as 'technical' barriers to trade (as distinct from physical barriers such as Customs controls or fiscal barriers) have been high on the list of these obstacles. Typically, they result from differences between national tradition and practice in the areas of regulations, standards, testing and certification.

Such barriers fall into two categories: those which result from national legislation and those which are the result of purely commercial activities, *i.e.* the requirements of private parties, notably individual customers and insurers, that products should meet certain national or industry technical specifications or standards.

Whichever category these technical requirements fall into—the regulated or voluntary areas respectively—the effect of such national differences is that a manufacturer wishing to sell the same product in a number of different national markets in the Union may have to meet a number of different national requirements.

In the field of testing and certification this may mean having to have products re-tested or re-certified every time they cross a national boundary in order to fulfil local commercial or regulatory requirements. This takes time and costs money and so a European policy to overcome trade barriers has been developed. The EU, in co-operation with the countries of the European Free Trade Association, has developed a policy on the testing and certification of products which is intended to overcome these barriers to trade.

The Regulated Area

In the regulated field this is being achieved through a series of directives. Products within a defined (fairly broad) scope which meet set criteria are to be entitled to circulate freely anywhere within the Union. Except where a manufacturer is entitled simply to declare the conformity of his products and chooses to do so, appropriate tests or certification procedures must be carried out by bodies approved and notified to the European Commission.

The advantage of this development is that the authorities in all Member States will be required to accept the findings of bodies notified in other Member States in the same way as they would those of bodies they had notified themselves; there is therefore no need for a manufacturer to have products checked more than once within the Union.

The Voluntary Area

Overcoming barriers in the area covered by EU legislation can only be partially effective so long as there remain national differences with regard to the identifying mark of the test laboratory or certification body which individual customers prefer to see.

A manufacturer's products may be legally entitled to circulate freely throughout the Union, but this will be of little comfort if the realities of the market place are such that the manufacturer cannot, in practice, sell those products without going through onerous re-checking procedures simply because products were tested by this rather than that laboratory or certificated by one rather than another certification body.

The European Commission's proposals were set out in their 1989 'Global Approach' document,[2] and were broadly endorsed by a Resolution of the Union's Council of Ministers adopted in December of that year. Creating a suitable environment within which to encourage the development of mutual recognition agreements between those most nearly concerned or the creation of pan-European certification agreements was seen as the best way to overcome technical barriers in this purely voluntary area. The idea is to build on existing systems and institutions to get them to work together according to European standards.

4 Legislative Requirements

It is essential that customers demand of providers of analytical data that their data meet established quality requirements. Formal quality requirements have now been developed and adopted on an international basis by both the EU and the Codex Alimentarius Commission (CAC); these are described below. Customers will be appreciative of these requirements so it is advisable that the laboratory fully complies with them to ensure that it meets its customer demands.

The European Union

For analytical laboratories in the food sector there are legislative requirements regarding analytical data which have been adopted by the European Union. In particular, methods of analysis have been prescribed by legislation for a number of foodstuffs since the UK acceded to the European Community in 1972. However, the Union now recognises that the competency of a laboratory (*i.e.* how well it can use a method) is equally as important as the 'quality' of the method used to obtain results. This is best illustrated by consideration of the Council Directive on the Official Control of Foodstuffs (OCF) which was adopted by the Community in June, 1989.[3]

In Article 13 it is stated:

'In order to ensure that the application of this Directive is uniform throughout the Member States, the Commission shall, within one year of its adoption, make a report to the European Parliament and to the Council on the possibility of establishing Community quality standards for all laboratories involved in inspection and sampling under this Directive.'

Following that the Commission, in September 1990, produced a Report which recommended establishing Community quality standards for all laboratories involved in inspections and sampling under the OCF Directive. Proposals on this have now been adopted by the Community in the Directive on Additional Measures Concerning the Food Control of Foodstuffs (AMFC).[4]

The relevant Articles are:

Article 3, which states:

'1. Member States shall take all measures necessary to ensure that the laboratories referred to in Article 7 of Directive 89/397/EEC[3] comply with the general criteria for the operation of testing laboratories laid down in European standard EN 45001[5] supplemented by Standard Operating Procedures and the random audit of their compliance by quality assurance personnel, in accordance with the OECD principles Nos. 2 and 7 of good laboratory practice as set out in Section II of Annex 2 of the Decision of the Council of the OECD of 12 Mar 1981 concerning the mutual acceptance of data in the assessment of chemicals.[6]

2. In assessing the laboratories referred to in Article 7 of Directive 89/397/EEC Member States shall:

   (a) apply the criteria laid down in European standard EN 45002;[7] and

   (b) require the use of proficiency testing schemes as far as appropriate.

   Laboratories meeting the assessment criteria shall be presumed to fulfil the criteria referred to in paragraph 1.

   Laboratories which do not meet the assessment criteria shall not be considered as laboratories referred to in Article 7 of the said Directive.

3. Member States shall designate bodies responsible for the assessment of laboratories as referred to in Article 7 of Directive 89/397/EEC. These bodies shall comply with the general criteria for laboratory accreditation bodies laid down in European Standard EN 45003.[8]

4. The accreditation and assessment of testing laboratories referred to in this article may relate to individual tests or groups of tests. Any appropriate deviation in the way in which the standards referred to in paragraphs 1, 2 and 3 are applied shall be adopted in accordance with the procedure laid down in Article 8.'

and Article 4, which states:

'Member States shall ensure that the validation of methods of analysis used within the context of official control of foodstuffs by the laboratories referred to in Article 7 of Directive 89/397/EEC comply whenever possible with the provisions of paragraphs 1 and 2 of the Annex to Council Directive 85/591/EEC of 23 December 1985 concerning the introduction of Community methods of sampling and analysis for the monitoring of foodstuffs intended for human consumption.'[9]

As a result of the adoption of the above directives, legislation is now in place to ensure that there is confidence not only in national laboratories but also those of the other Member States, thus facilitating the so-called 'mutual recognition' aspects which were described above.

The effect of the AMFC Directive is that organisations must consider the following aspects within the laboratory:

- the organisation of the laboratory,

- how well the laboratory actually carries out analyses, and
- the methods of analysis used in the laboratory.

All these aspects are inter-related, but in simple terms may be thought of as:

- becoming accredited to an internationally recognised standard; such accreditation is aided by the use of internal quality control procedures,
- participating in proficiency schemes, and
- using validated methods.

In addition it is important that there is 'co-operation with customers' as this is also required by virtue of the EN 45001 Standard at paragraph 6, and will be emphasised even more in future revised versions of EN 45001 and ISO/IEC Guide 25.[10]

Although the legislative requirements apply only to food control laboratories, the effect of their adoption is that other food laboratories will be advised to achieve the same standard in order for their results to be recognised as equivalent and accepted for 'due diligence' purposes.

The AMFC Directive requires that food control laboratories should be accredited to the EN 45000 series of Standards as supplemented by some of the OECD GLP principles. In the UK, Government Departments will nominate the United Kingdom Accreditation Service (UKAS) to carry out the accreditation of official food control laboratories for all the aspects prescribed in the Directive. However, as the accreditation agency will also be required to comply with the EN 45003 Standard and to carry out assessments in accordance with the EN 45002 Standard, all accreditation agencies that are members of the European Co-operation for Accreditation of Laboratories (EAL) may be asked to carry out the accreditation of a food control laboratory within the UK. Similar procedures will be followed in the other Member States, all having or developing equivalent organisations to UKAS. Details of the UK draft requirements for food control laboratories are described later in this Chapter.

Requirements of the Codex Alimentarius Commission: Guidelines for the Assessment of the Competence of Testing Laboratories Involved in the Import and Export Control of Food

The requirements of the Codex Alimentarius Commission (CAC) are becoming of increasing importance because of the acceptance of Codex Standards in the World Trade Organisation agreements. They may be regarded as being semi-legal in status. Thus, on a world-wide level, the establishment of the World Trade Organisation (WTO) and the formal acceptance of the Agreements on the Application of Sanitary and Phytosanitary Measures (SPS Agreement) and Technical Barriers to Trade (TBT Agreement) has dramatically increased the status of Codex as a body. As a result, Codex Standards are now seen as *de facto* international standards and are increasingly being adopted by reference into the food law of both developed and developing countries. In spite of the increased status of Codex Standards, funding constraints within the United Nations Organisation and associated agencies such as FAO and UNDP mean that the capacity of such agencies to support initiatives that may assist developing countries to implement Codex Standards is limited. It should also be noted that international trade in all commodities, including meat and meat products, is increasingly occurring under country-to-country certification agreements.

Because of the status of the CAC described above, the work that it has carried out in the area of laboratory quality assurance must be carefully considered. One of the CAC Committees, the Codex Committee on Methods of Analysis and Sampling (CCMAS), has developed criteria for assessing the competence of testing laboratories involved in the official import and export control of foods. These were recommended by the Committee at its 21st Session in March 1997[11] and adopted by the CAC at its 22nd Session in June 1997;[12] they mirror the EU recommendations for laboratory quality standards and methods of analysis.

The guidelines provide a framework for the implementation of quality assurance measures to ensure the competence of testing laboratories involved in the import and export control of foods. They are intended to assist countries in their fair trade in foodstuffs and to protect consumers.

The criteria for laboratories involved in the import and export control of foods, now adopted by the CAC, are:

- to comply with the general criteria for testing laboratories laid down in ISO/IEC Guide 25: 1990 'General requirements for the competence of calibration and testing laboratories';[10] (*i.e.* effectively accreditation),
- to participate in appropriate proficiency testing schemes for food analysis which conform to the requirements laid down in 'The International Harmonised Protocol for the Proficiency Testing of (Chemical) Analytical Laboratories'[13] (already adopted for Codex purposes by the CAC at its 21st Session in July 1995),
- to use, whenever available, methods of analysis which have been validated according to the principles laid down by the CAC, and
- to use internal quality control procedures, such as those described in the 'Harmonised Guidelines for Internal Quality Control in Analytical Chemistry Laboratories'.[14]

In addition, the bodies assessing the laboratories should comply with the general criteria for laboratory accreditation, such as those laid down in the ISO/IEC Guide 58:1993: 'Calibration and testing laboratory accreditation systems—General requirements for operation and recognition'.[15]

As will be explained later in Chapter 2, ISO/IEC Guide 25:1990 and the EN 45000 series of standards are similar in intent. Thus, as for the European Union, the requirements are based on accreditation, proficiency testing, the use of validated methods of analysis and, in addition, the formal requirement to use internal quality control procedures which comply with the Harmonised Guidelines.

It is only through these measures that international trade will be facilitated and the requirements to allow mutual recognition to be fulfilled will be achieved.

Implementation of the Requirements of the Additional Measures Food Control Directive and of the Codex Alimentarius Commission in the UK

The achievement of the requirements of the laboratory quality standards of both the European Union and of the Codex Alimentarius Commission will be difficult. In the UK it has been the normal practice for UKAS to accredit the scope of laboratories on a method-by-method basis. In the case of official food control laboratories undertaking many non-routine or investigative chemical analyses, it is accepted that it is not practical to develop accredited fully documented methods in the conventional sense, *i.e.* specific methods which specify in detail the complete analytical procedure on a sample/analyte/matrix basis. An alternative procedure, whereby a laboratory must have a protocol defining the approach to be adopted which includes the requirements for validation and quality control, is now to be encouraged. Full details of procedures used, including instrumental parameters, must be recorded at the time of each analysis in order to enable the procedure to be repeated in the same manner at a later date. It is therefore recommended that, for official food control laboratories undertaking analysis, appropriate method protocols are accredited on a generic basis with such generic accreditation being underpinned where necessary by specific method accreditation. The same approach may be taken with respect to participation in proficiency testing schemes. By using the combination of specific method accreditation and generic protocol accreditation, it will be possible for laboratories to be accredited for all the analyses which they are capable and competent to undertake. Method performance validation data demonstrating that the method was fit-for-purpose should be obtained before the test result is released and method performance should be monitored by on-going quality control techniques where applicable.

Thus, in the UK, the recommended scope of method accreditation and proficiency testing for official control laboratories is expected to be:

Methods of Chemical Analysis

Food analysis laboratories seeking accreditation (for the purposes of the AMFC Directive) shall include, as a minimum, the following specific determinations: acidity, alcoholic strength, ash, chloride, colouring matters—quantitative, colouring matters—qualitative, dietary fibre, fat, hydroxyproline, moisture, nitrite, nitrogen, non-meat protein, peroxide value, speciation of meat and/or fish, sulfur dioxide, sugars and total volatile nitrogen. Accreditation for other specific methods may be sought if the laboratory so wishes.

The laboratory shall also develop generic protocols for the following techniques: atomic absorption and/or ICP, GC, HPLC, microscopy, UV/visible spectrophotometry and sample preparation procedures (including digestion and solvent dissolution procedures). Other generic protocols which are acceptable to UKAS may also be developed. Full details of methods used, including any instrumental parameters affecting performance, must be recorded at the time of each analysis to enable the method to be repeated in the same manner at a later date.

Methods of Microbiological Examination

Official food control laboratories undertaking examination will be accredited on a method-by-method basis to detect and enumerate, where appropriate, pathogenic organisms and indicators routinely determined in food, including as a minimum aerobic plate count, Enterobacteriaceae or coliforms, *E. coli* (including serotype O157), *Staphylococcus aureus, Bacillus cereus, Clostridium perfringens, Salmonella* species, *Listeria monocytogenes and Campylobacter* species.

However, examinations which do not fall in the above list may be undertaken if the laboratory has in place a series of specific methods and protocols dealing with, for example: sample preparation, plate counting, impedimetric techniques, immunological procedures, gene probes methods and electron microscopy techniques.

Proficiency Testing for Laboratories Undertaking Chemical Analysis for Official Control Purposes

The scope of potential participation in proficiency testing schemes in the food sector is very wide. However, laboratories must participate in an appropriate proficiency testing scheme (*i.e.* one complying with the International Harmonised Protocol[13]). They should undertake analyses which are specific in nature and which are within the range of analyses offered by the laboratory but also be underpinned by participation in generic series within the proficiency test scheme. Thus, it may be expected that official food control laboratories will be required to participate, and achieve a satisfactory performance, in relevant series dealing with the testing of GC, HPLC, atomic absorption and/or ICP and of proximate analysis procedures. These series will be supplemented by participation in specific series where regulations are in force and the analytical techniques applied are judged to be sufficiently unique or specialised, *e.g.* aflatoxins or overall and specific migration, to require an independent demonstration of competence. Food enforcement analysis laboratories will be expected to participate in every round of a particular series of the proficiency testing scheme.

The above series (*i.e.* GC, HPLC, atomic absorption and/or ICP and proximate analysis) may be considered to be sufficiently representative of generic techniques without the need for further participation in specific series even if such series become available (*e.g.* a generic series for HPLC is considered representative of all HPLC determinations not of a unique or specialised nature even if further such series employing HPLC are developed). However, laboratories will be expected to demonstrate proficiency in extraction of components from different matrices.

Proficiency Testing for Laboratories Undertaking Microbiological Examination for Official Control Purposes

Proficiency test samples should mirror routine situations likely to be encountered when examining foods under the AMFC Directive in the UK. At least 12 samples should be distributed per year. Each sample may contain a single organism, a mixture of organisms or no organisms of significance at all, but the contents shall not be disclosed in advance. Determination of each of the pathogenic organisms and indicators listed in the methods of microbiological examination paragraph above should be required to be included in the proficiency testing scheme at least once each year. For purposes of recognition as an Official Control Laboratory, laboratories must participate in a proficiency testing scheme incorporating the parameters in the methods of microbiological examination paragraph above as a minimum.

Where quantitative determinations are assessed, schemes should treat the results statistically to determine whether performance is satisfactory, for example by converting counts to $\log_{10}$ values and then applying the procedures which have been developed in the International Harmonised Protocol for Proficiency Testing of (Chemical) Analytical Laboratories.[13]

As there are no nationally or internationally recognised protocols for assessing satisfactory performance in qualitative (presence/absence) food examinations, it is proposed that, in assessing this, schemes should take due account of false positive and false negative rates.

In addition it will be necessary for laboratories to be able to demonstrate quality control procedures to ensure compliance with the EN 45001 Standard, an example of which would be compliance with the ISO/AOAC/IUPAC Guidelines on Internal Quality Control in Analytical Chemistry Laboratories.[14]

5 Needs of Different Laboratories/Customers

Notwithstanding the legislative requirements described above, it is important to recognise that different laboratories and customers will have different requirements with regard to the production and use of analytical data. Many of the considerations given in this book refer to procedures and considerations which may be used by official laboratories when generating results which may be used for control purposes. Similar procedures and considerations will have to be made by trade ('manufacturers' or 'retailers') laboratories if they wish to be able to claim full due diligence using their analytical results. However, other users and providers of analytical data may wish to use data meeting less stringent quality requirements, but this should be done only in the full recognition that such data are frequently less reliable. Typically such results will be used only for screening purposes.

6 Quality Assurance Measures that may be Introduced by Laboratories

All reported data should be traceable to reliable and well documented standard materials, preferably certified reference materials. Accreditation of the laboratory by the appropriate national accreditation scheme indicates that the laboratory is applying sound quality assurance principles. Although those requirements are now well appreciated by the food analytical community, this book aims to provide further information on all aspects of quality assurance in the laboratory and highlights particular aspects of the accreditation process that analysts need to appreciate before they introduce quality assurance into their laboratory. Thus, as has been described above, it is now recognised that amongst the quality assurance measures to be employed by a laboratory is the need for it to demonstrate that it is in statistical control, to participate in proficiency testing schemes which provide an objective means of assessing and documenting the reliability of the data it is producing and to use methods of analysis which are 'fit-for-purpose'. These and other quality assurance measures must be considered for introduction by a laboratory which wishes to report reliable and appropriate data; such measures may be underpinned by the introduction of a formal quality system. The considerations that the food analysis laboratory must make when choosing the quality assurance systems and measures it should introduce are briefly described below in the order in which they will be considered. This commences with the choice of formal quality model followed by considerations on methods of analysis, aspects on the reporting of results (and in particular recovery and measurement uncertainty), internal quality control procedures that the laboratory should adopt and the participation in proficiency testing schemes. The assessment of the reported result against a limit is then considered together with experiences in the implementation of quality assurance and accreditation in the food analysis laboratory, primarily from a non-UK perspective.

Formal Quality System

It will be seen that it has now become necessary for a laboratory to introduce a formal quality system which is recognised through third party assessment either as a result of legislation or customer demands. The various choices are briefly described but, as will be seen from the formal legislative requirements now being placed on food analysis laboratories, it is the requirements stemming from the ISO/IEC Guide 25 that is of prime importance for these laboratories.

Analytical Methods

These must be validated before use, often by such methods having been tested through collaborative trials which conform to a recognised protocol. The methods must be well documented, staff adequately trained in their use and control charts should be established to ensure proper statistical control.

The following aspects of methods of analysis are therefore described:

- requirements and procedures for fully validated methods of analysis, *i.e.* methods which have been validated by collaborative trial;
- requirements for the in-house validation and verification of methods of analysis.

Use of Recovery Corrections When Reporting Results

Recovery studies are an essential component of the validation and use of analytical methods. It is important that all concerned with the production and interpretation of analytical results are aware of the problems and the basis on which the result is being reported. At present, however, there is no single well-defined approach to estimating, expressing and applying recovery information. In the absence of consistent strategies for the estimation and use of recovery, it is difficult to make valid comparisons between results produced in different laboratories or to verify the suitability of that data for the intended purpose. The International Analytical Community is now addressing these problems and so advice is given regarding recovery adjustments.

Measurement Uncertainty/Measurement Reliability

In quantitative chemical analysis, many important decisions are based on the results obtained and so it is therefore very important that an indication of the quality of the results reported is available. Analytical chemists are now more than ever coming under increased pressure to be able to demonstrate the quality of their results by giving a measure of the confidence placed on a particular result to demonstrate its fitness for purpose. This includes the level that the result would be expected to agree with other results irrespective of the method used. 'Measurement Uncertainty' is a useful parameter which gives this information.

In addition, measurement uncertainty has also become an important issue to accreditation agencies [*e.g.* to the United Kingdom Accreditation Service (UKAS) in the UK]. In document M10 of UKAS the instructions to accredited laboratories regarding uncertainty of measurement read as follows:[16]

'The laboratory is required to produce an estimate of the uncertainty of its measurements, to include the estimation of uncertainty in its methods and procedures for calibration and testing, and to report the uncertainty of measurement in calibration certificates and in test certificates and test reports, where relevant.

Estimates of uncertainty of measurement shall take into account all significant identified uncertainties in the measurement and testing processes, including those attributable to measuring equipment, reference measurement standards (including material used as a reference standard), staff using or operating equipment, measurement procedures, sampling and environmental conditions.

In estimating uncertainties of measurement, the Laboratory shall take account of data obtained from internal quality control schemes and other relevant sources. The Laboratory shall also ensure that any requirements for the estimation of uncertainty and for the determination of compliance with specified requirements as stated in relevant publications are complied with at all times.'

In view of the importance of measurement uncertainty the problems associated with its estimation and terminology in the food sector are considered.

Internal Quality Control

Internal quality control (IQC) is one of a number of concerted measures that analytical chemists can take to ensure that the data produced in the laboratory are of known quality and uncertainty. In practice this is determined by comparing the results achieved in the laboratory at a given time with a standard. IQC therefore comprises the routine practical procedures that enable the analyst to accept a result or group of results or reject the results and repeat the analysis. IQC is undertaken by the inclusion of particular reference materials, 'control materials', into the analytical sequence and by duplicate analysis.

Proficiency Testing

Participation in proficiency testing schemes provides laboratories with an objective means of assessing and documenting the reliability of the data they are producing. Although there are several types of proficiency testing schemes, they all share a common feature: test results obtained by one laboratory are compared with those obtained by one or more testing laboratories. Proficiency testing schemes must provide a transparent interpretation and assessment of results. Laboratories wishing to demonstrate their proficiency should seek and participate in proficiency testing schemes relevant to their area of work. The requirements for proficiency testing schemes to be able to meet international standards for proficiency testing schemes will be described.

Respecting a Limit

There is often confusion as to the actions that can be taken once an analytical result is generated. This chapter will consider how a result should be assessed, particularly with respect to a specified limit.

Experiences in the Implementation of Quality Assurance and Accreditation into the Food Analysis Laboratory

Some experiences of the authors on the implementation of quality assurance and accreditation into the food analysis laboratory will be discussed in the last few chapters of the book, and specifically:
Laboratory Aspects:

Part A: Requirements on the Management and Organisation of the Laboratory
Part B: Requirements on the Staff in the Laboratory
Part C: Requirements on the Laboratory Premises

Sampling, Sample Handing and Sample Preparation
Equipment:

Part A: Apparatus and Equipment
Part B: Calibration
Part C: Computers

Administrative Aspects:

Part A: Reports and the Chain of Documentation
Part B: Audits
Part C: Importance of Continuous Improvement

Specific Applications: Sensory Analysis

'Co-operation with Customers'

It must be appreciated that the requirements of the customer are the starting point for all quality work A correct choice of sampling plan, sample transportation and handling, analytical method, reporting format, *etc.* are prerequisites for this. If an analytical food laboratory is to carry out all of the required steps then the laboratory must give its customers its co-operation to enable the laboratory to determine and possibly clarify the requirements of its customers. Customer co-operation will therefore be discussed in all chapters of this book.

7 Sampling

Sampling is a critical component of the procedures whereby an analytical result is achieved. However, in this book, the considerations described only apply to the sample as received by the laboratory. Nevertheless, it is important that the customer of the analytical laboratory fully appreciates that the variability due to sampling error must also be assessed to ensure that the methods of analysis used are not unnecessarily precise, *i.e.* that the sampling and analysis errors are in a reasonable proportion to each other. It is important that undue attention is not given to reducing analytical errors at the expense of sampling errors; it is the most overall cost effective system that is required.

References

1 Food Safety Act, 1990, HMSO, London, 1990, **16**.
2 A Global Approach to Certification and Testing Quality Measures for Industrial Products (COM/89/209), *O. J.*, 1989, C267.
3 Council Directive 89/397/EEC on the Official Control of Foodstuffs, *O. J.,* 1989, L186.
4 Council Directive 93/99/EEC on the Subject of Additional Measures Concerning the Official Control of Foodstuffs, *O. J.*, 1993, L290.
5 General Criteria for the Operation of Testing Laboratories, European Standard EN 45001, CEN/CENELEC, Brussels, 1989.
6 Decision of the Council of the OECD of 12 Mar 1981 concerning the mutual acceptance of data in the assessment of chemicals, OECD, Paris, 1981.
7 General Criteria for the Assessment of Testing Laboratories, European Standard EN 45002, CEN/CENELEC, Brussels, 1989.
8 General Criteria for Laboratory Accreditation Bodies, European Standard EN 45003, CEN/CENELEC, Brussels, 1989.
9 Council Directive 85/591/EEC Concerning the Introduction of Community Methods of Sampling and Analysis for the Monitoring of Foodstuffs Intended for Human Consumption, *O. J.,*1985, L372.
10 General Requirements for the Competence of Calibration and Testing Laboratories, ISO/IEC Guide 25, ISO, Geneva, 1990,
11 Report of the 21st Session of the Codex Committee on Methods of Analysis and Sampling, FAO, Rome, 1997, ALINORM 97/23A.
12 Report of the 22nd Session of the Codex Alimentarius Commission, FAO, Rome, 1997, ALINORM 97/37.
13 The International Harmonised Protocol for the Proficiency Testing of (Chemical) Analytical Laboratories, ed. M. Thompson and R. Wood, *Pure Appl. Chem.*, 1993, **65,** 2123 (also published in *J. AOAC Int.*, 1993, **76,** 926).
14 Guidelines on Internal Quality Control in Analytical Chemistry Laboratories, ed. M. Thompson and R. Wood, *Pure Appl. Chem.,* 1995, **67,** 649.
15 Calibration and Testing Laboratory Accreditation Systems—General Requirements for Operation and Recognition, ISO/IEC Guide 58, ISO, Geneva, 1993.
16 General Criteria of Competence for Calibration and Testing Laboratories, UKAS, Queens Road, Teddington, TW11 0NA, NAMAS Accreditation Standard M 10, 1992, and Supplement, 1993.

CHAPTER 2

A Comparison of the Different Quality Models Available

1 Introduction

The introduction and continuous improvement of a quality system in a food laboratory gives assurance that all its analyses are carried out to the 'required quality'. However, it is not sufficient for only the laboratory to know that its results are of that quality; its customers also have to be assured that all specifications for quality are satisfied, these then making the analytical result 'fit for purpose'. As a consequence, the quality system has to be designed both to give analytical results with the 'right quality', and to give the customer confidence in their reliability.

When developing and introducing a quality system, different 'models' to achieve this can be used. There are models described in international standards and guidelines, and in particular:

- Standard EN 45001 (1989), General criteria for the operation of testing laboratories;
- ISO/IEC Guide 25 (1990), General requirements for the competence of calibration and testing laboratories;
- OECD environment monograph no. 45; the OECD Principles of Good Laboratory Practice (GLP);
- Standard EN ISO 9001 (1994), Quality systems—Model for quality assurance in design/development, production, installation and servicing;
- Standard EN ISO 14001 (1996), Environmental management systems—Specifications with guidance for use;
- Rules for European Quality Award (EQA); and
- Rules for British Quality Award (BQA), Swedish Quality Award (Utmärkelsen Svensk Kvalitet; USK) and other national total quality management (TQM) models.

All are of potential importance and relevance to an analytical food laboratory.

The list of standards, *etc.* on quality assurance that can be used by laboratories can be extended further by consideration of:

- ISO/IEC Guide 43, Development and operation of laboratory proficiency testing;
- ISO/IEC Guide 58, Calibration and testing laboratory accreditation systems—General requirements for operation and recognition;
- EN 45002, General criteria for the assessment of testing laboratories;
- EN 45003, General criteria for laboratory accreditation bodies;
- OECD environment monographs 45, 110, 111, *etc.* on GLP principles and their application;
- EN ISO 9002, Quality systems—Model for quality assurance in production, installation and servicing;
- EN ISO 9003, Quality systems—Model for quality assurance in final inspection and test;
- EN ISO 9004–1, Quality management and quality system elements—Part 1: Guidelines;
- EN ISO 9004–2, Quality management and quality system elements—Part 2: Guidelines for services;
- ISO 10011, Parts 1–3, Guidelines for auditing quality systems;
- EN 45011, General criteria for certification bodies operating product certification;
- EN 45012, General criteria for certification bodies operating Quality System certification; and
- EN 45013, General criteria for certification bodies operating certification of personnel.

However, as can be seen from the above, many standards/guidelines have been developed in this area. In fact there are so many that those wishing to introduce quality assurance into their laboratories are frequently discouraged! This chapter outlines the most important models which may be used by analytical food laboratories when developing their own quality system.

The Intent of the Available Models

The available models are developed for different purposes and as a result of different customer needs. Thus the history of the model influences which an analytical food laboratory should choose to use when developing a quality system. Sometimes the model to be chosen is prescribed by legislation as described in Chapter 1.

The intent of each model that the laboratory is most likely to choose is described below:

- EN 45001 and ISO/IEC Guide 25 are developed for the operation of *testing laboratories that sell their analytical services*;
- OECD GLP principles are developed to promote quality in *toxicological studies on chemicals and pharmaceuticals*;
- EN 45002 and 45003, ISO 10011, OECD Environment Monographs 110

and 111 give information on subjects such as *assessment, monitoring, inspection and accreditation* of different types of laboratories;
- EN ISO 9001–9003 are quality system standards developed to *facilitate the dialogue in between producers and their customers* (they also influence laboratories that are an integral part of a ISO 9000 certified company);
- EN ISO 9000–1, 9000–3, 9004 and 9004–2 are *guidelines for the application of ISO 9001–9003 standards* in different situations;
- EQA, BQA, USK, *etc.* are *satisfaction* (TQM) models supplying information on how *the management can improve quality, reduce cost and increase customer satisfaction.*

Models such as EN 45001 give information on 'technical specifications' for the quality of the analysis at a laboratory. Standards such as EN ISO 9001 present requirements on how a quality system should be developed without giving sufficient information on analytical specifications for laboratories. However, those standards are also 'technical' in their approach. TQM models, on the other hand, emphasise the satisfaction of customers and the functional quality of their products. TQM models supply no information on technical and organisational concerns.

By choosing well-known and internationally accepted models the laboratory does not have to develop 'quality assurance' procedures by relying solely on in-house expertise. In addition, the introduction of a defined model makes contact between laboratories and their customers easier as universally known and recognised quality concepts and quality words are used. However, there are also other reasons why laboratories should choose a specific quality model.

Laboratories which are an integral part of companies that operate to the EN ISO 9001 or EQA models must have a quality system that harmonises with the models chosen by the 'parent' organisation. In some analytical sectors, laboratories have to comply with legal requirements that certain standards or models should be used, *e.g.* as described previously for food control laboratories in the European Union. In these cases the quality assurance model must meet the demands of the EN 45001–45003 standards to be acceptable. Another example is the requirements included in the Organisation for Economic Co-operation and Development (OECD) Environment Monograph 45 on the procedures for carrying out toxicological studies on new chemicals and pharmaceuticals.

A laboratory that is simultaneously active in many analytical sectors has to consider how it should develop a quality system that meets the demands of the different quality models that it has to follow (*e.g.* when a laboratory works within the food, environmental monitoring and the pharmaceutical sectors). It must be stressed, however, that a laboratory always develops a quality system that is unique to it, no matter to what external standard it is nominally operating.

The different models from which an analytical food laboratory is most likely to chose when introducing an assurance system are briefly outlined below.

2 Laboratory Accreditation and Certification Models

There are three quality model schemes of relevance to food analysis laboratories, the choice depending on whether such laboratories undertake contract work for third parties or are an integral part of the food manufacturing process. They are, in Europe, the EN 45001 Standard (as used by National Accreditation Agencies and which is based on the ISO/IEC Guide 25), the BS 5750/ISO 9000 (now known as BS EN ISO 9000 in the UK) or the principles of Good Laboratory Practice (GLP). All models aim to achieve quality standards for laboratories and have many common features.

Accreditation as defined by ISO/IEC Guide 25 is generally the model required of food analysis laboratories by legislation so this model will be described in most detail.

Accreditation of laboratories carrying out analysis of foodstuffs or items associated with food (*e.g.* packaging materials) rests on the use of well documented analytical methods, adequately trained staff, proper sample identification procedures, regular quality audit and equipment calibration, well defined measurement traceability, an adequate system of result reporting and a maintenance of a system of records such that all information of practical relevance to the tests it has carried out is readily available. These requirements, however, are also an integral part of all the models being considered.

ISO/IEC Guide 25 and EN 45001

The models ISO/IEC Guide 25, General requirements for the competence of calibration and testing laboratories,[1] and the EN 45001, General criteria for the operation of testing laboratories,[2] are directly related to each other with the European Standard having been derived from the ISO/IEC Guide. Both are continually being updated. They are particularly appropriate for laboratories which ‘sell’ their analytical services to a third party and undertake routine analytical services.

The models EN 45001 and ISO/IEC Guide 25 are also used as specifications for accreditation of analytical laboratories, thereby giving recognition that an analytical laboratory is competent to carry out specified tests. In addition it must also be emphasised that both models may be recommended to laboratories that have no intent to obtain formal accreditation. All analytical food laboratories which wish to implement quality assurance procedures can benefit from the purpose of these quality models.

Both EN 45001 and ISO/IEC Guide 25 supply detailed information on how a laboratory should quality assure its analytical work. The models provide technical information on the design and use of, for example, premises and (internal) environment, equipment, methods and instructions, analytical reports, sample handling, archives, traceability to international standards, *etc.* Information on the design and maintenance of a quality system is, however, less comprehensive. This also applies to the information on how to obtain customer satisfaction regarding a laboratory's analytical results (*i.e.* the 'products' produced by the laboratory).

EN 45001

EN 45001 was published in 1989. The standard presents general criteria for the technical competence of an analytical laboratory, irrespective of the sector involved. It is intended for the use of laboratories and accreditation bodies as well as other bodies concerned with recognising the competence of analytical laboratories. This is further specified in the later part of the standard where requirements are given on 'co-operation with bodies granting accreditation' and on 'duties resulting from accreditation'. These requirements are not included in the 1990 draft of the ISO/IEC Guide 25.

The criteria in EN 45001 occasionally have to be supplemented when applied to a particular sector. This is best illustrated by the inclusion of the two OECD GLP principles (numbers 2 and 7 in Section II) in the laboratory quality requirements specified in the Additional Measures on Food Control (AMFC) Directive.[3] Because the text of EN 45001 is 'brief', many European accreditation bodies have supplemented this standard with paragraphs from ISO/IEC Guide 25.

Typically the following are included:

- Deputies shall be nominated in case of absence of technical manager or quality co-ordinator (ISO/IEC guide paragraph 4.2 h);
- Where audit findings cast doubts on the correctness of the analytical results, the laboratory shall take immediate corrective action and inform, in writing, clients whose work may have been affected (paragraph 5.3);
- Calibration status shall be labelled or marked on equipment and reference materials, where appropriate (paragraph 8.3);
- Requirements of calibration or verification before putting instruments into service are specified (paragraph 9.1);
- Requirements on instructions for evaluation of analytical results are specified (paragraph 10.2);
- Requirements on personal computers and automatic recording equipment are specified (paragraphs 10.7 and 10.8);
- The condition of the test sample shall be recorded upon receipt (paragraph 11.2); and
- Requirements on the transmission of analytical results *via* fax, e-mail, *etc.* are specified (paragraph 13.7).

The above examples are taken from the Swedish Board for Accreditation and Conformity Assessment Agency's (SWEDAC) directions/interpretation of EN 45001. Other accreditation agencies have similar requirements.

ISO/IEC Guide 25

The present (third edition) of ISO/IEC Guide 25 was published in 1990. A fourth, revised edition, is under preparation by the ISO/CASCO Working Group 10 and, after publication, will serve as the basis for the next revision of EN 45001.

ISO/IEC Guide 25 is much more clearly defined than EN 45001. If an analytical laboratory is able to choose its quality model to follow, it is recommended that the ISO/IEC Guide 25 is used. This also has the additional advantage in that the ISO/IEC Guide is the model that is accepted on a world-wide basis whereas EN 45001 is the European Standard. This is particularly important now that the Codex Alimentarius Commission has adopted standards for the assessment of the competence of testing laboratories involved in the import and export control of food.

Laboratories that fulfil the criteria in ISO/IEC Guide 25 also fulfil the quality requirements on calibration and testing activities stated in the ISO 9000 standards. As a consequence, an analytical food laboratory accredited according to the ISO/IEC Guide 25 also fulfils the requirements on a quality system when a company where the laboratory is active and certified according to the EN ISO 9001, 9002 or 9003 standards.

The ISO/IEC Guide 25 is used as the specifications for accreditation of analytical laboratories. However, it lacks information on co-operation with accreditation bodies. Neither does the ISO/IEC guide present any requirements on co-operation with clients to thus help them clarify their demands (see paragraph 6.1 in EN 45001; contract reviews, however, will be emphasised in future versions of the EN 45001 Standard). On the other hand the ISO/IEC Guide gives more detailed information on a laboratory quality system (ISO/IEC Guide paragraph 5.1), internal quality control, *etc.* (paragraph 5.6), internal audits (paragraph 5.3) and co-operation with sub-contractors (paragraph 15.1). Furthermore, the ISO/IEC Guide requires that the laboratory has to use documented procedures and appropriate statistical techniques to select samples, when sampling is carried out by the laboratory (paragraph 10.4). OECD environment monograph 45 (OECD GLP principles) also presents requirements on sampling, *etc.* but EN 45001 does not.

Certification of Laboratories Which are an Integral Part of the Manufacturing Process—BS EN ISO 9000

The BS EN ISO 9000 series is the national standard which promulgates, for use by UK suppliers and purchasers, the ISO 9000 series international standards for quality systems. However, this series was not designed for the food industry although the British Standards Institution has developed guidelines (available from the British Food Manufacturing Industry Research Association)[4] to help the UK food industry interpret the standard (see below). The principles of the standard are applicable to an organisation of any size and specify basic disciplines and procedures. As an example, a drinks bottling factory may work to the BS EN ISO 9000 standard. If there is a small quality control analytical laboratory attached to the factory, it is most probable that it has been decided that the laboratory will also work to the same standard.

Laboratories certified to BS EN ISO 9000 can only work as part of the manufacturing process, *i.e.* they are not accredited to take on third party work.

In the UK, a Committee of Representatives from organisations concerned with the manufacture and sale of food and drinks put together the Guide 'Quality systems for the food and drink industries—Guidelines for the use of BS 5750: Part 2:1987 in the manufacture of food and drink (ISO 9002:1987; EN 29002:1987)'. The document was published in 1989,[4] and is undergoing some revision at present. It makes extensive reference to 'IFST Food and drink manufacture—Good manufacturing practice: a guide to its responsible management'.[5]

Laboratory Quality Assurance within a Company's Quality System

Laboratories that carry out their analysis as a part of internal quality control procedures have to adapt to the over-arching company quality system. Quality models applicable for this type of quality systems are the EN ISO 9000 and the EN ISO 14000 standards. If the company is working with quality management models such as the European Quality Award (EQA), British Quality Award (BQA) or Utmärkelsen Svensk Kvalitet (USK), then the laboratory quality system has to be compatible with those TQM models.

Sometimes accreditation is recommended or even mandatory for internal process control laboratories, *e.g.* when directives prescribe that final inspection shall be performed at such laboratories. Independent of which quality model a company chooses, the laboratory models EN 45001 and ISO/IEC Guide 25 may be recommended for internal process control laboratories even if there is no formal reporting of results outside the organisation or acceptance of work for a third party.

Quality Systems Standards EN ISO 9001–9003

The quality systems standards in the ISO 9000 series have been developed to facilitate the dialogue between a producer and its customers. The standards specify quality system requirements that are aimed primarily at achieving customer satisfaction by preventing nonconformity at all stages in the production process. No product specifications are given in those standards, as is the case in the laboratory models EN 45001 and ISO/IEC Guide 25 (where the analytical result may be considered as the laboratory product).

An often heard comment on the EN ISO 9000 standards is that they 'guarantee that products are similar, not that they have quality'. To some extent this is a misunderstanding, as quality systems developed from the model

ISO 9000 ensure that the customer requirements on the product are specified within the company, and are met when the company's product is produced.

The ISO 9000 series comprises a number of standards of which three—EN ISO 9001, 9002 and 9003—deal with quality system requirements. New editions of these were published in 1994 under the titles:

- EN ISO 9001 (1994), Quality systems—Model for quality assurance in design/development, production, installation and servicing;
- EN ISO 9002 (1994), Quality systems—Model for quality assurance in production, installation and servicing; and
- EN ISO 9003 (1994), Quality systems—Model for quality assurance in final inspection and test.

The standard a company chooses depends on the scope of its business. EN ISO 9001 covers all types of activities from design development to production, installation and servicing. EN ISO 9002 is a suitable model for many food companies, as those generally 'only' produce and sell foodstuffs. The standard EN ISO 9003 is designed for companies working with final inspection and test (*e.g.* elevator riggers).

All three standards are now being discussed within the Codex Alimentarius Commission and are to be recommended as one of many quality systems models for official import and export control of foods. The standards are also discussed or referred to in other fora, *e.g.* in Directive 93/43/EEC, the Food Hygiene Directive.[6]

Other standards in the ISO 9000 Series include guidelines on how to implement EN ISO 9001–9003. Examples of such guidelines are:

- EN ISO 9000–1 (1994), Quality management and quality assurance standards—Part 1: Guidelines for selection and use;
- EN ISO 9004–1 (1994), Quality management and quality system elements—Part 1: Guidelines;
- EN ISO 9004–2 (1994), Quality management and quality system elements—Part 2: Guidelines for services; and
- ISO 10011 (1990), Parts 1–3 Guidelines for auditing quality systems.

An analytical food laboratory operating according to ISO/IEC Guide 25 (1990) fulfils the requirements on internal process control laboratories in the standards EN ISO 9001–9003.

Certification

'Certification of conformity' has been defined as:

Action by a third party, demonstrating that adequate confidence is provided that a duly identified product, process or service is in conformity with a specific standard or other normative document.

Certification can be given as quality system certification according to EN

ISO 9001, 9002 and 9003. Certification is also possible for specific products or services.

Quality system certifications according to EN ISO 9001, 9002 and 9003 are performed by organisations such as Lloyds Register Quality Assurance (LRQA), Bureau Veritas Quality Assurance (BVQA) and Den Norske Veritas (DNV). The company usually takes the initiative for certification but some product directives (*e.g.* for medical implants such as pacemakers) require that companies within specific sectors shall be certified according to EN ISO 9001, *etc.*

Audits of companies certified according to EN ISO 9001, 9002 and 9003 are performed as prescribed in the guidelines ISO 10011 (1990), Parts 1–3, Guidelines for auditing quality systems.

Accreditation of Bodies Operating Certification

Organisations granting certification have to implement quality systems of their own. The organisations can then obtain a formal recognition (*i.e.* an accreditation) that they are competent to carry out certification. Accreditation for certification bodies operating quality system certification (EN ISO 9001, 9002 and 9003) is given according to EN 45012. Certification bodies operating product certification are accredited according the EN 45011 standard, the bodies operating certification of personnel according to the EN 45013 standard.

Good Laboratory Practice (GLP)

This, as with the BS EN ISO 9000 standard, is an organisation-wide scheme and does not list individual analyses in the laboratory's scope of activity.

OECD GLP Principles for the Testing of New Chemicals and Pharmaceuticals

The OECD GLP principles were originally developed for laboratories involved in the toxicological testing of new chemicals and pharmaceuticals to obtain data on their properties and/or safety with respect to human health or the environment. GLP principles can be used for laboratories involved in *in vitro* or *in vivo* genetic toxicological studies, toxicity studies with mammals, environmental toxicity studies, analytical, pharmaco- and toxicokinetic studies.

The OECD GLP principles are based on rules established by the US Food and Drug Administration (FDA) in late 1970. They were adopted in 1981 and are published in different OECD 'environment monographs'. Of these, No. 45 describes 'the OECD principles of Good Laboratory Practice (GLP)'. Two paragraphs (2 and 7) in section II of this monograph have become mandatory to official food control laboratories, as they are prescribed in Directive 93/99/EEC, The Additional Measures Food Control Directive.[3]

Some other OECD environment monographs within the GLP sector are given below:

GLP Consensus Documents

- Quality assurance and GLP; Environment monograph No. 48;
- Compliance of laboratory suppliers with GLP principles; Environment monograph No. 49; and
- The application of the principles of GLP to computerised systems; Environment monograph No. 116.

Guidance for GLP Monitoring Authorities

- Guides for compliance monitoring procedures for good laboratory practice; Environment monograph No. 110; and
- Guidance for the conduct of laboratory inspections and study audits; Environment monograph No. 111.

In parallel to the OECD GLP principles, similar principles from the FDA and EPA [(US) Environmental Protection Agency] are in use (see below). However, the differences in content between those different types of GLP documents has largely been eliminated.

It should be noted that the wording 'GLP—Good Laboratory Practice' is sometimes used as being synonymous to quality assurance but expressed in a lower case acronym ('we have glp in our laboratory....'). This is the case, for example, in Directive 89/187/EEC on 'Powers and Conditions of Operation of the Community Reference Laboratories'[7] and Directive 86/469/EEC on the 'Examination of Animals and Fresh Meat for the Presence of Residues'.[8]

Purpose of OECD GLP Principles

The OECD GLP principles are documentary in nature. They cover the organisational process and the conditions under which laboratory studies are planned, performed, monitored, recorded and reported. The ultimate aim of the studies is usually submission of the preparation involved for registration by international licensing bodies. A registration can take place many years after the study has been completed. The model is not recommended for routine analytical food laboratories.

The OECD GLP principles are designed for studies on one substance (chemical, pharmaceutical or cosmetic) and set no limits for the number or types of methods to be used. The OECD model refers to the same quality elements as the accreditation models EN 45001 and ISO/IEC Guide 25. Some important differences exist, however, between these models.

The OECD GLP principles require the presence for each study of a study director, who is responsible for the overall conduct of the study and that this follows the study plan (monograph No. 45, section II, 1.2). Furthermore, a study plan has to be written and kept up to date under the responsibility of the study director (paragraph 8). Specific requirements are given for 'biological' (*e.g.* animal) test systems (paragraph 5.2).

The GLP principles stress to a greater degree the requirements on archives and quality assurance programs than is the case in the EN 45001 and ISO/IEC Guide 25 Standards, *e.g.* only personnel authorised by management should have access to the archives. Movement of material in and out of those shall be recorded (paragraph 10). Quality assurance personnel should not be involved in the conduct of the study being assured. Furthermore, the personnel should verify that the study has been performed according to the study plan and review final reports to confirm that methods, observation, *etc.* are accurately described, and they shall prepare and sign a statement with dates of inspections and dates of any findings (paragraph 2). It should be noted that the quality assurance functions do not carry out any scientific evaluation of the results from a GLP study. Such evaluations are carried out by the legislative authorities on the submission for approval of clinical studies or registration.

Supervision of GLP Compliance

The supervision of compliance to GLP principles is a task for governments and is not therefore a voluntary obligation. This is in contrast to accreditation according to EN 45001 and ISO/IEC Guide 25 in that it may be requested on a voluntary basis by a laboratory. OECD GLP principles were implemented in 1989 in the EU through the EC Directive 89/569/EEC 'on Acceptance of an OECD Decision/Recommendation on Compliance with Principles of Good Laboratory Practice'.[9] EU Member States then have to implement the GLP principles within their national legislation. 24 Member States of the OECD, including the EU countries, USA, Canada and Japan, have accepted the GLP principles.

The implementation of the OECD GLP principles differs between different countries. In Denmark, France, Italy and Sweden, the same authorities are responsible for both the supervision of GLP compliance and final registration. In other countries, *e.g.* Germany and Finland, different authorities on federal and state level are responsible for different tasks in the supervision. Assessments (inspections) of the GLP studies are carried out every second year, partly dependent on the result. GLP authorities in Europe provide 'statements of compliance with the OECD GLP Principles'. Authorities in the USA do not, on the other hand, sign or accept any documents that give laboratories or studies a GLP status.

In the UK, the GLP scheme is operated by the Department of Health (DoH) GLP Unit and is concerned with the organisational processes and the conditions under which laboratory studies are planned, performed, monitored, recorded and reported. The first inspections for compliance were conducted by inspectors in 1982 when the UK Health and Safety Executive established an inspectorate to monitor the testing of industrial chemicals. In 1983 this was expanded to the pharmaceutical, agrochemical, cosmetic and food additives fields. Major participants in this DoH scheme are medical laboratories.

Bilateral Agreements

All recognitions on mutual acceptance of information on GLP studies are based on bilateral agreements. No system exists today that guarantees a harmonised and equivalent implementation of the OECD GLP principle within the OECD countries. The OECD GLP panel, where member states are represented, discusses technical GLP aspects and produces new GLP documents but does not formally assess their use.

The EU Commission, however, has taken an initiative for a commission surveillance system on the implementation of OECD GLP principles within Europe.

3 Other Standards of Interest

There are other standards/guidelines prepared by the international standardising organisations which are of interest to the food analysis laboratory as they may directly influence the work of the laboratory. These include:

ISO/IEC Guide 43 on Proficiency Testing

ISO/IEC Guide 43, 'Development and operation of laboratory proficiency testing',[10] is a guide for those that arrange such testing schemes. ISO/IEC Guide 43 supplies information on the quality assurance of proficiency testing schemes.

Laboratories participating in proficiency testing schemes should be aware of and appreciate this guide. As participation normally requires a financial commitment from the 'customer' (the laboratory), the laboratory itself needs to be assured that the proficiency testing scheme is operating to an acceptable standard of quality, normally demonstrated by the scheme being accredited to an International Standard, and the scheme is thus able to look after the interests of its participants. The ISO/IEC guide provides information on requirements on sample homogeneity and sample stability, on evaluation and reporting procedures, *etc.*, information that can be especially useful as some organisers lack good procedures for planning and reporting.

The demand that organisers of proficiency testing schemes should also quality assure their work is becoming greater as such schemes have an increasingly large influence on the work of many laboratories. The customers should drive their requirements but in future it will be possible, and desirable, for accreditation bodies to formalise these requirements. In the food sector it is essential that any proficiency testing scheme not only complies with ISO/IEC Guide 43 but also meets the requirements of the International Harmonised

Protocol for the Proficiency Testing of (Chemical) Analytical Laboratories,[11] that protocol having been adopted by both the EU and the Codex Alimentarius Commission.

Accreditation According to EN 45002, EN 45003 and ISO/IEC Guide 58

In Europe, as in the rest of the world, there exists a large number of different accreditation programmes. Here it has to be stressed that the word 'accreditation' has slightly different meanings depending on which standard or context it is used; accreditation based on EN 45000 standards, ISO/IEC Guide 25 and ISO/IEC Guide 58 is discussed below.

The term 'accreditation' can, within the scope of EN 45001, be defined as:

A formal recognition that a testing laboratory is competent to carry out a specific test or specific types of tests.

An accreditation is based on requirements in the models EN 45001, parts of EN 45002 and ISO/IEC Guide 25. Consideration is also given to specific demands raised by legal authorities and other customers. Accreditation is given for one or many specified methods that can be used for the analysis of specific analytes in specific matrices. If the method specifications differ from one accredited laboratory to another, then the accreditations are not strictly comparable.

Accreditation is carried out by organisations that themselves have implemented appropriate quality assurance and quality systems according to the following models:

1. EN 45002, General criteria for the assessment of testing laboratories;
2. EN 45003, General criteria for laboratory accreditation bodies;
3. ISO/IEC Guide 58, Calibration and testing laboratory accreditation systems—General requirement for operation and recognition.

The EN ISO 14000 Series

A new series of environmental management standards is being prepared. The ISO 14000 standards will specify the requirements of an environmental management system. The overall aim is to support environmental protection and prevention of pollution in balance with socio-economic needs. At present the following standards exist:

- EN ISO 14001, Environmental management systems—Specifications with guidance for use;
- EN ISO 14010, Guidelines for environmental auditing—General principles;
- EN ISO 14011, Guidelines for environmental auditing—Audit procedures—Auditing of environmental management systems; and

- EN ISO 14012, Guidelines for environmental auditing—Qualification criteria for environmental auditing.

The EN ISO 14001 standard shares common management system principles with the ISO 9000 quality system standards. The environmental standard requires companies to identify areas of their work that impact on the environment, to prepare an environmental policy which makes commitments to continuous improvement and implement a management system for environmental activities. Companies that have an ISO 9000 certification may elect to use the management system within this environmental model when working for a certification according to the future ISO 14001. Analytical food laboratories which are an integral part of ISO 14001 certified companies will also have to fulfil the environmental requirements in that standard.

Laboratories carrying out emission control and emission surveillance according to legal requirements have to be accredited for the methods used according to EN 45001 in European countries such as Sweden and Norway.

EMAS

EMAS is an acronym for 'Eco Management and Audit Scheme' and is based on the EU Regulation 1836/93/EEC 'allowing voluntary participation by companies in the industrial sectors in EMAS'.[12] The intention of this EU Regulation is that responsibility for the environmental effects of a business is voluntary and an internal company matter. It requires a company to have chosen to follow EMAS; it then has to develop and implement a management system for environmental activities, formulate an environmental policy for each plant, carry out environmental audits, *etc.*, similar to the requirements in the ISO 14001 standard. As with ISO 14001, there are many similarities between the requirements of ISO 9001 and those of EMAS in respect of system structure, documentation, internal auditing, *etc.*

In the EMAS system, it is the environmental assessor who is responsible for assessment of a company's environmental work against the requirements of EMAS. This assessor also approves the company's environmental statement as to whether the requirements have been fulfilled. As a consequence, analytical food laboratories which are an integral part of EMAS certified companies also have to fulfil those environmental requirements, this mirroring the situation as with ISO 14001.

Total Quality Management (TQM) Models

The European Quality Award (EQA), the British Quality Award (BQA) and the Utmärkelsen Svensk Kvalitet (USK) are examples of national and international TQM models. The basic elements of these are:

- there is to be customer satisfaction and customer oriented management;
- there is to be active leadership for quality;
- there is to be employee involvement at all levels and teamwork;
- TQM begins with education and ends with education;
- TQM has to be given time and priority;
- TQM is process orientated;
- TQM is to be based on facts;
- TQM must enable fast reactions to be introduced; and
- TQM must emphasise and promote continuous quality improvements.

The various TQM models all put 'customer satisfaction' as the highest priority. No requirements are given in the models on how a business shall be organised, or on what sort of technical quality the products shall fulfil.

Companies certified according to EN ISO 9001, and laboratories that have an ISO/IEC Guide 25 and/or EN 45001 accreditation, can also benefit if they implement a TQM model. It is therefore recommended that the management of analytical food laboratories use TQM models to give their analytical results 'added value' with regards to their quality and customer satisfaction.

4 National Accreditation Agencies

Most European Union countries and a significant of other countries have designated a 'National Accreditation Agency', *i.e.* an accreditation agency that is a signatory of the mutual recognition agreements developed by the European Co-operation for Accreditation of Laboratories (EAL) [formerly known as the Western European Laboratory Accreditation Co-operation (WELAC)] and the International Laboratory Accreditation Co-operation (ILAC). This leads to the accreditation agency complying with the requirements of the ISO/IEC Guide 58:1993, Calibration and testing laboratory accreditation systems—General requirements for operation and recognition'.[13] It is as a result of the mutual recognition agreements of the accreditation agencies that each accreditation agency may be considered equivalent to any other signatory of the agreement. Thus, for a laboratory, accreditation by one signatory is as valid as accreditation by any other signatory.

The major National Accreditation Agencies in Europe are briefly described or listed below:

UK—United Kingdom Accreditation Service (UKAS)

UKAS was an agency of the UK Department of Trade and Industry but is now a private company; it was formed in 1985 by the amalgamation of the British Calibration Service and the National Testing Laboratory Accreditation Scheme. When a laboratory is accredited by UKAS it is normally for a number of individual analyses, the list of these being found in the laboratory's quality manual which describes in detail the quality policy of the laboratory (however, the concept of generic accreditation is now being introduced, as described for food analysis laboratories in Chapter 1). Individual sections in a laboratory are frequently accredited for specific analyses whilst in other sections of the laboratory there is no accreditation.

UKAS has negotiated mutual recognition agreements with equivalent organisations in other countries; thus, to the laboratory, one of the benefits of a laboratory being accredited by UKAS is that, as a result of these agreements, which recognise the equivalence of services and analytical certificates issued, the laboratory will be recognised by all the other signatories of the agreement.

How UKAS Operates

UKAS operates to the EN 45000 series of standards.

UKAS is run by an executive of professionals experienced in calibration and testing, together with an administrative support staff. The Executive originally formed a service within the National Physical Laboratory at Teddington, near London, but is now a privatised company. The UKAS Executive is responsible for all aspects of operation of the Service but because of the impracticability of directly employing experts in every field of testing and calibration, UKAS contracts individual experts in the fields concerned to assess laboratories on its behalf. These experts are trained by UKAS in the principles of quality assurance and the techniques of assessment and together with a UKAS Technical Officer form the team that assesses each laboratory. To ensure that its policies and practices meet the needs of laboratories and their users, UKAS takes advice from committees comprising representatives from laboratories and their users and other interested parties.

Information Available

UKAS produces, for laboratories wishing to become accredited, a number of publications. Some of the most important are:

- M10—General Criteria of Competence for Calibration and Testing Laboratories;[14]
- M11—Regulations to be met by Calibration and Testing Laboratories;[15] and
- M16—The Quality Manual: Guidance for Preparation.[16]

Swedish Board for Accreditation and Conformity Assessment (SWEDAC)

SWEDAC is Sweden's central administrative agency for accreditation and conformity assessment. Its overall function is to provide support for both industry and the public sector in conformity with international agreements. SWEDAC functions as Sweden's national authority for accreditation of laboratories and certification and inspection bodies. The agency is also responsible for the accreditation of organisations which meet the environmental requirements specified in the EC's Eco Management and Audit Scheme (EMAS). In addition, SWEDAC functions as Sweden's national body for approval of chemical laboratories in accordance with GLP regulations issued by the OECD.

SWEDAC works with European and internationally harmonised standards, primarily the EN 45000 series. Where a laboratory is to be accredited, ISO/IEC Guide 25[1] is also used. In addition, there are certain areas to which special regulations apply, *e.g.* in the case of official food control there are national criteria which have to be complied with, these having been developed and issued by regulatory authorities.

Information Available

SWEDAC has prepared for laboratories wishing to become accredited a number of publications, albeit in Swedish, to aid such laboratories. Some of the most important are:

- SWEDAC DOC 92:5—Kvalitetsmanual för ackrediterade laboratorier—vägledning (Quality manual for accredited laboratories—guidelines);
- SWEDAC DOC 95:5—Vägledning för skattning och angivande av mätosäkerhet (Guidelines on estimation and expression of measurement uncertainty); and
- SWEDAC DOC 96:20—Ackreditering av metoder avsedda för bestämning av flera grundämnen eller substanser (Accreditation of methods for determination of several elements or substances).

Accreditation Agencies Within Europe (Members of EAL)

Table 2.1 indicates the national members of the European Co-operation for Accreditation of Laboratories (EAL) for a number of European countries:

Table 2.1 *National members of EAL*

Country	*Agency (Acronym)*
Austria	BMwA
Belgium	BELTEST
Denmark	DANAK*
Finland	FINAS*
France	COFRAC*
Germany	DAR
Great Britain	UKAS*
Greece	ELOT
Iceland	ISAC
Ireland	ILAB*
Italy	SINAL*

The Netherlands	STERLAB*
Norway	NA*
Spain	RELE*
Sweden	SWEDAC*
Switzerland	SAS

Other Organisations with Bilateral Agreements with EAL

There is a multi-lateral agreement organised through the International Laboratory Accreditation Co-operation (ILAC) which covers non-European countries.

The main signatories are given in Table 2.2.

Table 2.2 *Non-European agencies having a bilateral agreement with EAL*

Country	*Agency (acronym)*
Australia	NATA
Hong Kong	HOKLAS
New Zealand	TELARC

Signatories

Multilateral agreements on mutual acceptance of analytical results from laboratories accredited according to EN 45001 and ISO/IEC Guide 25 have been signed by a number of EAL member countries (countries that have signed are marked with * in Table 2.1). As a consequence of this agreement, accreditation bodies themselves are assessed regularly by auditors from other member bodies within EAL. The purpose of this is to evaluate how the requirements in EN 45002, *etc.* are fulfilled. Furthermore, the implementation of accreditation within Europe is then harmonised; other positive effects of the audits are that the quality of accreditation is improved.

Bilateral agreements on mutual acceptance of analytical results from laboratories accredited according to EN 45001 and ISO/IEC Guide 25 have been signed between some EAL member countries and Australia, Hong Kong and New Zealand. Here the world-wide co-operation on accreditation is organised *via* ILAC.

Established accreditation bodies sometimes do not fulfil the requirements in EN 45002 regarding planning, carrying out the assessments, or reporting the results of assessments and this should be rectified. The requirements of EN 45002, EN 45003 and ISO/IEC Guide 58 provide information that can be of use when the implementation of the accreditation system has to be discussed by the customers (the laboratories) and the accreditation bodies.

5 Guidance on Laboratory Quality Assurance

Although the general information provided here can encourage a laboratory to choose and implement in a positive and developing way the appropriate model for it, analytical food laboratories which want to implement quality assurance according to EN 45001 and ISO/IEC Guide 25 require detailed guidance on the interpretation of the models. The general information given here has to be supplemented by such information.

Detailed information on quality assurance is given in the following publications:

- 'Food and drink laboratory accreditation—A practical approach', Sandra Wilson and Geoff Weir; Chapman & Hall (1995) (ISBN 0 412 59920 1);
- 'Quality assurance principles for analytical laboratories', Fredrick M. Garfield; AOAC International (1991) (ISBN 0 935584 46 3);
- 'Quality assurance guidelines—for microbiological laboratories', NMKL report No. 5, 2nd edition (1994) (ISSN 0281-5303);
- 'Guidelines for quality assurance at chemical food laboratories', NMKL report No. 8, 2nd edition (1997);
- 'Performance check and in-house calibration of analytical balances', NMKL Procedure No. 1 (1995);
- 'Performance check and in-house calibration of thermometers', NMKL Procedure No. 2 (1995);
- 'Control charts and control materials in internal quality control in food chemical laboratories' NMKL Procedure No. 3 (1996);
- 'Validation of chemical analytical methods,' NMKL Procedure No. 4 (1996); and
- 'Estimation and expression of measurement uncertainty in chemical analysis', NMKL Procedure No. 5 (1997).

In addition, EAL has also published a number of guidelines on how ISO/IEC Guide 25 and EN 45001 can be interpreted by chemical, microbiological or sensory laboratories. The purpose of the guidelines is to help both laboratories and accreditation bodies implement the models in the optimal way. It has to be emphasised that the guidelines *do not* replace the documents EN 45001 or ISO/IEC Guide 25.

EAL guidelines of interest for analytical food laboratories are:

- 'Accreditation for chemical laboratories—Guidance on the interpretation of the EN 45000 series of standards and ISO/IEC Guide 25'; WGD 2 (edition 1, 1993);
- 'Accreditation guide for laboratories performing microbiological testing', EAL-G18 (edition 1, 1996);
- 'Accreditation of sensory testing laboratories—Guidance on the inter-

pretation of the EN 45000 series of standards and ISO/IEC Guide 25'; EAL-G16 (edition 1, 1995); and

- 'Internal audits and management review for laboratories'; EAL-G3 (edition 2, 1996).

In addition, most of the above guidelines can be profitably studied and used by laboratories operating within an EN ISO 9000, EMAS or EN ISO 14000 certification environment.

References

1 General Requirements for the Competence of Calibration and Testing Laboratories, ISO/IEC Guide 25, ISO, Geneva.
2 General Criteria for the Operation of Testing Laboratories, European Standard EN 45001, CEN/CENELEC, Brussels, 1989.
3 Council Directive 93/99/EEC on the Subject of Additional Measures Concerning the Official Control of Foodstuffs, *O. J.*, 1993, L290.
4 Quality Systems for the Food and Drink Industries—Guidelines for the use of BS 5750 Part 2 1987 in the manufacture of food and drink (ISO 9002:1987; EN 29002:1987), BFMIRA, Leatherhead, 1989.
5 IFST Food and Drink Manufacture—Good Manufacturing Practice: a Guide to its Responsible Management, IFST, London, 1991.
6 Council Directive 93/43/EEC of 14 June 1993 on the Hygiene of Foodstuffs, *O. J.*, 1993, L.168/1.
7 Council Directive 89/187/EEC on Powers and Conditions of Operation of the Community Reference Laboratories, *O. J.*, 1989, L.066.
8 Council Directive 86/469/EEC on the Examination of Animals and Fresh Meat for the Presence of Residues, *O. J.*, 1986, L.275.
9 Council Decision 89/569/EEC on the Acceptance by the European Economic Community of an OECD Decision/Recommendation on Compliance with Principles of Good Laboratory Practice, *O. J.*, 1989, L.315/1.
10 Development and Operation of Laboratory Proficiency Testing, ISO/IEC Guide 43, ISO, Geneva, 1997.
11 The International Harmonised Protocol for the Proficiency Testing of (Chemical) Analytical Laboratories, ed. M. Thompson and R. Wood, *Pure Appl. Chem.*, 1993, **65**, 2123 (also published in *J. AOAC Int.*, 1993, **76**, 926).
12 Eco Management and Audit Scheme EU Regulation 1836/93/EEC Allowing Voluntary Participation by Companies in the Industrial Sectors in EMAS, *O. J.*, 1993, L.168/1.
13 Calibration and Testing Laboratory Accreditation Systems—General Requirements for Operation and Recognition, ISO/IEC Guide 58, ISO, Geneva, 1993.
14 General Criteria of Competence for Calibration and Testing Laboratories, UKAS, Queens Road, Teddington, TW11 0NA, NAMAS Accreditation Standard M 10, 1992 and Supplement, 1993.
15 Regulations to be met by Calibration and Testing Laboratories, UKAS, Queens Road, Teddington, TW11 0NA, 1992, NAMAS Accreditation Standard M 11.
16 UKAS 1989, The Quality Manual: Guidance for Preparation, UKAS, Queens Road, Teddington, TW11 0NA, 1989, NAMAS Accreditation Standard M 16.

CHAPTER 3

Methods of Analysis—Their Selection, Acceptability and Validation

The selection and development of methods of analysis has traditionally been a subject of importance to analytical laboratories working in the food sector, often to the extent that the practical application of the method is neglected. Possibly this greater emphasis is because most organisations, be they governmental or one of the international standardising organisations working in the foodstuffs area, develop methods of analysis and incorporate them into legislation or international standards, but then did not have any mechanism to assess how well such methods are being applied. However, this approach is becoming superseded with the need to demonstrate that the application of the method is also being successfully achieved. In addition, analysts will be allowed a greater freedom of choice of analytical method provided the method chosen meets certain pre-defined criteria. This so-called 'criteria approach' towards methods of analysis is being progressively adopted by legislative authorities. As has been stressed previously, the emphasis in the methods of analysis in the foodstuffs area is now changing with the formal introduction of accreditation, proficiency testing and defined internal quality control procedures into the laboratory. Nevertheless, it is essential that the quality of the method of analysis is fully recognised and appreciated, this particularly with the introduction of the criteria approach as analysts will increasing have to justify their choice of method.

Thus, notwithstanding other quality assurance requirements, laboratories must ensure the quality of all their analytical methods in use; they must be validated and verified. The various means of ensuring the quality of methods varies from one analytical discipline to the other and the objective of the analysis influences the extent of the quality assurance work. The requirements on methods used in process control are, for example, generally less demanding than the requirements on methods for end product control and official food control.

The requirement for food analysis laboratories to use a 'fully validated' method of analysis is now universally accepted or required within the food sector. A description of these requirements is therefore given below as food analysts will increasing be required to justify their choice of method in the light of these 'fully validated' requirements. The meaning of 'fully validated' within the food sector is illustrated by the requirements of the various international organisations which develop or adopt Standard Methods of Analysis; these are described in Part A of this chapter.

Most method validation guides start with discussions on how criteria such as specificity, accuracy and precision of the method should be validated. The analytical problem, requirements of the customers and choices of analytical principles are seldom mentioned in this context. The introduction of a 'criteria approach' can also be seen as a way of introducing a 'customer approach' into method validation discussion. Quality can be defined as 'fulfilling the customer requirement', something that is possible only when the method to be used is suitable for solving the analytical problem.

The first step in a 'full validation procedure' therefore should be to identify and document 'customer requirements' and the analytical problem, what is analytically and economically possible and other specific requirements on sampling (see Chapter 10), laboratory environment, external environment, *etc.* A 'validation plan' should be written that indicates the method criteria needed and addresses questions such as:

- when is the method going to be used (official food control and in-house process control methods may have to fulfil different criteria on, *e.g.* precision and accuracy)?
- what type of answer is required—qualitative or quantitative?
- in what state is the analyte, *i.e.* is it bound, free, *etc.*?

The importance of a validation (or verification) plan is discussed in the Nordic Committee on Food Analysis (NMKL) procedure no. 4 on 'In-House Validation of Chemical Analytical Methods' and the EURACHEM document on 'The fitness for purpose of analytical methods'. These are commented on in Part C of this chapter.

It cannot be over-stressed that analysts must appreciate the requirements for methods to be recognised as fully validated within the food sector (especially for official food control), the procedures for the validation of methods of analysis and the application of such procedures. Thus it is particularly important that analysts appreciate the requirements for and analysis of collaborative trial data and how the 'criteria approach' will directly affect their working practices.

This chapter therefore describes the requirements for methods of analysis, and in particular the extent of the validation required and the considerations to be made in their selection. It is divided into three parts, these being:

Part A: Introduction and the acceptability of methods of analysis, types of validation, and the criteria approach for methods of analysis.

Part B: The requirements for and procedures to obtain methods that have been fully validated through collaborative trials.

Part C: The requirements for and procedures to obtain methods that have been validated 'in-house'.

Part A: Introduction and the Acceptability of Methods of Analysis, Types of Validation and the Criteria Approach for Methods of Analysis

Analytical methods must be selected on the basis of customer needs. In many cases this is possible only if the laboratory assists the customer to select the method, since they often lack specific analytical competence. Prerequisite for a successful selection is that the purpose of the analysis or examination, together with other specific needs, is well documented in an analytical order/request or in a project or study plan.

Laboratories carrying out analytical work for official food control do not always have the freedom to select the method. In particular, microbiological methods are often laid down in the legislation, *e.g.* both EU and national legislation frequently specify which reference methods are to be used in cases of dispute.

In some cases, laboratories use internally (in-house) developed methods or significantly modify standard methods. Such methods must be validated at the relevant concentration ranges, before being taken into routine use.

If the method of choice is an established standard reference method from, for example, the AOAC International, the laboratory usually only needs to verify that it can achieve the performance characteristics given in the method, especially trueness and precision, and demonstrate that the method is suitable for the intended use (see Table 3.1). The extent and nature of such verification work depend on the needs of the customer. If, for example, the method can give 1% precision whilst only 5% is needed, it is usually sufficient if the laboratory is able to demonstrate that 5% precision is achieved.

It has to be appreciated that far from all international standard methods have been validated in full method–performance interlaboratory (collaborative) trials. Further information on validation and verification of analytical methods is given in Parts B and C of this chapter.

1 The Introduction of a New or Unfamiliar Method into the Laboratory

Responsibility for Carrying out Validation and Verification

When a laboratory intends to use a method with which it is unfamiliar, it is the responsibility of the laboratory to verify that it is competent to use the method. Usually national or international organisations, such AOAC International, NMKL, International Dairy Federation (IDF), *etc.* have undertaken the interlaboratory validation of the method in a method performance (collaborative) trial. The extent of laboratory internal validation and verification depends on the context in which the method is to be used. Some suggestions as to the extent of validation and verification measures are given in Table 3.1.

Table 3.1 *Laboratory internal validation and verification*

Existing validation	*Laboratory requirement*
Fully validated method (has been studied in a collaborative trial)	Verification that the laboratory is capable of achieving the performance characteristics of the method (or is able to fulfil the requirements of the analytical task)
Fully validated method, but new matrix or new instruments used	Verification of trueness and precision; possibly also the detection limit
Well-established, but not collaboratively studied method	Verification, supplemented with limited validation (*e.g.* with regard to reproducibility)
Method published in the scientific literature; characteristics given	Verification, supplemented with limited validation (*e.g.* with regard to repeatability and reproducibility)
Method published in the scientific literature; no characteristics given	Full validation and verification
Method developed in-house	Full validation and verification

Competence Requirements

The introduction of new analytical methods requires that both the parties developing and validating them, as well as the laboratory that will subsequently verify its ability to use them routinely, are sufficiently competent, *i.e.* are sufficiently knowledgeable, experienced and proficient. When new analytical techniques based on, for example, immunological or DNA properties are considered, the laboratory may have to employ new staff with specific competence.

It is not unusual for new rapid test kits to be marketed with the argument that they are 'so simple that anyone can use them'. Often this is not correct and it is necessary for the laboratory to acquire new knowledge or otherwise widen its competence before starting to use the kit. Manufacturers of rapid analysis kits sometimes make minor changes both in the documentation and in the composition of the kit; it is therefore necessary for the user to be knowledgeable and observant but also that any instructions accompanying the kits, *etc.* are carefully read. Knowledge of components included and inter-batch variation may be of critical importance when, for example, polyclonal antibodies are used in ELISA analyses, or Tris buffers are used in PCR analyses.

Evaluation of Published Validation and Verification Data

Validation results, *i.e.* the performance characteristics of analytical methods, are often given as repeatability and reproducibility data, normally as standard deviations or relative standard deviations.

The value of the published data varies depending on the extent of the validation or verification. In some cases, results are based on experiments carried out using synthetic aqueous standard solutions. The results in such cases are naturally not applicable to the same analyte present in a complex matrix. In other cases, method studies have been appropriately carried out in accordance with internationally established protocols (see Part B). In order to give the user the possibility of evaluating and utilising published results, it must be clearly described how the performance characteristics were estimated or determined.

Methods published in the scientific literature are often accompanied by insufficient information on the nature and extent of quality assurance measures applied during the validation of the method, such as a description of how important analytical parameters (temperature and pH) were controlled. It is also important to know whether the originating laboratory has participated in proficiency testing schemes or has used reference materials to support the development work, and indeed whether it had been subjected to any third party evaluation, *e.g.* through formal accreditation.

The absence of performance characteristics or lack of information on the general competence of the originating laboratory makes it difficult to assess the credibility of incomplete validation data published in methods.

Using a New or Unfamiliar Method

Before taking new or unfamiliar methods into routine use, authorised staff must decide if the requirements of the customer and of the analytical problem are being met. The judgement should be based both on published validation data and on experimental data resulting from the laboratory's own verification work. Methods should be formally approved in writing, and their fields of application clearly described.

Laboratory-specific Instruction (the 'SOP')

Standard methods from organisations which develop methods are often clearly worded and can therefore be used without modifications. In some cases, however, methods contain alternative procedures for extraction or other analytical steps. In such cases, authorised staff must specify which alternative is to be used. If it is necessary to introduce modifications, these must be clearly documented, dated and signed by authorised staff. Alterations and specifications must be written unambiguously; some laboratories permit minor alterations to be inserted directly into the texts of methods, provided they are clear, dated and signed. More commonly, such instructions are contained in a laboratory-specific instruction, which describes exactly the performance of the method. The OECD GLP principles (see Chapter 2) unconditionally require that laboratories have available written, detailed 'standard operating procedures' (SOP).

The laboratory shall maintain records of all authorised instructions, *etc.*, and the instructions must be included in the document control system of the laboratory. Document control means that:

- laboratory-specific instructions are in writing, preferably in a uniform format, and dated and signed by authorised staff;
- alterations, if any, are dated and signed by authorised staff;
- all documents are maintained up-to-date and readily available to staff; and
- laboratory-specific instructions are in a language readily understood by staff.

Measurement Uncertainty

Food laboratories are often required to provide information on the estimated measurement uncertainty of reported results. This is commented on in detail in Chapter 5. However, microbiological and sensory laboratories have more difficulties in fulfilling this requirement than chemical laboratories.

At present there are no comprehensive models for the estimation and expression of measurement uncertainty in microbiological and sensory work. The extreme inhomogeneity of micro-organisms in native samples makes it difficult to estimate the uncertainty of a numerical microbiological result.

Laboratories must be able to give information to the customer on how uncertainties are estimated, the number of measurements, type of analyte, concentration range, *etc.*

The quality models EN 45001 (in paragraph 5.4.3) and ISO/IEC Guide 25 (in paragraph 13.2) require that, where relevant, a statement of the estimated uncertainty of an analytical result is included in the test report. The models explicitly require that quantitative results are given together with calculated or estimated uncertainty. Experience shows that it is not always appropriate to include measurement uncertainty in all test reports as it may only confuse customers.

Whether or not results are reported together with estimated uncertainties, it is reasonable to require that laboratories are aware of at least the order of magnitude of the uncertainty associated with the result.

Some laboratories estimate measurement uncertainty using pure, ‘synthetic’ solutions without the presence of a matrix. Other laboratories report uncertainties based on more extensive, and more relevant, evaluations using real samples or reference materials. Such evaluations will normally lead to higher estimated uncertainties than those obtained using synthetic solutions.

Often no information is given on how the uncertainty was estimated. Ambitious and credible laboratories reporting realistic uncertainties estimated from results on real samples may lose customers to less conscientious laboratories reporting lower uncertainties estimated using synthetic solutions.

Different analytical tasks require the reporting of different types of measurement uncertainties. Results from studies for the determination of compliance of foods with legislation must have a high degree of reliability as the results may be presented as evidence in a court of law. In these cases, judgement is often based on the reproducibility (between-laboratory variation) at a 99% probability level. Analytical results to be used, for example for production control purposes within one and the same plant, may be evaluated on the basis of repeatability or reproducibility at lower probability levels.

Information Needed on Methods for Accreditation Purposes

A laboratory seeking to be accredited must ensure the quality of the methods included in its scope of accreditation. Naturally this requires a more extensive effort than the effort required to ensure the quality of a method used, for example, in internal production control. For accreditation it is now often required that the method be validated in a method–performance (collaborative) trial and that the laboratory has introduced suitable quality control measures, such as the regular use of reference and control materials and participates in appropriate proficiency testing schemes.

Accreditation according to EN 45001 and ISO/IEC Guide 25 is a formal recognition that a laboratory is competent to carry out specific tests or types of tests. In most European countries, analytical results reported by accredited laboratories are accepted unconditionally across borders. Clients requiring analytical data that will be globally accepted usually purchase analytical services only from accredited laboratories.

The technical assessments are often based on preselected critical control points in the analytical chain. The assessment should cover the entire chain from receipt of a sample to the reporting of the analytical results and filing of the report. If the laboratory's scope of accreditation covers a large number of methods, only a representative selection of methods are assessed. The EAL publication 'Consideration of Methods and Criteria for the Assessment of the Scope of Accreditation' (EAL-G14) is used by accreditation bodies as a guide for this selection. The publication presents examples of criteria such as:

- evidence of the implementation of the quality system, experience and capability of modification/innovation of analytical methods;
- technical complexity;
- consequence of errors;
- frequency of use (high or low);
- balance between standard methods (normalised) and non-standard methods (customer specifications, in-house procedures, *etc.*); and
- balance between complete monitoring of the how the analytical work is

carried out, checks of analytical reports and inspection of analytical facilities.

Examples of Requirements Regarding Analytical Methods from an Accreditation Body

Laboratories seeking accreditation for chemical, microbiological and sensory food analyses must be able to demonstrate that they can competently use the methods included in the scope of the accreditation. If a method is to be used for the official control of foods there are extensive requirements on internal verification, *i.e.* that the laboratory is able to demonstrate that it can use the method in a way which enables the analytical task to be solved. The following requirements are examples of factors which laboratories seeking accreditation should pay attention to, since they often are included in a competence assessment:

- the laboratory must have information on the method: is it based on a standard or reference method, or has it been internally developed?;
- any deviation in a method as compared to a reference method is fully described and the effects of the deviation have been investigated;
- the method has been verified, *e.g.* by analysing spiked samples of relevant matrices;
- the laboratory's own written method text is available;
- the method has been in use in the laboratory for a time period of a minimum of three months during which a number of 'real' samples of relevant types have been analysed;
- quality control procedures are in place, *e.g.* analysis of reference or control materials, or control strains (see Chapter 6);
- if possible, the laboratory participates in proficiency testing schemes and evaluates, on a continuous basis, the results (see Chapter 7);
- where relevant, the measurement uncertainty has been estimated (see Chapter 5); and
- if a sensory laboratory, it monitors the performance of individual sensory assessors and of panels.

Documentation showing that the laboratory complies with the requirements presented above must normally be available to the accreditation body and their technical assessors three to four weeks before the assessment. This information is a useful tool for the assessors when they select which parts of an analytical chain are to be assessed.

The evaluation of a laboratory's results on the basis of the elements listed above is carried out in order to assess the analytical activities and capabilities of a laboratory to obtain an overall impression of the laboratory. The result should demonstrate whether the laboratory is competent and proficient in the use of the methods for which accreditation is sought.

2 Requirements of Legislation and International Standardising Agents for Methods of Analysis

The development of methods of analysis for incorporation into international standards or into foodstuff legislation was, until comparatively recently, not systematic. As a consequence, not all present standard methods have been collaboratively validated; even if collaboratively validated, the method performance characteristics are not always published as part of the method, thus making it difficult for the analyst to verify his or her capability to use the method in an appropriate way.

In general, it is more common for chemical standard methods to be 'fully validated' and supplied with information on, *e.g.* precision (reproducibility and repeatability), than is the case for microbiological methods. One reason why there are few 'fully validated' microbiological methods is that there are no dedicated guidelines on method validation for microbiological methods in the food sector; in practice the chemical protocols are adapted by the microbiological sector but this is not being carried out in a systematic manner.

Most international organisations now develop their own methods in a defined way or stipulate conditions to which their methods should comply; the most important of these, for the AOAC International (AOACI), the Codex Alimentarius Commission, European Union and the European Committee for Standardization (CEN), are described below, as is the future ISO guidance on validation of microbiological methods (ISO/TC 147/SC 4/WG 12).

The essential requirement of all these organisations is that the method has been subjected to a collaborative trial before it is adopted or recommended by that organisation. There are other possibilities for the standardisation of a method, *i.e.* through the results of proficiency testing schemes or through an in-house validation, but the first and 'easiest' procedure for a food laboratory seeking a method for any particular analyte/matrix combination is to use a method published by one of the international standardising organisations.

The requirements of the four main non-commodity-specific method standardising organisations in the food sector are:

AOAC International (AOACI)

This organisation has required for many years that all of its methods, which are used on a world-wide basis, be collaboratively tested before being accepted for publication in the organisation's 'Official Methods of Analysis' book; it now adopts only methods which have been validated in a collaborative trial. It was probably the first international organisation working in the food sector which laid down principles for the establishment of its methods. The procedure for carrying out the necessary collaborative trials is well defined and has formed part of the internal requirements of the organisation for many years. In addition, methods are adopted by the organisation on a preliminary/tentative basis (as 'Official First Action Methods') before being adopted as 'Official Final Action Methods' by it after at least two years in use by analysts on a routine basis. This procedure ensures that the methods are satisfactory on a day-to-day basis before being finally accepted by the AOACI.

The AOACI has also developed two other procedures for 'method validation' which, though not dealing directly with the adoption of methods of analysis which have been assessed by collaborative trial, do result in some validation. These are:

- peer validated methods—in which a method of analysis is assessed in a very limited number of laboratories (usually three) and an assessment made of its performance characteristics, and
- methods assessed by the AOAC Research Institute—such methods are normally proprietary products (*e.g.* 'test kits') for which a 'certificate' of conformity with specification is issued by the Research Institute.

The Codex Alimentarius Commission

This was the first international organisation working at the government level in the food sector which laid down principles for the establishment of its (chemical) methods of analysis. That it was necessary for such guidelines and principles to be laid down reflects the confused and unsatisfactory situation in the development of legislative methods of analysis that existed until the early 1980s in the food sector.

The 'Principles For The Establishment Of Codex Methods Of Analysis'[1] are given below; other organisations which subsequently laid down procedures for the development of (chemical) methods of analysis in their particular sector followed these principles to a significant degree. Some of the principles, such as the definitions of types of methods below, can be used within both microbiology and chemistry. The general criteria, on the other hand, have to be modified to be usable within the microbiological field.

Principles For The Establishment Of Codex Chemical Methods Of Analysis

Purpose of Codex methods of analysis

The methods are primarily intended as international methods for the verification of provisions in Codex standards. They should be used for reference, in calibration of methods in use or introduced for routine examination and control purposes.

Methods of analysis

(A) *Definition of types of methods of analysis (note: though developed for chemical methods, they are also appropriate to microbiological methods).*

(a) Defining Methods (Type I)

Definition: a method which determines a value that can only be arrived at in terms of the method *per se* and serves by definition as the only method for establishing the accepted value of the item measured.

Examples: Howard Mould Count, Reichert–Meissl value, loss on drying, salt in brine by density (note: Coliform bacteria on VRB agar would be an example from the microbiological sector).

(b) Reference Methods (Type II)

Definition: a Type II method is the one designated reference method where Type I methods do not apply. It should be selected from Type III methods (as defined below). It should be recommended for use in cases of dispute and for calibration purposes.

Example: potentiometric method for halides (note: *Brochotrix thermosphacta* and *Escherichia coli*, AOAC 1984 method, would be examples from the microbiological sector).

(c) Alternative Approved Methods (Type III)

Definition: a Type III method is one which meets the criteria required by the Codex Committee on Methods of Analysis and Sampling for methods that may be used for control, inspection or regulatory purposes.

Example: Volhard Method or Mohr Method for chlorides (note: Salmonella in food, Entis 1995 method, would be an example from the microbiological sector).

(d) Tentative Method (Type IV)

Definition: a Type IV method is a method which has been used traditionally or else has been recently introduced but for which the criteria required for acceptance by the Codex Committee on Methods of Analysis and Sampling have not yet been determined.

Examples: chlorine by X-ray fluorescence, estimation of synthetic colours in foods (note: Yeasts and Moulds, NMKL no. 98, 1995 method, and *Shigella* bacteria would be examples from the microbiological sector).

(B) *General criteria for the selection of methods of analysis*

(a) Official methods of analysis elaborated by international organisations occupying themselves with a food or group of foods should be preferred.

(b) Preference should be given to methods of analysis the reliability of which have been established in respect of the following criteria, selected as appropriate:

specificity;

accuracy;

precision, repeatability intra-laboratory (within laboratory), reproducibility inter-laboratory (within laboratory and between laboratories);

limit of detection;

sensitivity;

practicability and applicability under normal laboratory conditions;

other criteria which may be selected as required.
(Note: addition criteria to be chosen when validating microbiological methods are discussed below, under ISO document 'Validation of microbiological methods'.)

(c) The method selected should be chosen on the basis of practicability and preference should be given to methods which have applicability for routine use.

(d) All proposed methods of analysis must have direct pertinence to the Codex Standard to which they are directed.

(e) Methods of analysis which are applicable uniformly to various groups of commodities should be given preference over methods which apply only to individual commodities.

The European Union

The European Union (EU) is attempting to harmonise sampling and analysis procedures to meet the current demands of the national and international enforcement agencies and the likely increased problems that the open market will bring. To aid this, the EU issued a Directive on Sampling and Methods of Analysis.[2] The Directive contains a technical annex, in which the need to carry out a collaborative trial on a method before it can be adopted by the Community is emphasised.

The criteria to which Community methods of analysis for foodstuffs should now conform are as stringent as those recommended by any international organisation following adoption of the Directive. The requirements follow those described for Codex above, and are given in the Annex to the Directive. They are:

'1. Methods of analysis which are to be considered for adoption under the provisions of the Directive shall be examined with respect to the following criteria:
   (i) specificity
   (ii) accuracy
   (iii) precision; repeatability intra-laboratory (within laboratory), reproducibility inter-laboratory (within laboratory and between laboratories)
   (iv) limit of detection
   (v) sensitivity
   (vi) practicability and applicability under normal laboratory conditions
   (vii) other criteria which may be selected as required.

2. The precision values referred to in 1 (iii) shall be obtained from a collaborative trial which has been conducted in accordance with an internationally recognised protocol on collaborative trials (*e.g.* international organisation of Standardization 'Precision of Test Methods' (ISO

5725/1981)). The repeatability and reproducibility values shall be expressed in an internationally recognised form (*e.g.* the 95% confidence intervals as defined by ISO 5725/1981). The results from the collaborative trial shall be published or be freely available.
3. Methods of analysis which are applicable uniformly to various groups of commodities should be given preference over methods which apply to individual commodities.
4. Methods of analysis adopted under this Directive should be edited in the standard layout for methods of analysis recommended by the International Organisations for Standardisation.'

European Committee for Standardisation (CEN)

In Europe, the international organisation which is now developing most standardised methods of analysis in the food additive and contaminant areas and in some commodity specific areas is the European Committee for Standardization (CEN). Although CEN methods are not prescribed by legislation, the European Commission does place considerable importance on the work that CEN carries out in the development of specific methods in the food sector; CEN has been given direct mandates by the Commission to publish particular methods, *e.g.* those for the detection of food irradiation. CEN, like the other organisations described above, has adopted a set of guidelines to which its methods technical committees should conform when developing a method of analysis. The guidelines are:

'Details of the interlaboratory test on the precision of the method are to be summarised in an annex to the method. It is to be stated that the values derived from the interlaboratory test may not be applicable to analyte concentration ranges and matrices other than given in the annex.'

The precision clauses shall be worded as follows:

Repeatability: 'The absolute difference between two single test results found on identical test materials by one operator using the same apparatus within the shortest feasible time interval will exceed the repeatability value r in not more than 5% of the cases.

The value(s) is (are): . . . .'

Reproducibility: 'The absolute difference between two single test results on identical test material reported by two laboratories will exceed the reproducibility value R in not more than 5% of the cases.

The value(s) is (are): ....'

There shall be minimum requirements regarding the information to be given in an informative annex, these being:

- year of interlaboratory test and reference to the test report (if available);
- number of samples;

- number of laboratories retained after eliminating outliers;
- number of outliers (laboratories);
- number of accepted results;
- mean value (with the respective unit);
- repeatability standard deviation (s_r) (with the respective unit);
- repeatability relative standard deviation (RSD_r) (%);
- repeatability limit (r) (with the respective units);
- reproducibility standard deviation (s_R) (with the respective unit);
- reproducibility relative standard deviation (RSD_R) (%);
- reproducibility limit (R) (with the respective unit);
- sample types clearly described;
- notes if further information is to be given.

In addition, CEN publishes its methods as either finalised standards or as prestandards, the latter being required to be reviewed after two years and then either withdrawn or converted to a full standard. Thus, CEN follows the same procedure as the AOACI in this regard.

Requirements of Official Bodies—The Valid Analytical Method

Consideration of the above requirements means that all legislative methods from the European Union and Codex, as well as methods of analysis stemming from the main international standards organisations, must be fully validated—*i.e.* have been subjected to a collaborative trial conforming to an internationally recognised protocol. These requirements are now adopted by most other international bodies. Thus it is essential that food laboratories use methods which comply with the requirements in order to be able to ensure acceptance of their analytical methodology by their customers.

The concept of the valid analytical method in the food sector is such that the following essential attributes of the method are determined:

Accuracy

Accuracy is defined as the closeness of the agreement between the result of a measurement and a true value of the measureand.[3] It may be assessed by the use of reference materials.

However, in food analysis, there is a particular problem. In many instances, though not normally for food additives and contaminants, the numerical value of a characteristic (or criterion) in a standard is dependent on the procedures used to ascertain its value. This illustrates the need for the (sampling and) analysis provisions in a standard to be developed at the same time as the numerical value of the characteristics in the standard are negotiated, to ensure that the characteristics are related to the methodological procedures prescribed.

Precision

Precision is defined as the closeness of agreement between independent test results obtained under prescribed conditions.[4]

In a standard method, the precision characteristics are obtained from a properly organised collaborative trial, *i.e.* a trial conforming to the requirements of an international standard (*e.g.* the AOAC/ISO/IUPAC Harmonised Protocol[5,6] or the ISO 5725 Standard[7]). Because of the importance of collaborative trials, the fact that Codex and European Union methods must have been subjected to a collaborative trial before acceptance, and the resource that is now being devoted to the assessment of precision characteristics of analytical methods, they are described in detail in Part B of this chapter.

3 Future ISO Guidance on the Validation of Microbiological Methods

Guidance on the validation of microbiological methods is now being developed within the ISO Working Group ISO/TC 147/SC 4/WG 12. In the present draft (1997), there is emphasis on selective methods, dealing both with the background and principles of those methods, and the introduction of guidance on validation practices, statistical designs and numerical validation criteria.

Microbiological methods are divided into three types:

1. presence/absence (P/A);
2. most probable number (MPN); and
3. colony culture.

Many of the so-called rapid or modern microbiological methods also fit within the definition. Performance characteristics for which specifications and validation procedures are addressed are classified as: robustness, linearity, working limits, selectivity/specificity and recovery.

The ISO guidance also discusses limitations and characteristic features of analytical microbiological methods. Examples are the unavoidable problem with random variation due to particle (micro-organism) distribution between perfectly mixed parallel sample solutions, viability problems and the human influence on the analytical result when reading plates, *etc.*

The introduction of an internationally accepted guidance on validation of microbiological methods will have a large impact on the credibility of microbiological results.

4 NMKL Guidelines on Validation of Chemical Analytical Methods

The Nordic Committee on Food Analysis (NMKL) has published a guide to be used by NMKL referees, 'Guidelines for Referees on Validation of Chemical Analytical Methods'.[8] This provides information on how to develop chemical methods, on the organisation of method performance studies and the statistical evaluation of the result from those, and information on further references.

Referees appointed by NMKL are required to use these guidelines for organising collaborative trials. The guidelines can be of use to others who would like to develop and evaluate chemical methods in collaborative trials.

5 Future Requirements for Methods of Analysis—Criteria of Methods of Analysis

Notwithstanding the above, there is now an increasing tendency for the laboratory to be allowed freedom of choice of the analytical method. This is because it is now recognised that for a laboratory to report satisfactory (and acceptable) analytical results it must undertake a number of measures, only one of which is to choose and use an appropriate or prescribed method of analysis. This is best illustrated by consideration of activities in:

1. GATT, where the agreement on the Application of Sanitary and Phytosanitary Measures recommends mutual recognition between governments. For the Codex Alimentarius Commission, this means that the concept of equivalence should be adopted;
2. the Codex Alimentarius Commission, which has endorsed the IUPAC/ISO/AOAC Harmonised Protocol for the Proficiency Testing of (Chemical) Analytical Laboratories,[9] and the CCMAS which is recommending for endorsement by the Codex Alimentarius Commission the IUPAC/ISO/AOAC Harmonised Guidelines for Internal Quality Control in Analytical Chemistry Laboratories[10] and which, at its 21st Session, recommended laboratory quality standards for laboratories involved in import/export certification work;[11]
3. the EU, where as a result of the adoption of the Additional Measures Food Control Directive,[12] food control analytical laboratories will be required to become accredited (to the European equivalent of ISO Guide 25), participate in proficiency testing schemes and to use fully validated methods whenever such methods are available. Such validation for methods equates to the requirements for methods of analysis outlined in the Codex Alimentarius Commission's Procedural Manual.[1]

As a result of such activities the focus of attention is moving towards how well the analyst (laboratory) carries out his analytical procedures and away from just ensuring that he is using a prescribed analytical method.

Prescriptive Analytical Methodology?

At the 19th to 21st Sessions of the Codex Committee on Methods of Analysis and Sampling, papers were discussed in which the arguments were given for amending the present Codex procedure whereby the prescribed numerical values in Codex standards are determined using prescribed methods of sampling and analysis elaborated through defined Codex procedures. It was stated that there were a number of criticisms to be made of the present Codex procedure in regard to the elaboration of methods of analysis. These were:

1. the analyst is denied freedom of choice and thus may be required to use an inappropriate method in some situations;
2. the procedure inhibits the use of advanced methodology especially if the 'official' method is dated;
3. the procedure inhibits the use of automation; and
4. it is administratively difficult to change a method found to be unsatisfactory or inferior to another currently available.

The Committee then discussed an alternative approach whereby a defined set of criteria to which methods should comply without specifically endorsing specific methods should be adopted. It was stated that:

1. This approach incorporates greater flexibility than the present procedure adopted by Codex, and in the case of non-defining methods, eliminates the need to consider and endorse as a series of Type III methods (*i.e.* methods for the analysis of a specific chemical entity). The Committee recognised that the endorsement of many Type III methods for any specific determination does, in practice, rarely occur; that this reduces the effectiveness of the present Codex system for the endorsement of methods was appreciated.
2. In some areas of food analysis there are many methods of analysis which are available, which meet Codex requirements as regards method characteristics, but which are not considered by CCMAS and the Commission because of time constraints on the Committee.
3. The adoption of a more generalised approach would ensure that such methods are brought into the Codex system and not disadvantage developments being undertaken elsewhere in the analytical community.
4. It may be necessary to continue to prescribe a single Type II reference method but that the criteria approach could certainly be applied to the present Type III methods.

Many delegates to the recent sessions of CCMAS reacted favourably to the new approach being suggested.

CCMAS has therefore developed examples of this new criteria-based approach which is also being adopted by the European Union.

Three examples are given below in order to emphasise to the practical food analyst that he is being given increased freedom of choice of methods of analysis. However, with the adoption of this approach it is now essential that the analyst appreciates whether his method fully complies with the recognised requirements of being 'validated' as that laboratory now has to be able to justify the choice of analytical procedure made.

Example 1: Codex Generalised Criteria for Lead

For the determination of lead, the criteria to be used for the selection of appropriate methods, as discussed at the 21st Session of the Codex Committee on Methods of Analysis and Sampling,[13] are as given in Table 3.2, *i.e.* any method may be used provided it meets these criteria.

Table 3.2 *Criteria to be used for the selection of method for determining lead*

Parameter	*Value/Comment*
Applicability	All foods
Detection limit	No more than one twentieth of the value of specification
Determination limit	No more than one tenth of the value of specification
Precision	$HORRAT_r$ and $HORRAT_R$ values of less than two in the validation collaborative trial
Recovery	80–105%
Specificity	No cross interferences permitted

Example 2: Outline of the Guidelines for Laboratories Carrying out the Determination of Nitrate in Lettuce and Spinach: EC Monitoring Programme—Method of Analysis to be Used by Monitoring Laboratories

The European Commission Working Group has decided it is not appropriate to prescribe a specific method of analysis for nitrate to the laboratories undertaking the monitoring exercise, but that laboratories may use whatever method of analysis they so wish provided the method chosen complies with the general criteria given in Table 3.3 and that the laboratory itself meets certain quality standards.

It has therefore developed and accepted the following criteria for methods of analysis:

Methods of analysis used for food control purposes must comply whenever possible with the provisions of paragraphs 1 and 2 of the Annex to Council

Directive 85/591/EEC.[14] The provisions of that Directive should apply to methods used by laboratories undertaking the nitrate monitoring.

The effect of the criteria of Table 3.3 is that laboratories may use any method of analysis provided it has been validated as being capable of performing to the above standards. A number of specific methods have been shown to meet the required precision.

Table 3.3 *Specific criteria for methods of analysis used in monitoring nitrate levels*

Criterion	*Concentration range*	*Recommended value*	*Maximum permitted value*
Blanks	All	Negligible	—
Recovery	All	90–110%	
Precision RSD_R	All	As derived from Horwitz equation	Two times value derived from Horwitz equation

Specific values for precision RSD_R at concentrations of interest in the monitoring programme derived from the Horwitz equation are:

Precision RSD_R	1000 mg/kg	5.6	11.2
Precision RSD_R	2000 mg/kg	5.1	10.2
Precision RSD_R	3000 mg/kg	4.8	9.6
Precision RSD_R	4000 mg/kg	4.6	9.2
Precision RSD_R	5000 mg/kg	4.4	8.8
Precision RSD_R	6000 mg/kg	4.3	8.6

(Precision RSD_r may be calculated as 0.66 times Precision RSD_R at the concentration of interest)

Example 3: Outline of the Guidelines for Laboratories Carrying Out the Determination of Mycotoxins—Method of Analysis to be Used by Control Laboratories

Methods of analysis used for food control purposes must comply whenever possible with the provisions of paragraphs 1 and 2 of the Annex to Council Directive 85/591/EEC.[14] The provisions of that Directive should apply to methods used by laboratories undertaking aflatoxin control analysis.

The specific criteria for methods of analysis used for the control of aflatoxin levels in some foods and milk are given in Table 3.4.

1. Values to apply to both B1 and sum of B1 + B2 + G1 + G2.
2. If sum of individual aflatoxins B1 + B2 + G1 + G2 are to be reported, then response of each to the analytical system must be either known or equivalent.
3. The detection limits of the methods used are not stated as the precision values are given at the concentrations of interest.

Table 3.4 *Specific criteria for methods of analysis used in monitoring aflatoxin levels*

Criterion	*Concentration range*	*Recommended value*	*Maximum permitted value*
Blanks	All	Negligible	—
Rccovcry—milks	0.01–0.05 μg/kg	60–120%	
	> 0.05 μg/kg	70–110%	
Recovery—foods*	< 1.0 μg/kg	50–120%	
	1–10 μg/kg	70–110%	
	> 10 μg/kg	80–110%	
Precision RSD_R	All	As derived from Horwitz equation	Two times value derived from Horwitz equation

Specific values for precision RSD_R at concentrations of interest for aflatoxin control purposes derived from the Horwitz equation are:

Precision RSD_R	0.03 μg/kg	76.7	153.4
Precision RSD_R	0.05 μg/kg	71.0	142.0
Precision RSD_R	0.1 μg/kg	64.0	128.0
Precision RSD_R	0.2 μg/kg	57.7	115.4
Precision RSD_R	0.3 μg/kg	54.2	108.4
Precision RSD_R	1.0 μg/kg	45.3	90.6
Precision RSD_R	2.0 μg/kg	40.8	81.6
Precision RSD_R	4.0 μg/kg	36.7	73.4
Precision RSD_R	5.0 μg/kg	35.5	71.0
Precision RSD_R	10.0 μg/kg	32.0	64.0
Precision RSD_R	20.0 μg/kg	28.8	57.6
Precision RSD_R	50.0 μg/kg	25.1	50.2

(Precision RSD_r may be calculated as 0.66 times Precision RSD_R at the concentration of interest)

* See note 1 on p. 54

4. The precision values are calculated from the Horwitz equation, *i.e.*:
   $RSD_R = 2^{(1-0.5 \log C)}$
   where:
   RSD_R is the relative standard deviation calculated from results generated under reproducibility conditions $[(s_R/\bar{x}) \times 100]$;
   C is the concentration ratio (*i.e.* 1 = 100g/100g, 0.001 = 1000 mg/kg)
   This is a generalised precision equation which has found to be independent of analyte and matrix but solely dependent on concentration for most 'routine' methods of analysis.

The effect of the criteria of Table 3.4 is that laboratories may use any method of analysis provided it has been validated as being capable of performing to these standards. A number of specific methods have been shown to meet the required precision.

Recovery Calculation

The analytical result shall be reported corrected for recovery provided the recovery obtained falls within the range given above. Due note of the (draft) Harmonised Guidelines for the Use of Recovery Information in Analytical Measurement (see Chapter 4) developed under the auspices of IUPAC/ISO/AOAC shall be taken.

Part B: The Requirements for and Procedures to Obtain Methods that have been Fully Validated Through Collaborative Trials

As seen from the above, all 'official' methods of analysis are required to include precision data; such data can only be obtained through a collaborative trial and hence the stress that is given to collaboratively tested and validated methods in the food sector.

6 What is a Collaborative Trial?

A collaborative trial is a procedure whereby the precision of a method of analysis may be assessed and quantified. The precision of a method is usually expressed in terms of repeatability and reproducibility values. Accuracy is not the objective. Because of the importance of collaborative trials they are described in greater detail below using the Outline and Recommendations from the IUPAC/ISO/AOAC Harmonised Protocols.

IUPAC/ISO/AOAC International Harmonisation Protocol

Recently there has been progress towards a universal acceptance of collaboratively tested methods and collaborative trial results and methods, no matter by whom these trials are organised. This has been aided by the publication of the IUPAC/ISO/AOAC International Harmonisation Protocols on Collaborative Studies.[5,10] These Protocols were developed under the auspices of the International Union of Pure and Applied Chemistry (IUPAC) aided by representatives from the major organisations interested in conducting collaborative studies. In particular, from the food sector, the AOAC International, the International Organisation for Standardisation (ISO), the International Dairy Federation (IDF), the Collaborative International Analytical Council for Pesticides (CIPAC), the Nordic Analytical Committee (NMKL), the Codex Committee on Methods of Analysis and Sampling and the International Office of Cocoa and Chocolate were involved.

The Protocols give a series of recommendations, the most important of which are described below:

The Components That Make Up A Collaborative Trial

Participants

It is of paramount importance that all the participants taking part in the trial are competent *i.e.* they are fully conversant and experienced in the techniques used in the particular trial, and that they can be relied upon to act responsibly in following the method/protocol. It is the precision of the method that is being assessed, not the performance of the trial participants. However, to use only laboratories who are 'experts' in the use of the method being considered but which will not ultimately be involved in the routine use of the method could give an exaggerated precision performance for the method. Thus it is not expected for participants to become involved in a trial unless they are likely to have to use the method on a subsequent occasion.

The number of participants must be at least eight. Only when it is impossible to obtain this number (*e.g.* if very expensive instrumentation or specialised laboratories are required) may the trial be conducted with fewer, but with an absolute minimum of five, laboratories. Ideally, if the method being assessed is intended for international use, laboratories from different countries should participate in the collaborative trial.

Sample type

These should, if possible, be normal materials containing the required levels of analyte. If this is not possible then alternative laboratory prepared samples must be used. These have the obvious disadvantage of being a different analyte/sample matrix from that of the normal materials. The types of samples used should cover as wide a range as possible of sample types for which the method is to be used. Ideally for collaborative trial purposes, the individual sample/sample types should be identical in appearance so as to preserve sample anonymity.

Sample homogeneity

The bulk sample, from which all sub-samples which form the collaborative trial samples are taken, should be homogenous and the sub-sampling procedure such that the resulting sub-samples have an equivalent composition to that of the original bulk sample. Tests should also be carried out to determine the stability of the sample over the intended time period of the trial. The precision of the method can only be as good as the homogeneity of the samples allows it to be; there must be little or no variability due to sample heterogeneity.

Sample plan—Number of materials (material—analyte/concentration level/ matrix combination)

For a trial of method performance, at least five samples covering the range of analyte concentration representative of the commodity should be used. These samples should be duplicated, making a minimum of ten samples in all. The samples should be visually identical and given individual sample codes to preserve sample anonymity as far as possible. Only when a single-level specification is involved for a single matrix may this minimum required number of materials be as low as three.

In the case of blind duplicates, it may be possible to link the pairs together. This possibility may be reduced by the use of:

Split level samples, where the samples differ only slightly in concentration of analyte and now any artificial attempt to draw these 'duplicate' samples closer together will in fact give decreased precision as it is the difference *between* the split level samples that is involved in the calculation of the precision parameter, and this will be a quantifiable amount.

Sometimes it is not feasible to prepare blind duplicates or split level samples and laboratories are then requested to analyse samples as known duplicates. This is the least favoured option for collaborative trial design.

Number of Replicates

The repeatability precision parameters must be estimated by using one of the following set of designs (listed in approximate order of desirability):

Split level (single or double)

For each level which is split and which constitutes only a single *material* for purposes of design and statistical analysis, use two nearly identical materials that differ only slightly in analyte concentration. For the single split level, each laboratory is to make only one determination on each (split) level (total for two per material); for the double split level, two known (non-blind) determinations are made on each (split) level (total for four per material). Alternatively, for the double split level, the two replicates for each (split) level may be submitted as blind replicates [one determination on each portion submitted (total for four per material)].

Note: the double split level procedure is no longer recommended following a revision of the requirements in the 1995 Harmonised Protocol.

Combination blind replicates and split level

Use split levels for some materials and blind replicates for other materials in the same trial (single values from each submitted portion).

Blind replicates

For each material, use blind identical replicates; when data censoring is impossible (*e.g.* automatic input, calculation and printout), non-blind identical replicates may be used.

Known replicates

For each material, use known replicates (two or more analyses of portions from the same test sample), but only when it is not practical to use one of the preceding designs.

Independent replicate analyses

Use only a single portion from each material (*i.e.* do not replicate) in the collaborative trial, but rectify the inability to calculate repeatability parameters by quality control parameters and other within-laboratory data obtained independently of the collaborative trial.

The method(s) to be tested

The method(s) to be tested should have been tested for robustness (*i.e.* how susceptible the procedure is to small variations in method protocol/instructions) and optimised before circulation to participants. A thorough evaluation of the method at this stage can often eliminate the need for a retrial at a later stage.

Pilot trial/pre-trial

These are invaluable exercises if resources permit. A pilot trial involves at least three laboratories testing the method, checking the written version, highlighting any hidden problems that may be encountered during the trial and giving an initial estimation of the precision of the method. A pre-trial can be used in place or to supplement the pilot trial; it is particularly useful where a new method or technique is to be used. Participants are sent one or two reference samples with which to familiarise themselves with the method before starting the trial proper. A successful pre-trial is usually a prerequisite to starting the trial proper and can be used to rectify any individual problems that the laboratories are having with the method.

The trial proper

Participants should be given clear instructions on the protocol of the trial, the time limit for return of results, the number of determinations to be carried out per sample, how to report results, to how many decimal places, as received or on a dry matter basis, corrected or uncorrected for recovery, *etc.*

Statistical analysis

It is important to appreciate that the statistical significance of the results is wholly dependent on the quality of the data obtained from the trial. Data which contain obvious gross errors should be removed prior to statistical analysis. It is essential that participants inform the trial co-ordinator of any gross error that they know has occurred during the analysis and also if any deviation from the method as written has taken place. The statistical parameters calculated and the outlier tests carried out are those used in the internationally agreed Protocol for the Design, Conduct and Interpretation of Collaborative Studies.[5,10]

In the statistical analysis of the collaborative trial data, the required statistical procedures listed below must be performed and the results reported. Supplemental, additional procedures are not precluded.

The statistical analysis recommended in the Harmonised Protocols is outlined in the flowchart given as Figure A1.

A detailed description of the exact procedure to be used for the analysis of data from a collaborative trial is given in Appendix IV to this chapter.

Valid data

Only valid data should be reported and subjected to statistical treatment. Valid data are those data that would be reported as resulting from the normal performance of laboratory analyses; they do not include results subject to method deviations, instrument malfunctions, unexpected occurrences during performance, or by clerical or typographical errors, *i.e.* aberrant data.

One-way analyses of variance

One-way analyses of variance and outlier treatments must be applied separately to each material to estimate the components of variance and repeatability and reproducibility parameters.

Initial estimation

Calculate the mean, $\bar{x}$ (average of laboratory averages), repeatability relative standard deviation, RSD_r, and reproducibility relative standard deviation, RSD_R, with no outliers removed, but using only valid data.

Outlier treatment

More time has been devoted by the analytical community to the procedures to be used for the removal of outliers than is warranted. However, the procedure prescribed in the Harmonised Protocol is now accepted within the food sector. The estimated precision parameters that must be reported are based on the initial valid data with all outliers identified by the outlier procedure described in the 1988 Harmonised Protocol. This procedure essentially consists of sequential application of the Cochran and Grubbs tests [at 1% probability (P) level, 1-tail for Cochran, 2-tail for single Grubbs, overall for paired Grubbs] until no further outliers are flagged or until a drop of more than 22.2% ($= \frac{2}{9}$) in the original number of laboratories would occur.

The Grubbs tests are to be applied to one material at a time to the set of replicate means from all laboratories, and not to individual values from replicated designs, because their differences from the overall mean for that material are not independent.

The 1995 Protocol recommends the same outlier tests but at a 2.5% probability level. This has resulted in different probability levels for outlier tests now being recommended by different organisations, *i.e.* either 1% (*e.g.* ISO) or 2.5% (*e.g.* AOAC International and Codex).

Analysis of data for outliers—Cochran test

First apply the Cochran outlier test (1-tail test at $P = 1\%$ or 2.5%) and remove any laboratory whose critical value exceeds the tabulated value given in the Harmonised Protocol but stop removal if more than 22.2% (2 of 9 laboratories) would be removed.

This measure compares the ratio of the largest variance between individual replicates with the sum of the individual variances. Because collaborative trials are only concerned with duplicates, then squared differences can be used instead of variances. The calculated Cochran's value is compared with the tabulated value (1-tail $P < 1\%$, but the value of $P < 2.5\%$ will probably become generally accepted in future). If the calculated value is greater than the tabulated value, then the duplicate results from that laboratory for that sample are outliers and are not used in the calculation of the precision parameters.

If the Cochran's test is positive then further Cochran's tests are carried out on the remaining data until no outliers are detected or until 22.2% of the data are removed.

After a negative Cochran's test, the data are analysed for differences in the means of the duplicates from the different laboratories using Grubbs's test.[5] This may be either Grubbs's Single Value Test (initially) and then Grubbs's Paired Value Test or 'Double Grubbs'.

Analysis of data for outliers—Grubbs's test

Apply the single-value Grubbs test (2-tail) and remove any outlying laboratory; if no laboratory is identified, then apply the pair-value test (two values at the same end and one value at each end, $P = 1\%$ or 2.5% overall). Remove any laboratory(ies) identified by these tests, using the values given in the Harmonised Protocol, but stop removal if more than 22.2% (2 of 9 laboratories) would be removed by both outlier tests.

In the Grubbs's Single Value Test the means of the duplicates of each laboratory are calculated and the standard deviation of these means(s) calculated. Standard deviations are also calculated for the means omitting the lowest mean (s_l) and the highest mean (s_h). The percentage decrease in standard deviation is calculated using s_l and s_h respectively. The value which gives the greatest percentage decrease is then compared with the tabulated value.

In the Grubbs's Paired Value Test, the % decrease in standard deviation is calculated for the means excluding the two lowest results (s_{ll}) and the two highest results (s_{hh}), and the highest and lowest results (s_{hl}). The value which gives the greatest % decrease is compared with the tabulated value at the 1% or 2.5% level.

When an outlier is identified, return to the start of the scheme and re-test for Cochran's and Grubbs's outliers as before.

It is essential that no more than 22.2% of the original data should be removed as outliers; stop the outlier procedure as soon as that point is reached.

Repeatability and reproducibility parameters are then calculated from the remaining data.

Precision parameters

The most commonly quoted precision parameters are the repeatability and reproducibility. Typical definitions of these are given in Chapter 14. These are then calculated using the procedures described in Appendix IV to this chapter.

7 Final Report

The final report should be published and should include all valid data.

8 Other Points of Note

When s_r is negative

By definition, s_R is greater than or equal to s_r in collaborative trials; occasionally the estimate of s_r is greater than the estimate of s_R (the range of replicates is greater than the range of laboratory averages and the calculated s_L^2 is then negative). When this occurs, set $s_L = 0$ and $s_R = s_r$.

9 Other Procedures for the Validation of a Method of Analysis

International organisations are now considering other procedures for the validation of methods of analysis besides a complete collaborative trial. Instructions on the use of results from proficiency testing schemes have been developed by the UK Ministry of Agriculture, Fisheries and Food[15] and there is now a procedure for the validation of test kit methods developed under the 'MicroVal' project.

These are outlined below:

UK Ministry of Agriculture, Fisheries and Food Validated Methods of Analysis: Results from Proficiency Testing Schemes

Introduction

The purpose of a proficiency testing scheme is to test the competence of the laboratory and not to validate a method of analysis. In most proficiency testing schemes, participants have a free choice of method of analysis and so there is no opportunity to validate a method formally using a proficiency testing scheme, *i.e.* a multiplicity of methods may be used by participants. However, in some situations, there is the possibility of validating a method of analysis if:

1. there are sufficient participants in the proficiency testing scheme who choose to use the same defined method of analysis; or
2. a method of analysis is prescribed by the scheme co-ordinators.

For most proficiency testing schemes, it is the former situation which will predominate, but in the case of microbiological proficiency testing schemes in particular, it is very likely that a method of analysis will be prescribed. The former situation will tend to occur when a very empirical determination is being assessed in the proficiency testing scheme. The procedure to be adopted is outlined in the introduction to the MAFF Validated Methods for the Analysis of Foodstuffs Series.[15]

Organisation of Proficiency Testing Scheme

The proficiency testing scheme whose results are to be used must conform to the AOAC/ISO/IUPAC International Protocol on the organisation of proficiency testing schemes for the results to be recognised by MAFF.

The AOAC/ISO/IUPAC Protocol stipulates the procedures that must be incorporated in any proficiency testing scheme, and in particular the work that must be carried out by the scheme co-ordinators to ensure that samples received by the participants in the scheme are homogeneous. At present there are no internationally agreed stipulations to ensure sufficient homogeneity of samples used for collaborative trial exercises, whereas there are for proficiency test exercises.

Number of Samples

In the case of a collaborative trial conforming in design to the Harmonised Protocol a minimum of five test materials are required to be prepared. However, in most proficiency testing schemes there are insufficient test materials sent out in any one round (*i.e.* the dispatch of test material at a specific time) to meet the minimum requirements for the number of materials as specified in the Harmonised Guidelines for Collaborative Studies. Because of that it is necessary to 'build up' the number of samples used to validate a method over a period of time. This may mean that the time taken to accumulate sufficient results to ensure validation of a method may extend over one or two years, depending upon the test materials which are being used in the proficiency testing scheme.

Replication of Results

In most cases the validation of a method by a collaborative trial results in both within- and between-laboratory precision characteristics (*i.e.* repeatability and reproducibility). Most collaborative trials are dispatched as either blind duplicates or as split level test materials. This means that one of the aims of a collaborative trial, that of determining the within-laboratory variability, is readily achievable. However, because the aim of a proficiency test is different from a collaborative trial, the results of replicate analyses of any particular test material are normally not reported to the proficiency test co-ordinator, *i.e.* it is only the single result, as reported to the customer, that is returned. Because of that, it is frequently the case that it is not possible to obtain the within-laboratory variability of the method. In such cases only the overall precision of the method *i.e.* the between-laboratory precision, will be quoted and not the within-laboratory precision in the MAFF Validated Method Series.

Laboratories

Because of the nature of a proficiency testing scheme the same laboratories will not necessarily participate in each round of the scheme. Thus it may be expected that more variability is to be obtained through assessment by using proficiency testing results.

An example of a method which has been standardised by this approach is given in the MAFF method for the enumeration of listeria monocytogenes.[16]

MicroVal Protocol for the Validation of Alternative Microbiological Methods

MicroVal—Validation of Microbiological Methods for Food and Drink in a European Perspective—is a project which commenced in June 1993. It was initially started as a Eureka project with the objective to develop a European standard (EN) giving the general principles for the validation of alternative methods in the field of microbiological analysis of food. Technical rules for validation criteria and general rules for the certification of microbiological alternative methods will be provided.

The principle aim of the project is to discuss new, alternative types of microbiological methods based on recent developments within the field of biotechnology, automation and micro-electronics. Such methods are often more rapid than traditional procedures. The scope of the project extends to the analysis of micro-organisms, bacterial toxins and mycotoxins with emphasis on the two first areas. Guidance will be given for both quantitative and qualitative methods. The MicroVal protocols that are being developed are to be completed in 1998.

10 Assessment of the Acceptability of the Precision Characteristics of a Method of Analysis: Calculation of HORRAT Values

There is no formal requirement in the European Union or in Codex as to the acceptability of the precision characteristics of any particular method. However, the calculated repeatability and reproducibility values can be compared with existing methods and a comparison made. If these are satisfactory then the method can be used as a validated method. If there is no method with which to compare the precision parameters then theoretical repeatability and reproducibility values can be calculated from the Horwitz equation.[17] This is best achieved by the use of HORRAT values to give a measure of the acceptability of the precision characteristics of a method.

The HORRAT value is:

(RSD_R derived from the collaborative trial)/(RSD_R predicted from the Horwitz equation).

Thus, Ho_R, the HORRAT value for reproducibility, is the observed RSD_R value divided by the RSD_R value calculated from the Horwitz equation at the concentration of interest.

Interpretation

If the HORRAT values are two or less, then the method may be assumed to have satisfactory reproducibility values. Laboratories should ensure that the methods which they employ meet this criterion.

Calculation of the Horwitz Value

The Horwitz value is derived from the Horwitz trumpet and equation, which states that for any method:

$$RSD_R = 2^{(1-0.5 \log C)}$$

and that the value is independent of matrix/analyte.

The major values are given in Table 3.5.

Table 3.5 *Horwitz values*

Concentration ratio	*RSD_R*
1 (100%)	2
10^{-1}	2.8
10^{-2} (1%)	4
10^{-3}	5.6
10^{-4}	8
10^{-5}	11
10^{-6} (ppm)	16
10^{-7}	23
10^{-8}	32
10^{-9} (ppb)	45

Horwitz has derived the equation after assessing the results from many (~3000) collaborative trials. Although it represents the average RSD_R values and is an approximation of the possible precision that can be achieved, the data points from 'acceptable' collaborative trials are less than twice the predicted RSD_R values at the concentrations of interest. This idealised smoothed curve is found to be independent of the nature of the analyte or of the analytical technique that was used to make the measurement. In general, the values taken from this curve are indicative of the precision that is achievable and acceptable of an analytical method by different laboratories. Its use provides a satisfactory and simple means of assessing method precision acceptability.

This procedure is increasingly being used by organisations to assess the acceptability of the precision characteristics of their methods.

Part C: The Requirements for and Procedures to Obtain Methods that have been Validated 'In-house'

It has been stipulated in Chapter 1 that methods of analysis used in the food sector should, wherever possible, be fully validated, *i.e.* have been subjected to a collaborative trial. There are, however, many situations where this is not feasible or such methods are not available. As a result, the need for laboratories to develop and use their own in-house methods of analysis is well recognised in the analytical community. Until recently, such in-house method validation has been undertaken on an *ad hoc* basis. It has become recognised that such validation should be carried out on a more formal basis and a number of organisations have developed procedures and protocols which meet such needs. The Codex Alimentarius Commission has agreed that the topic should be included in its formal work programme.[18] In particular, it is recognised within the Codex system that there are some sectors of food analysis, *e.g.* the veterinary residues in food sector, where it is very unlikely that fully validated methods are, or are likely to become, available. The situation within the Veterinary Drug Residues in Food Codex Committee has been described by the Australian Delegation to that Committee.[19]

Validation can be defined as the process of determining the suitability of a measurement system for providing useful analytical data. The term interlaboratory comparison (*i.e.* collaborative trial) is often taken to be synonymous with method validation in the food sector. According to ISO/IEC Guide 25, this is but one of a number of ways of validating analytical methods. The others include one or more of the following:

- calibration using references or reference materials;
- comparison of results achieved with other methods;
- systematic assessment of the factors influencing the result; and
- assessment of the uncertainty of the results based on scientific knowledge and practical experience.

It is the current procedures which are being developed for formal In-House Method Validation which are described in Part C of this chapter.

11 Protocols for the In-house Validation of Analytical Methods

There are a number of authoritative texts which have been developed for the in-house validation of analytical methods, and in particular:

- a Protocol on the Validation of Chemical Analytical Methods developed by the Nordic Committee on Food Analysis;[20]
- a generic laboratory guide developed by EURACHEM produced by the

UK Laboratory of the Government Chemist with the support of the UK Department of Trade and Industry Valid Analytical Measurement Initiative;[21]

- an Interlaboratory Analytical Method Validation Short Course developed by the AOAC International;[22]
- a guide to the Validation of Methods developed by the Dutch Inspectorate for Health Protection;[23] and
- a guide to Analytical Quality Assurance in Public Analyst Laboratories prepared by the UK Association of Public Analysts.[24]

The NMKL protocol is described in detail in Appendix I of this chapter as it serves as a good reference procedure for those food analysts who have to validate their methods to an internationally recognised protocol. Elements of the EURACHEM Guide (see Appendix II) is given as additional aid to analysts who have to develop their own in-house validation procedures.

The contents of both the AOAC International Short Course and the Dutch Guide to the Validation of Methods are given in Appendix III.

APPENDIX I

Validation of Chemical Analytical Methods as Adopted and Published by the Nordic Committee on Food Analysis as NMKL Procedure No. 4, 1996

Foreword

This procedure was elaborated by a working group established by the Nordic Committee on Food Analysis at its Annual General Meeting on the Faroe Islands in 1995.

The following persons were members of the working group:

Denmark:	Inge Meyland, The National Food Agency
	Lisbeth Lund, DANAK (observer)
Finland:	Harriet Wallin, VTT Biotechnology and Food Research
Iceland:	Arngrmur Thorlacius, Agricultural Research Institute
Norway:	Kåre Julshamn, Institute of Nutrition, Directorate of Fisheries (chairman)
Sweden:	Anders Nilsson, The National Food Administration

The working group extends its thanks to the Institute of Nutrition, Directorate of Fisheries in Bergen and to the National Food Administration in Uppsala for making available to the working group their internal procedures as discussion papers.

The working group also extends its thanks to the Nordic accreditation bodies DANAK (Denmark), FINAS (Finland), Norsk akkreditering, NA (Norway) and SWEDAC (Sweden) which forwarded valuable comments to this procedure.

This procedure is available from the General Secretariat of NMKL, c/o VTT Biotechnology and Food Research, PB 1500, FIN-02044 VTT, Finland, tel. +358 9 4565164, fax +358 9 4552103, e-mail: harriet.wallin@vtt.fi

NMKL invites all readers and users of this procedure to submit comments and views. Suggestions and comments should be sent to the General Secretariat of NMKL (address above).

Contents

1. Introduction
2. What is understood by validation of a chemical analytical method?
3. Description of a validation and verification procedure
   3.1 Protocol for a validation and verification
   3.2 Specificity and standard curve
   3.3 Precision
   3.4 Trueness
   3.5 Concentration or measurement range
   3.6 Limit of detection
   3.7 Limit of quantification/limit of determination
   3.8 Robustness
   3.9 Sensitivity
   3.10 Evaluation of validation results
4. Documentation of validation or verification
5. Monitoring
   5.1 Continuous monitoring
   5.2 Monitoring as a result of altered circumstances in the performance of the method
6. References

1 Introduction

In order to create confidence in the analytical results reported by a laboratory, the analytical methods employed must fulfil certain quality requirements. By validation is here understood the measures taken in order to test and describe whether a method in respect to its accuracy, use, implementation and sources of errors operates at all times in accordance with expectations and laid down requirements.

The regulations of the European Union require that, whenever possible, validated methods are employed in official food control. This requirement is also applicable in laboratories which are accredited or intend to seek for accreditation, irrespective of whether they are associated with official food control or not. In this connection it is emphasised that this procedure in no way should replace method–performance studies where such studies are possible. Laboratories should always give priority to methods which have been tested in method–performance studies if such are available, unless the laboratory has specific reasons for selecting other methods. When studied methods are not available, other methods must be used, which must be validated internally. A method tested in a method–performance study must not be taken into routine use before the laboratory has demonstrated and documented (verified) that the method is suitable for the analytical task. In other words it should be ensured that the method is 'fit for its intended purpose', not only that the laboratory can achieve the performance characteristics stated in the method.

All laboratories of a certain size develop methods and thereby have considered quality requirements of analytical methods. It is a prerequisite that laboratories working with internal validation of methods employ competent staff and have access to adequate equipment required for the validation. The aim should be that this important work, as carried out in different laboratories, should be performed as similarly as possible and according to defined guidelines. This procedure describes the parameters to be investigated, their definitions, the way in which the work should be carried out and criteria for acceptance of methods.

The users of this procedure will represent laboratories of various sizes having different resources. Laboratories thus have different prerequisites and possibilities, and the requirements on the performance of the validation of analytical methods will naturally vary. The recommended acceptance criteria can and must be adjusted to the needs of the individual laboratory, which means that laboratories may deviate from the recommended criteria. Such changes should be documented in the form of laboratory-specific standard operating procedures which are made available to the staff.

2 What is Understood by Validation of a Chemical Analytical Method?

The working group found it suitable to include six different categories of analytical methods in use in chemical laboratories (Table A1). The categories

Table A1 *Different categories of analytical methods according to the degree of validation and recommended further work*

Degree of external validation	*Recommended internal validation*
The method is externally validated in a method–performance study	Verification of trueness and precision
The method is externally validated but is used on a new matrix or using new instrument(s)	Verification of trueness and precision, possibly also detection limit
Well established, but not tested method	Verification, possibly a more extensive validation
The method is published in the scientific literature and states important performance characteristics	Verification, possibly a more extensive validation
The method is published in the scientific literature without presentation of performance characteristics	The method needs to be fully validated
The method was internally developed	The method needs to be fully validated

are based on the available documentation on the degree of validation of the method. For the various categories, the working group recommends validation work to be performed in accordance with Table A1.

Explanations to terms used in Table A1:

The classification 'The method is externally validated in a method–performance study' indicates methods which have been tested in interlaboratory method–performance studies (previously designated collaborative studies) with an acceptable result, and which include information on central performance characteristics, such as working range, trueness, precision and limit of quantification. The study must have been conducted, and its results interpreted in accordance with internationally accepted protocols, such as ISO 5725 or the ISO/IUPAC/AOAC harmonised protocol.

It must be kept in mind that ISO and other organisations have published well-established standard methods, which have not been tested in interlaboratory studies.

By 'verification' is understood that the laboratory, prior to taking a method into routine use, tests and documents that it is competent to use the method. This means that the laboratory investigates and documents that it, for example for trueness and precision, obtains results corresponding to the performance characteristics of the method. Above all, laboratories must be able to demonstrate that the method is suitable to solve the analytical task in question.

By 'a more extensive validation' is understood that the laboratory examines and documents, fully or partly (see Section 3), the characteristics of the method before taking it into routine use.

By 'full validation' is understood a study including all the aspects given in Section 3, where relevant and possible.

3 Description of a Validation and Verification Procedure

This section contains recommendations on the conduct and sequence of the various steps of a validation and verification procedure. However, the work should always be adjusted to the possibilities of the laboratory and to the requirements of the method in question. It is not always relevant to study all the steps given in these recommendations.

Recommended procedures:

1. Design a validation or verification protocol (3.1)
2. Determine specificity and standard curve (3.2)
3. Determine precision, expressed as repeatability and reproducibility (3.3)
4. Determine trueness (3.4)
5. Determine working range/measuring range (3.5)
6. Determine detection limit (3.6)
7. Determine limit of quantification/Limit of determination (3.7)
8. Determine robustness (3.8)

9. Determine sensitivity (3.9)
10. Evaluate the results from steps 2–8 against the protocol (3.10)
11. Document the work in a report (see Section 4)

The field of application of the method will dictate the extent of the validation work. Methods to be used in the official control of food will, for example, require full validation in accordance with Table A1, since the analytical results must be equal, irrespective of from which laboratory they are reported. On the other hand, a simple verification may be sufficient in connection with in-house methods used in process control in industry. Analytical methods used to determine concentrations which are close to the detection limit of the method require more extensive validation than methods used to determine concentrations well above the detection limit. The extent and the emphasis of the work depends on whether the method is a quantitative or a qualitative method as well as on the principle of the method (chromatography, determinations using antibodies, *etc.*).

Validation or verification must always be carried out before taking a new method into routine use. Such work should be repeated in full or in part when the results of the first validation or verification indicate that the method needs to be modified, for example when a laboratory is unable to demonstrate that the method will solve the analytical problem.

Verification should also be repeated, for example, when:

- major instruments are replaced;
- new batches of major reagents are taken into use (for example, new batches of polyclonal antibodies);
- changes made in the laboratory premises may influence the results;
- methods are used for the first time by new staff;
- the laboratory takes an already validated and verified method into use after it has been out of use for a long time.

Validation and verification work should be clearly distinguished from the experimental work conducted in connection with the development of a method. During validation and verification work, it is strictly forbidden to introduce any changes into the method. If changes are made, the validation and verification must be repeated. After a method has been taken into routine use, its performance should be monitored according to the quality control measures as described in the quality manual of the laboratory, for example using suitable control charts [see NMKL Proedure No. 3 (1996)].

Performed validation and verification work must always be documented in such a way that the results can be checked later, and the work if necessary be repeated. The rules in section 5.4.4 of the EN 45001 standard must be observed.

The required documentation may be divided into four categories:

1. protocol and planning (see Section 3.1);
2. documentation of raw data from experimental work;

3. documentation of the evaluation of the results;
4. compilation of the results of validation or verification (see Section 3.10) into a report describing the results of the work (see Section 4).

It is the responsibility of the laboratory carrying out validation work to archive all pertinent documentation for as long as the method is in use and for the period of time the laboratory refers to the analytical results obtained using the method. This time period normally stretches over several decades.

3.1 Protocol for a Validation or Verification

Before starting the experimental work in the laboratory, a protocol or plan for the validation or verification should be established. The protocol should contain a compilation of all the requirements which the method must fulfil. The protocol may be regarded as a 'work order' concerning the tasks to be performed, and should be based on the following elements:

- the needs of the client;
- what is analytically achievable;
- conditions of the laboratory which have an impact on validation or verification, such as, for example, work environment and available equipment.

Concrete requirements should form the starting point of the protocol and should give answers to the following questions:

- *In what circumstances is the method to be used?* For example, official food control or production control. This has a bearing on the choice of analytical principle, and places requirements on reproducibility.
- *Is a qualitative or an exact quantitative result required?* This consideration affects requirements on specificity, limits of detection and quantification, *etc.*
- *In which chemical forms does the analyte occur?* Bound or free? As various chemical compounds? This has a bearing on the choice of sample matrices selected for the validation and on the design of recovery tests.
- *What is the matrix?*
- *Is there a risk of interference from the sample matrix or other components?* Influences selectivity and specificity, *etc.*
- *How much sample is usually available?* Are the samples, for example, homogenous? Influences requirements on reproducibility and repeatability, sensitivity, *etc.*
- *In which working range is the method to be used?* Close to the limit of determination or at higher concentrations? Influences the requirements on linearity, detection and quantification limits, reproducibility and repeatability, *etc.*

- *Can results be verified by using reference materials or by participating in proficiency testing schemes?* This has a bearing on the requirements of trueness, reproducibility, robustness, *etc.*
- *Which environmental requirements must be fulfilled?* Personal safety? Effects of the environment, such as extreme temperatures, particles in the air, *etc.* Use of chemicals toxic to the environment, such as ozone-depleting solvents, *etc.*? Are the requirements on robustness influenced, and thereby indirect requirements on staff, equipment, reagents, premises, *etc.*?
- *Which economical requirements must be fulfilled?* Sets limits for the entire validation and verification work.

The validation or verification plan should, in a concise manner, describe which validation elements are to be evaluated and the order in which this is to be done. The protocol should also state which results the laboratory wishes to achieve on the various elements. If there are no specified requirements on, for example, precision, the laboratory should state in its protocol what is analytically reasonable to require of the method in question.

If the laboratory is developing a totally new method, the protocol should also include the development work and aspects of the choice of the analytical principle, purchase of reasonable equipment, *etc.* In many cases these factors are already determined, and if the laboratory selects a method which has already been tested in an interlaboratory method–performance study, the protocol will only describe the steps of the verification work.

It is important that the protocol and the results of the experimental work are compared at all times. Have the required specifications for linearity, trueness, *etc.* been fulfilled? If the laboratory during the work detects considerable deviations in method performance as compared to specifications, the method must be modified, and the validation or verification work repeated.

The field of application is an important characteristic of most analytical methods. The evaluation of the field of application will depend on the type of method. For example, a method for the determination of vitamin D would require much more work in order to establish its suitability for all types of foods than would an atomic absorption spectroscopy method for the determination of calcium. Often work may be saved by allowing the needs of the laboratory, not the limitations of the method, to dictate the field of application. If a method is used solely for the analysis of meat products, it is unnecessary to investigate its applicability to, for example, dairy products. However, it is important that the final report of the validation or verification describes the nature of the samples tested, and the basis for the determined field of application.

Some analytical methods are self-defining and their field of application implicit. Thus, for example, the amount of Kjeldahl nitrogen equals the amount of nitrogen determined in accordance with a Kjeldahl determination, and the dry matter content and ash will usually depend on the method (drying at 105 °C, vacuum drying at 70 °C, freeze drying, ashing at 550 °C, *etc.*).

Sections 3.2 to 3.9 describe the elements which should be included in a full internal analytical method. If it is decided to omit some of these elements in an internal validation, the reason for doing so should be documented. Section 3.3 (precision) and 3.4 (trueness) describe the elements to be included in a verification. These should possibly be supplemented with elements from Section 3.6 (limit of detection); see Table A1.

3.2 Specificity and Standard Curve

Specificity

Definition
The ability of an analytical method to distinguish the analyte to be determined from other substances present in the sample.

Procedure
A blank sample and a sample to which a known amount of the analyte has been added may be analysed in order to check that there are no inerferences with the analyte from any expected compounds in the sample, degradation products, metabolites or known additives. In some cases, *e.g.* in the analysis of pesticides, a more concentrated extract of the blank may be analysed in order to demonstrate that no signals occur.

Specificity may also be examined by carrying out determinations in the presence of substances suspected of interfering with the analyte. The analyst should be aware that the analyte might be present in the sample in more than one chemical form.

By experience, the analyst will often be aware of the kind as well as the nature of interferences to be expected in connection with specific methods. For example, spectral inerferences occur in inductively coupled plasma (ICP) determinations, whereas in chromatographic measurements, several substances may have the same retention time. Such problems may have unfortunate consequences for the analytical results, and these techniques therefore require more extensive examination of specificity than techniques associated with none or only a few known interferences.

Self-defining methods need not be investigated for specificity; for example, certain substances will be lost when drying at 105 °C, but that fraction is of no interest in cases where dry matter is defined as the matter remaining after such treatment.

Standard Curve

Definition
The standard curve reflects the ratio between amount/content of the analyte in a sample solution and the resulting measurement response.

Procedure

Measurement response should be determined using at least six different measurement points. Determinations should be made on reference samples or on blank samples to which the analyte has been added in evenly distributed concentrations covering the entire working range. The concentrations should be selected on both sides of the active working range.

The experiment should be repeated at least once. The results may be graphically presented, and the equation should be given for the linear regression as well as the correlation coefficient as a measure of the distribution ($R > 0.999$).

When linearity cannot be achieved, the standard curve in the relevant concentration range should be based on sufficient points to determine accurately the non-linear response function. The so-called curvature coefficient is a better measure of the linearity than the correlation coefficient. The curvature is estimated by adjusting the results to the function:

$$R = k \times c^n$$

where R is the response corrected for the blank, c is the concentration, k is the estimate of the sensitivity, and n is the curvature coefficient.

The function is converted to the logarithmic form:

$$\log(R) = \log(k) + n \log(c)$$

A linear regression of $\log(R)$ against $\log(c)$ will then give n as the angular coefficient. The value of n should be above 0.9 and below 1.1.

3.3 Precision

Definition

Precision is the degree of agreement between independent analytical results obtained under specific circumstances. Precision depends only on the distribution of random errors, and is not associated with the true value. Precision is usually expressed as the standard deviation of the analytical result. Low precision gives a high standard deviation. Precision is an important characteristic in the evaluation of all quantitative methods.

The precision of a method depends very much on the conditions under which it has been estimated. Repeatability and reproducibility conditions represent clearly different conditions, while internal reproducibility lies in between these two extremes.

Repeatability or reproducibility

An estimate of the repeatability of a method is acquired when analytical results are obtained on identical test portions, in the same laboratory, using the same equipment and within a short period of time. The reproducibility of the method can be estimated on the basis of results obtained when the method has been used to analyse identical test portions in different laboratories using different equipment. By internal reproducibility is understood the agreement of results when the analysis has been carried out in the same laboratory, using the same method, but performed at different times by different analysts using, for example, different batches of reagents.

It is important that it is exactly documented how, and on what materials, repeatability and internal reproducibility have been estimated (reference materials, control materials, authentic samples, synthetic solutions). In cases where the method under study is intended to be used over a large concentration range, the precision should be estimated at several concentration levels, *e.g.* at low, medium and high levels. It is recommended that the estimation of precision is repeated in whole or partly during validation work.

It should be kept in mind that the precision depends very much on the concentration of the analyte and on the analytical technique.

Procedure

Make at least ten determinations from the same sample material. The conditions may be those of repeatability, reproducibility or something in between. Calculate the standard deviation either from results of replicate single determinations or from duplicate determinations.

The standard deviation of single determinations is an estimate of the distribution around the average, whereas the standard deviation derived from duplicate determinations is an estimate of the variation between two single determinations.

Single determinations: Carry out for example at least ten determinations, and calculate the standard deviation, s, from the obtained individual results, x_i, according to the formula:

$$s = \sqrt{\frac{\sum(x_i - \bar{x})^2}{n - 1}}$$

where $\bar{x}$ is the average, i is 1, 2, ..., n, and n is the number of determinations.

Duplicate determinations: A duplicate determination consists of two determinations a_i and b_i. Carry out two determinations and calculate the standard deviation, s, based on d differences $(a_i - b_i)$ according to the formula:

$$s = \sqrt{\frac{\sum(a_i - b_i)^2}{2d}}$$

where i is 1, 2, ..., d, and d is the number of duplicate determinations.

Precision may be expressed either directly as the standard deviaion, as the relative standard deviation, RSD, or as the confidence interval.

The relative standard deviation is given by: RSD (%) = $(s/\bar{x}) \times 100$

The confidence interval is given by: $x \pm D$ where $D = 3s$

The coefficient 3 in the formula of the confidence interval is a good approximation for 10 to 15 measurements and a confidence interval of 99%. If the precision estimate is based on a different number of measurements, it is recommended that, instead of the coefficient of 3, a value from a two-sided t-test table is used, especially if the number of observations is fewer than 10.

For information, Table A2 contains typical values of acceptable relative standard deviations for repeatability, based on the concentrations of the analyte (*Pure Appl. Chem.*, 1990, **62**, 149).

Table A2 *Recommended acceptable relative standard deviation for repeatability for different analyte concentrations*

Analyte concentration	*RSD (%)*
100 g/kg	2
10 g/kg	3
1 g/kg	4
100 mg/kg	5
10 mg/kg	7
1 mg/kg	11
100 μg/kg	15
10 μg/kg	21
1 μg/kg	30
0.1 μg/kg	43

The above concerns quantitative determinations. In the case of qualitative methods, the precision must be approached in a different manner since a qualitative analysis in practice is a yes/no measurement on a given threshold concentration of the analyte. The precision of a qualitative analysis may be expressed as the ratio of true or false positive or negative results, respectively. These ratios should be determined at different concentrations of the analyte, at and above the threshold concentrations of the method. The result should be verified by using a separate reliable method, if available. If such an alternative reliable method is not available for comparison, a recovery test using suitable 'spiked/not spiked samples' of suitable matrices may be used.

% false positive = (false positive × 100)/sum of the known negative

% false negative = (false negative × 100)/sum of the known positive

3.4 Trueness

Definition

Trueness is the agreement between the sample's true content of a specific analyte and the result of the analysis. In this context a result is to be understood as either the average of duplicate determinations or a single result, depending on what is normal for the method. A distinction should be made between different analytical methods, on the basis of whether or not there are available reference methods and/or certified reference materials as well as proficiency testing schemes.

Alternative definition

The degree of agreement of the true content of the analyte in the sample and the average obtained as a result of several determinations.

The true content of the sample will always be unknown. In order to evaluate the trueness of a method, laboratories should depend on accepted norms such as a certified content of a reference material, results obtained using a validated method or a result obtained in a proficiency test. In all three cases, several laboratories, analysts and instruments should be involved in order to minimise the errors.

Determinations for which Certified or Other Reference Material is Available

Procedure

Investigate the trueness of the method by analysing a certified reference material. If a certified reference material is not available, use other reference materials, or a material prepared in the laboratory (in-house material) containing a known amount of the analyte. The content of the analyte in in-house control materials must be thoroughly investigated, preferably using two or more analytical methods based on different physical chemical principles and, if possible, based on determinations carried out in different laboratories. An in-house material should, whenever possible, be calibrated against a certified reference material. Carelessness in the assessment of the analyte concentration of an in-house control material is a serious loophole in the quality assurance of the laboratory.

Commentary

A certified reference material or an in-house control material must have an analyte concentration which is similar to the concentration level occurring in authentic samples. Certified reference materials and other control materials provide an estimate of the performance of the method in the examined range only. The analysis of reference materials is of little value if the reference material represents a different analyte level. For example, it is not possible to test the trueness of a method in the range from 0.1 μg/kg to 0.1 mg/kg using a reference material having an analyte concentration of 1 mg/kg.

Attention should also be paid to the fact that certified reference materials and other control materials are not always typical materials. Compared to authentic samples they are easier to handle, for example to extract and to ash. They are also more homogeneous than ordinary food samples. This may mean that results obtained from reference materials are often 'better' than those obtained from unknown samples, resulting in false security regarding the results. It is therefore not recommended to use results obtained on reference materials uncritically as evidence that the analysis of an unknown food sample has the same trueness. Often a reference sample included in an analytical run does not go through the same homogenisation steps as the unknown samples. This means that it is difficult to detect whether the samples have been contaminated during homogenisation.

Finally, it must be recognised that not all materials designated as 'certified reference materials' have the same high quality. Some caution is recommended.

With reference to what has been said above, the analysis of certified reference materials does not alone verify the trueness of an analytical method. The analysis of reference materials should be supplemented with other quality criteria, such as, for example, recovery tests (spiking).

Determinations for which Reference Methods are Available

Definition

A reference method is a method which has been studied in a method–performance study.

Procedure

Examine the trueness of the method by analysing the same samples with the method to be verified or validated and the reference method. If the reference method is not in routine use in the laboratory, it is not justified to introduce the method only in order to evaluate a new method. In such cases it is recommended that samples are sent to a laboratory having the necessary competence regarding the reference method, preferably an accredited laboratory.

Determinations for which Proficiency Tests are Available

Procedure

Examine the trueness of the analytical method by using the method when participating in a proficiency test including samples corresponding to those for which the candidate method is intended. Be aware that documened trueness holds only for relevant analyte levels and matrices.

Determinations for which No Reference Materials or Proficiency Tests are Available

Procedure

Determine the recovery by spiking samples with suitable amounts of a chemical containing the analyte. Recovery tests offer a limited control of the systematic error by checking the recovery of the analysis. The technique is especially useful in connection with unstable analytes, or when only a limited number of determinations are to be made. It is recommended that recovery tests are carried out in the relevant concentration ranges. If the method is used on several levels, recovery tests should be carried out on at least two levels. A known amount of the analyte is added to a test portion of a sample having a known concentration of the analyte. The added amount should correspond to the level normally present in the sample material. The test portion to which the analyte has been added is analysed along with the original sample with no added analyte. It is recommended to carry out ten replicates weighings, five of the material without the added analyte, and five with the added analyte. The averages of each series are calculated. The recovery (%) of the added analyte is calculated as 100 times the difference in averages divided by the amount of the added amount, *i.e.*:

$$[(\text{Analysed amount} - \text{Original amount in the sample})/(\text{Added amount})] \times 100$$

The obvious advantage of using recovery tests is that the matrix is representative. The technique may be used for all analytes and for most matrices, provided that the analyte is available in the laboratory as a stable synthetic compound. The greatest limitation is that there may be a difference in chemical form between the analyte in the authentic sample and in the added synthetic compound. Furthermore, it is almost impossible to check all relevant concentration ranges.

Where recoveries in the range of 80–110% are obtained, it is often sufficient to test for recovery as described above three times. The determinations should be carried out during a limited period of time. It generally holds that more determinations should be made the lower the analyte concentration is, because random errors increase with decreasing concentrations. Calculate the average and the standard deviation of the results obtained. Possibly apply the t-test in order to demonstrate whether or not the obtained recoveries are significantly different from 100.

It is important to evaluate the trueness of all quantitative methods. This also holds for self-defining methods, such as fat, *etc.* It will always be important that laboratories obtain results comparable to those obtained by other laboratories using the same method.

The requirements on trueness and precision will, in both cases, depend on the concentration level and on the aim of the determination. In the case of trace analysis, a deviation in trueness and precision of over 10%

is often acceptable, whereas such deviations will normally be totally unacceptable in determinations of, for example, Kjeldahl nitrogen or dry matter.

3.5 Concentration or Measurement Range

Definition
The range of the analyte concentration which has been experimentally demonstrated to fulfil the quality requirements of the method (see also precision and trueness).

All methods have a limited sensitivity which restricts the concentration range in which the method is applicable. The lower limit for reliable quantification is defined as the limit of quantification and is discussed in Section 3.7. The upper limit of quantification is especially important when using methods including standardisation. This is best illustrated by some examples. Spectrometric methods usually have a linear working range up to a certain concentration. At higher concentrations, the curve will bend towards the concentration axis. Unless the method is to be used for very limited purposes in a known concentration range, it is necesary to study the extent of the linear range. In addition, it is worthwhile to determine to what extent the requirements regarding precision and trueness are satisfied in this range. In practice, a limitation concerning the lowest part of the standard curve will be obtained in the form of a limit of quantification.

Ion-selective electrodes show a response which against the logarithm of the concentration exhibits a linear relationship over a wide concentration range (often from mg/kg to molar concentrations), but the curve will bend towards the y-axis when the concentration approaches zero. The limit of quantification may then be defined as the lowest region of the linear standard curve. Samples with very low concentrations may be lifted up to the linear range by adding known amounts of the analyte. During validation and verification it should also be ensured that requirements on precision and trueness are met within the linear range.

In some cases it may be practical to use a non-linear standard curve in order to expand the working range, or on the basis of the use of a detector having an implicit non-linear response (for example a flame photometric sulfur detector in gas chromatography). The problem will resemble a linear standardisation since it should be investigated in which concentration range the applied standardisation relation is valid and also ensured that other quality requirements of the method are fulfilled.

Gravimetric and titrimetric methods (among these some important analytical methods for the determination of dry matter, ash, fat and Kjeldahl nitrogen) do not require evaluations of linearity, but the limit of quantification should be determined if the methods are used for the determination of very low concentrations.

3.6 Limit of Detection

Definition
Quantitative determinations: The amount or the content of an analyte corresponding to the lowest measurement signal which with a certain statistical confidence may be interpreted as indicating that the analyte is present in the solution, but not necessarily allowing exact quantification.

Qualitative determinations: The threshold concentration below which positive identification is unreliable.

Procedure
Quantitative determinations: Determine the limiting amount or concentration of an analyte from the analysis of a number (> 20) of blank samples. The limit of detection is calculated as three times the standard deviation of the obtained average of the blank results (three standard deviations correspond to a confidence of 99%). Calculate the amount or content in the sample using a standard curve.

Some analytical methods will give no certain signal in the blank sample. In such cases it can be attempted to enlarge the instrumental noise and state the limit of detection as three times the standard deviation of the noise.

When enlarging instrument noise, extracts of the blank sample (concentrated blank samples) should be used. This extract should in addition be significantly more concentrated than extracts from normal samples in order to establish that there is no interference from any component.

The limit of detection is very important in trace elemental analysis. On the other hand, it will often be unnecessary to estimate this limit when evaluating methods for the determination of the principal components of foods.

Qualitative determinations: When evaluating qualitative methods, it is recommended that the range in which the method gives correct results (responses) is investigated. This may be carried out by analysing a series of samples, *i.e.* a blank sample and samples containing the analyte at various concentration levels. It is recommended that about ten replicate determinations are carried out at each level. A response curve for the method should be constructed by plotting the ratio of positive results against the conentration. The threshold concentration, *i.e.* the concentration at which the method becomes unreliable, can be read from the curve. In the example in Table A3, the reliability of a qualitative method becomes less than 100% at conentrations below 100 μg/g.

Table A3 *Example of results obtained for a qualitative method*

Concentration $\mu g/g$	*Number of determinations*	*Pos./neg. results*
200	10	10/0
100	10	10/0
75	10	5/5
50	10	1/9
25	10	0/10

3.7 Limit of Quantification/Limit of Determination

Definition

The limit of quantification of an analytical procedure is the lowest amount of analyte in a sample which can be quantitatively determined with a certain confidence. The limit can also be called the limit of determination.

Procedure

Using the method, analyse a number of blank samples (> 20). The limit of quantification is calculated as 10 times the standard deviation of the average of the blank samples. The amount or the content of the samples is determined from the standard curve.

In cases where the noise of the instrument is used as the basis of the determination of the limit of determination, see Section 3.5.

When evaluating quantitative methods, it is important to define and estimate the lowest concentrations which can be measured with a sufficiently defined precision. The procedure recommended above is based on the assumption that the precision at the limit of determination is equal to a blank analysis (background noise will dominate). A measurement corresponding to 10 standard deviations will then have a relative precision of 30% since the absolute precision (99% confidence, 10–15 measurements) is estimated to be three standard deviations.

3.8 Robustness

Definition

Robustness is defined as meaning the sensitivity of an analytical method to minor deviations in the experimental conditions of the method. A method which is not appreciably influenced is said to be robust.

Procedure

Some form of testing of robustness should be included in the evaluation of analytical methods elaborated in the laboratory. In the literature it is recommended to perform testing of robustness prior to the conduct of the method–performance testing. The nature of the analytical method will define which parameters of the methods need to be studied. The most frequently occurring parameters tested, and those which may be critical for an analytical method are: the composition of the sample (matrix), the batch of chemicals, pH, extraction time, temperature, flow rate and the volatility of the analyte. Blank samples may be used in robustness tests since these will reveal effects from, for example, the matrix and new batches of chemicals. Information gained in robustness testing may be used to specify the conditions under which a method is applicable.

3.9 Sensitivity

Definition
The sensitivity of a method is a measure of the magnitude of the response caused by a certain amount of analyte. It is represented by the angular coefficient of the standard curve.

Procedure
Determine the standard curve of the method using at least six concentrations of the standard. Calculate the angular coefficient by the least squares method.

3.10 Evaluation of Validation Results

After having carried out verification or validation work, the results obtained should be carefully compared against the verification or validation protocol as outlined under Section 3.1. The laboratory should at this stage evaluate whether or not the studied method meets the criteria of the protocol, for example in respect to working range, precision and trueness. If requirements are not met, the method should not be taken into routine use. Only after all predetermined criteria are met should the method be used.

4 Documentation of Validation or Verification

After completed verification or validation, a report of the work should be written. It is of particular importance that the report includes all raw data from the experimental work, or references to where such data can be found. It should be accurately specified which characteristics were studied, and if some aspects have not been included in the studies, then the underlying reasons should be documented. The report should contain the calculated results and it must be clear how the results have been calculated. It is important to describe all details, for example, the matrices and analyte levels included in the studies on precision, trueness and limits of detection and quantification.

The report should describe how the results obtained were evaluated against the analytical requirements laid down in the verification or validation protocol. Finally, the report should, in an unambiguous manner, conclude for which analytical problems the verified or validated method is appliable. If the method has been found not to be applicable to certain matrices or certain analyte levels, then this should be clearly stated in the report.

The validation or verification report should be filed together with the protocol for at least the period of time that the laboratory keeps the method in question on file. It goes without saying that the report, as well as reports from periodic follow-up of the validation/verification (see Section 5), must be kept available to all users of the method.

5 Monitoring

When a validated/verified method is taken into routine use in the laboratory, it is important that a competent analyst is made responsible for monitoring that the method continuously performs in accordance with the results obtained in the validation/verification. The follow-up may be carried out according to the guidelines given below.

5.1 Continuous Monitoring

The trueness and the precision of the method should be monitored continuously. Obtained results must agree with the validation/verification results. The responsible analyst should decide on the frequency of the monitoring and document this in the method or elsewhere.

5.2 Monitoring as a Result of Altered Circumstances in the Performance of the Method

Alterations in the circumstances under which the method is used may cause changes in the performance of the method. In such cases it may be necessary to repeat the validation of the method, partly or in full.

Minor modifications of the method: If the method has been modified, however slightly, it must be ensured that, after modification, it gives comparable or better results (*e.g.* limit of detection, specificity, trueness, precision) compared to those related to the validation/verification. All relevant matrices must be studied.

Method used on new matrix: If the method is to be used on a matrix which was not included in the validation/verification, the applicability of the method to the new matrix must be ensured by checking specificity, trueness and precision.

New chemicals: If important chemicals from, for example, a new supplier are taken into use, or large batch variations are expected, as is the case with, for example, polyclonal antibodies, it must be checked that parameters such as detection limit and sensitivity are unchanged.

New instruments: When new instruments are taken into use, it should be checked that there are no changes in working range, linearity, limit of quantification and precision.

New premises: If the carrying-out of the method is moved to new premises, it may be necessary to check that the environment has no effect, for example, on blanks, limit of quantification, linearity and precision. The steps to take depend on the analyte, the matrix and the instruments used.

New analyst: When new staff are to be authorised to perform the determinations, it should be ensured that the persons are sufficiently well versed for the task. This may be done by checking, for example, limit of quantification, trueness and precision.

6 References

Accreditation for Chemical Laboratories: Guidance on the Interpretation of the EN 45000 series of Standards and ISO/IEC Guide 25; Joint working group of WELAC (Western European Laboratory Accreditaion Co-operation) and EURACHEM. EURACHEM Guidance Docuent No. 1/WELAC Guidance Document No. WGD 2.

Harmonized Guidelines for Internal Quality Control in Analytical Chemistry Laboratories, *Pure Appl. Chem.*, 1995, **67**, 649.

Harmonized Protocols for the Adoption of Standardized Analytical Methods and for the Presentation of their Performance Characteristics, *Pure Appl. Chem.*, 1990, **62**, 149.

Method Validation—A Laboratory Guide, Eurachem Draft 1.0–3/96. Produced by LGC with the support of the UK DTI VAM initiative.

NMKL Procedure No. 3 'Control charts and control materials in the internal quality control in chemical food laboratories', The Nordic Committee on Food Analysis, Espoo (1996) pp 22, 5 appendices.

NMKL Report No. 8 'Quality Assurance Principles for the Chemical Food Laboratory', The Nordic Council of Ministers, Copenhagen, (1990) pp 65.

Precision of test methods—Determination of repeatability and reproducibility for a standard test method by inter-laboratory tests, ISO 5725, ISO, Geneva, 1986.

Protocol for the Design, Conduct and Interpretation of Method–Performance Studies, *Pure Appl. Chem.*, 1995, **67**, 331.

Quality manual of the Institute of Nutrition, Directorate of Fisheries, Bergen, Appendix 9:2 (1996) (in Norwegian).

APPENDIX II

Method Validation—A Laboratory Guide (EURACHEM)

The EURACHEM draft document is a laboratory guide to method validation, produced by LGC with the support of the UK DTI VAM initiative. Whilst it is generic in nature and not specific to the food sector, much of its content is relevant to the food sector. The aim of the document is described as being 'to explain some of the issues around method validation and increase the reader's understanding of what is involved and why it is important'. This information is presented in the context of the six principles of analytical practice identified in the Valid Analytical Measurement programme and in particular the principle that 'analytical measurements should be made using methods and equipment which have been tested to ensure they are fit for purpose'. It is intended to give guidance on methods testing.

The document addresses in turn the following points:

- what method validation is;
- why method validation is necessary;
- when methods should be validated;
- how methods should be validated;
- using and documenting validated methods; and
- using validation data to design QC.

The guide describes method validation as the set of tests used to establish that a method is fit for a particular purpose.

In presenting method validation within the context of good working practice, the guide stresses the need for the analyst to be competent and for equipment to be used appropriately. The interrelationship of the process of method validation to the development of the method itself is also highlighted, this relationship being described as 'an iterative process' that continues until specified performance requirements have been met.

Methods must be validated, and used, appropriately in order to ensure that reliable analytical results are produced. The validation of a method is required when it is necessary to verify that its performance characteristics are adequate for use for a particular problem. It is recommended that the reliability of a result, and by implication the adequacy of the performance characteristics, is expressed in terms of the uncertainty of measurement, 'quoted in a way that is widely accepted, internally consistent and easy to interpret'. Such a statement of the measurement uncertainty is said 'to not only provide information on the confidence that can be placed on the result but also provides a basis for the intercomparability with other results of the measurement of the same quantity'.

The guide suggests that the question of what constitutes 'appropriate' validation may be addressed by illustrating the different situations of Table A4 where method validation is necessary.

Table A4 *Appropriate validation levels*

Situation	*Degree of validation or revalidation required*
New method developed for particular problem	Full
Existing method tested for applicability to a particular problem	Full
Established method revised to incorporate improvements	Partial or Full
When quality control indicates an established method is changing with time	Partial or Full
Established method in a different laboratory	Partial
Established method with different instrumentation	Partial
Established method with different operator	Partial

In this way an attempt has been made to draw a distinction between the varying degrees of validation which may be considered appropriate in different situations. However, the use, in this generic document, of the term 'full' to mean the highest degree of within-laboratory validation does not correspond with existing practice in the food sector, where it is generally understood that 'full validation' is that achieved by collaborative trial.

The question of deciding what degree of validation is required is returned to in the section of the guide 'how methods should be validated.' It is suggested that starting with a carefully considered analytical specification provides a good base on which to plan the validation process. Only general guidance is otherwise provided on this point however, it merely being suggested that the analyst takes into account 'past experience, customer requirements and the need for comparability with other similar methods already in use'.

Nevertheless the responsibility for ensuring that a method is adequately validated is stated as resting with the laboratory that wishes to use a particular method for a particular specification.

The document, however, does raise some pertinent questions regarding the situation where it is inconvenient or impossible for a laboratory to enter into collaborative study to test a method:

- can laboratories validate methods on their own, and if so how?
- will methods validated in this way be recognised by other laboratories?

- what sort of recognition can be expected for 'in-house' methods used in a regulatory environment?

While not directly answering these questions, the guide does suggest that in the absence of collaborative testing it may be feasible to get some idea of the comparability of the method with others used elsewhere by analysing certified reference materials or by benchmarking the method under examination against one for which the validation is well characterised.

The guide gives details of the validation characteristics typically studied during within-laboratory validation and gives, without accompanying in-depth statistical treatments, some limited advice on how these characteristic parameters may be determined practically, addressing both qualitative and quantitative analysis as appropriate.

Notably, ruggedness is included as one of the method validation characteristics. By its inclusion, the process of method validation, in addition to giving an indication of a method's performance capabilities and limitations in routine use, will indicate what might cause the method to go out of control. Validation data can therefore be used to design appropriate quality control.

In summary the validation characteristics presented are:

- selectivity/specificity;
- limit of detection;
- limit of quantification;
- recovery;
- working and linear ranges;
- accuracy/trueness;
- repeatability precision;
- reproducibility precision; and
- ruggedness/robustness.

Once again the information contained in this generic document, at times, appears to conflict with existing practice in the food sector, where for example, reproducibility precision is expressed as that derived by collaborative trial.

The proper documentation of validated methods, however, which is recognised as essential in this document, does conform to established practice in the food area.

The main information given for each of the validation characteristics or parameters is as follows:

Selectivity

If interferences are present, which either are not separated from the analyte of interest or are not known to be present, they may:

- inhibit analyte confirmation;
- enhance or suppress the concentration of the analyte.

In these cases, carrying out further development may overcome these effects, thereby raising the confidence of the analysis.

Limit of detection (LOD)

It is recognised that a number of conventions exist for determining 'the lowest concentration of the analyte that can be conclusively detected by the method', based on various statistical justifications. The approaches suggested involve either of the following:

- carrying out 10 independent analyses of sample blanks. The LOD is expressed as the analyte concentration corresponding to the sample blank plus three sample standard deviations.
- carrying out 10 independent analyses of a sample blank fortified at the lowest concentration for which an acceptable degree of uncertainty can be achieved. The LOD is expressed as the analyte concentration corresponding to 4.65 sample standard deviations.

For qualitative measurements, the concentration threshold below which positive identification becomes unreliable can be ascertained by carrying out approximately 10 replicate assessments at each of a number of concentration levels across the range.

Limit of quantification

The limit of quantification is the lowest concentration of analyte that can be determined with an acceptable level of uncertainty, or alternatively it is set by various conventions to be 5, 6 or 10 standard deviations of the blank mean. It is also sometimes known as the limit of determination.

Sensitivity

This is the measure of the change in instrument response which corresponds to a change in analyte concentration. Where the response has been established as linear with respect to concentration, sensitivity corresponds to the gradient of the response curve.

Working and linear ranges

For any quantitative method there is a range of analyte concentrations over which the method may be applied. At the lower end of the concentration range, the limiting factor is the value of the limit of detection and/or limit of quantification. At the upper end of the concentration range, limitations will be imposed by various effects depending on the detection mechanism.

Within this working range there may exist a linear range, within which the detection response will have a linear response to analyte concentration. Note that the working and linear range may be different for different sample types according to the effect of interferences arising from the sample matrix.

It is recommended that in the first instance the response relationship should be examined over the working range by carrying out a single assessment of the response levels at six concentration levels at least. To determine the response relationship within the linear range, it is recommended that three replicates are carried out at each of at least six concentration levels.

Accuracy

Whether or not a method is accurate depends on the systematic errors which are inherent either within the method itself, in the way the method is used, or in the environment in which the method is used. These systematic errors cause bias in the method. Bias can be investigated using alternative analytical techniques, the analysis of suitable (matrix-matched) certified reference materials or where a suitable material is not available by carrying out analyses on spiked test portions.

Ideally the absolute value of bias can be determined by traceability of the measurand to the true value of an internationally accepted standard.

It is recommended that a reagent blank and standard (or reference material) are each analysed 10 times and the (mean analyte–mean blank) value compared with the true/acceptable value or the standard (or reference material). A comparison with similar measurements carried out using independent/primary methods will indicate the method's relative bias.

Precision

The two most useful precision measures in quantitative analyses are repeatability and reproducibility. Both are expressed in terms of standard deviation and are generally dependent on analyte concentration, so should be determined at a number of concentrations across the working range.

For qualitative analyses, precision cannot be stated as a standard deviation but may be expressed as true and false positive rates. These rates should be determined at a number of concentrations, below, at and above the threshold level.

It is recommended that 10 repeat determinations are carried out at each concentration level, both for repeatability and within-laboratory reproducibility.

Ruggedness

This is a measure of how effectively the performance of the analytical method stands up to less-than-perfect implementation. In any method there will be certain parts which, if not carried out sufficiently carefully, will have a severe effect on method performance. These aspects should be identified and, if possible, their influence on method performance evaluated using ruggedness tests, sometimes also called robustness tests.

An example of eight combinations of seven variable factors used to test the ruggedness of the method is given.

APPENDIX III

Intralaboratory Analytical Method Validation

Both the AOAC International, through its short course, and the Dutch Inspectorate for Health Protection are interested in this area. The contents of the short course and the Validation Guide are given below:

1 AOAC International Short Course

The AOAC International has developed a short course manual devoted to the subject of 'intralaboratory analytical method validation'. In the manual, intralaboratory method validation is introduced by presenting the topic within the broader context of method validation generally.

Validation is defined as the process of determining the suitability of methodology for providing useful analytical data. The process involves the production of a set of quantitative and experimentally determined values, known as the performance characteristics of the method, for a number of fundamental parameters of importance in assessing the suitability of the method.

Overall, the objective of method validation is stated as being 'to have properly designed, adequately studied and well documented analytical methods capable of producing useful data for the intended purpose'.

The AOACI manual indicates the benefits of conducting method validation studies and describes when methods need to be validated and suggests standardised terminology for this purpose.

The manual suggests that the benefits of method validation, in addition to the assessment, characterisation and optimisation of method performance, include:

demonstration of ability to meet project requirements;
provision of information to assist the design of QA/QC strategy;
generation of baseline data for comparison to long-term quality control;
identification of instrument, equipment and supply needs and availability;
identification of personnel skill requirements and availability;
consideration of safety and hazardous waste issues;
facilitation of operational efficiency when the method is implemented.

The course suggests that intralaboratory validation of a method is required to be carried out, not only in the first instance by the laboratory which has developed the methodology for a particular purpose, but in addition, subsequently in other laboratories deploying the methodology. Intralaboratory validation of the method will also be required, dependent on the intended use of the data and the customer's willingness to accept qualitative and/or quantitative uncertainties, when changes are made to the circumstances of the previous validation, for example when a new application is undertaken for a currently validated method.

The course covers the following aspects of method validation:

Method Validation Studies: Introduction and Background Information

Methods and Method Validation: Definitions
Types of Method Validation Studies
Benefits of Conducting Method Validation Studies
Determining When a Validation Study is Needed
Scientific Validity
Legal Defensibility
Standardised Terminology

Method Scope and Applicability

Data Quality Objectives and Measurement Goals
Examining a Method for Applicability
Qualitative Requirements
Quantitative Requirements

Preparing for a Method Validation Study

Standards, Samples, Chemicals, Supplies, Equipment, and Other Resources
Planning the Conduct of the Study
SOP for a Method Validation Study: Example
An Approach for Planning a Method Validation Study: Example

Technical Components of a Method Validation Study—Overview
Errors in Analytical Measurements

Systematic and Random Errors, Mistakes
Bias, Variability, Accuracy, and Precision
Sources of Bias and Variability from the Laboratory
Overview of Approaches for Assessing Bias and Precision

Sensitivity and Calibration

Definitions
Simple Linear Regression (Method of Least Squares)
Weighted vs. Unweighted Least Squares Regression

- Single Point Calibrations
- Multiple Point Calibration Example
- Justifying a Single Point Calibration: Example

Selectivity and Specificity

- Definitions
- Blanks
- Demonstrating Selectivity

Assessing Bias and Precision When Reference Materials or Methods are Available

- Comparison to an Accepted Reference Value
  - One-Sample t Procedure
    - One-Sample to Procedure Example
    - Confidence Interval Calculation Example
- Comparison to Another Method
  - Two-Sample t Procedure
    - Two-Sample t Procedure Example
  - Comparing Standard Deviations—F Test
    - F Test Example
  - Two-Sample t Procedure when Variances are Not Equal
  - Matched Pairs t Procedure
    - Matched Pairs t Procedure Example
  - Least Squares Regression Analysis

Assessing Bias and Precision When Reference Materials or Methods are Not Available

- Blanks and Spikes
- Spike Recovery Assumptions
- Analysing Recovery Data
  - Comparing Groups of Recovery Data—Example
  - Least Squares Regression
    - Least Squares Regression Example
- Recovery Data from Spiked Positive Samples

Within- and Between-run Variability, Ruggedness Testing

- Designing a Within- and Between-run Study
  - One-way ANOVA Example
- Ruggedness Testing: Benefits and Approaches
  - Seven Factor Test Design
  - Ruggedness Test Example

Level of Detection and Quantitation

Practical Considerations
American Chemical Society Definitions
Hypothesis Testing and Decision Errors
Defining a Detection and Quantitation Level
EPA Detection Level Definition
Signal-to-Noise Ratio Calculations
Estimates from Calibration Curves
40 CFR Part 136, Appendix B

Method Review and Documentation

Components of the Review Process
Preparing a Method for Internal Use
Distributing and Controlling Analytical Methods
SOP for Method Format and Content: Example

Tables

Method Review and Planning
Data for Multiple Point Calibration Example
Regression Analysis Summary for Multiple Point Calibration
Data and Results for Single Point Calibration Example
Student's t Distribution Critical Values
F Distribution Critical Values ($\frac{1}{2}$ = 9 only)
Data for Matched Pairs t Procedure Example
Data for Spike Recovery Examples
ANOVA Summary for Recovery Comparison Example
Regression Analysis Summary for Unweighted Recovery Data
Regression Analysis Summary for Weighted Recovery Data
Within- and Between-run Data for Example
ANOVA Summaries for Within- and Between-run Data
Summary Table for Within- and Between-run Standard Deviations
Seven Factor Ruggedness Test Design
Data for Ruggedness Test Example
Ruggedness Test Result Summary
Decision Error Table for Detection and Non-detection
(Type I and Type II Errors)

Figures

Bias, Variability, Accuracy, Precision
X-Y and Residual Plots for Calibration Standards
Graphical Presentations of Recovery Data
X-Y Plot for Unweighted Recovery Data

Residuals Plot for Unweighted Recovery Data
X-Y and Residuals Plots for Weighted Recovery Data
Normal Distributions of Blank and Sample Data
Detection Level Defined by 3 Standard Deviations of the Blank
Relationship Between Peak-to-Peak
Noise and the Standard Deviation of a Blank Signal

2 Inspectorate for Health Protection, Rijswijk, The Netherlands

The Dutch Inspectorate for Health Protection has also addressed the area, and the contents of their Guide[23] are given below:

1. Introduction
2. Terms/parameters used for validation
   - 2.1 Terms
   - 2.2 Parameters
   - 2.3 Ruggedness test
3. Validation
   - 3.1 General
     - 3.1.1 (Physico) chemical methods
     - 3.1.2 Mechanical methods
   - 3.2 Methods developed and validated in other laboratories
   - 3.3 Methods for screening
   - 3.4 Research and development work
4. Reporting validation data
5. Literature

Annex 1: Important points for each analysis technique
Annex 2: Method validation form

APPENDIX IV

Procedure for the Statistical Analysis of Collaborative Trial Data

1. Arrangement of Data and Cochran's Test Calculation

1 Discard aberrant data. Set out data to be used in the statistical analysis of the collaborative trial results in the form shown in Table A5 and calculate the w^2 values:

Table A5 *Data tabulation and Cochran's test*

	Sample 1		*Difference in duplicates (w) [(1) − (2)]*	
Laboratory	*(Duplicate 1) mg/kg*	*(Duplicate 2) mg/kg*	w	w^2
1	46	41	5	25
2	42	44	−2	4
3	47	20	27	729
4	48	41	7	49
5	45	45	0	0
6	60	62	−2	4
7	44	47	−3	9
8	48	47	1	1
9	43	45	−2	4
10	46	42	4	16
				$\sum w^2 = 841$

2. Calculate the Cochran critical variance ratio, C

$$C = \frac{w^2(\max)}{\sum w^2} = \frac{729}{841} = 0.867$$

3. Decide on the probability level to be used for the outlier tests. Compare value with critical variance ratio:

Table A6 *Critical values for Cochran maximum variance test 1-tail, at P = 1% level, expressed as critical variance ratio for 10 laboratories (abstracted from Table A5)*

Number of laboratories	*Number of replicates from each laboratory*				
	2	3	4	5	6
10	0.718	0.536	0.447	0.393	0.357

As the calculated C (0.867) > critical value (0.718) for two replicates, the results for sample 1 from laboratory 3 are outliers.

The full critical values are given in Table A9.

2 Grubbs's Test Calculation for Single Values

1. Set out data to be used in the statistical analysis of the collaborative trial results in the form shown in Table A7, but without values which have been identified as outliers by the Cochran's Test.

Table A7 *Grubbs's outlier test*

Laboratory	*Sample 1* (Duplicate 1) mg/kg	(Duplicate 2) mg/kg	*Mean of duplicates*	*Standard deviations [(s) from all means, (s_l) excluding lowest mean, (s_h) excluding highest mean]*
1	46	41	43.5	
2	42	44	43	(s_l) = 5.842
3	Values identified as outlying by Cochran's test			
4	48	41	44.5	
5	45	45	45	
6	60	62	61	(s_h) = 1.408
7	44	47	45.5	
8	48	47	47.5	
9	43	45	44	
10	46	42	44	
				(s) = 5.61

2. Calculate the Grubbs's critical values. For the single value test, calculate the percentage decrease in standard deviation with the lowest (s_l) and the highest (s_h) values removed, *i.e.* % decrease = 100 × $(1 - s_l/s)$.

Compare whichever value gives the greatest percentage decrease with the tabulated values.

Single value: % decrease (s_l) = 100 × [1 − (5.842/5.610)] = −4.135

% decrease (s_h) = 100 × [1 − (1.408/5.610)] = 74.902

3. Compare value with Grubbs's test critical value:

Table A8 *Critical value for Grubbs's test at P = 1% level, expressed as the percent reduction in standard deviation caused by removal of the suspect value(s) for nine laboratories (abstracted from Table A5)*

Number of Laboratories	*Grubbs's test*	
	Single values	*Double values*
9	52.3	69.4

As the critical value s_h (74.902) > critical value (52.3), the results for sample 1 from laboratory 6 are outliers.
The full critical values are given in Table A9.

3 Grubbs's Test Calculation for Paired Values

1. The Grubbs's test for paired values is calculated as for the single value test, except that the percentage decrease in standard deviation is calculated for the two lowest values (s_{ly}) and the two highest values (s_{hh}),

$$i.e.\ \%\ \text{decrease} = 100 \times (1-(s_{hh})/s)$$

2. Compare whichever value gives the greatest percentage decrease with the tabulated values.

4 Critical Values and Outlier Scheme

The critical values for the outlier test to be used are given in Tables A9–A11 and the outlier scheme to be followed is given in Figure A1.

5 Precision Parameters

1. The precision parameters to be calculated are:
   s_r: the standard deviation of the repeatability
   RSD_r: the relative standard deviation of the repeatability
   r: the repeatability
   s_R: the standard deviation of the reproducibility
   RSD_R: the relative standard deviation of the reproducibility
   R: the reproducibility
2. Calculations for repeatability parameters

The following are to be calculated:

$$s_r = \sqrt{\frac{\sum w^2}{2n}}$$

where w is the difference between duplicates, and n is the number of laboratories.

$$\mathrm{RSD}_r = \frac{s_r}{m} \times 100$$

where m is the overall mean value for sample.

$$r = 2.83 \times s_r$$

3. Calculations for reproducibility parameters
The following are to be calculated:

$$s_R = \sqrt{(s_r)^2 + (s_1)^2}$$

$$\left(s_1 = \sqrt{\frac{n\sum y^2 - (\sum y)^2}{n(n-1)} \quad \frac{-\sum w^2}{4n}} \right)$$

where y is the mean of each duplicate.

$$\mathrm{RSD}_R = \frac{s_R}{m} \times 100$$

$$R = 2.83 \times s_R$$

4. Conversion of s_r/s_R into r/R values
The value of $2.83 = 2\sqrt{2}s$
where $2 \times s$ represents the approximate 95% confidence limits provided the number of laboratories is relatively large ($n = \infty$, figure = 1.96).
$\sqrt{2}$ arises from the fact that both repeatability and reproducibility are defined as differences between two sets of results.
Assuming the two sets have similar standard deviations,

$$i.e.\ r = 2\sqrt{(s_1)^2 + (s_2)^2} = \sqrt{2\,s^2} = \sqrt{2} \times s$$

where $s_1 = s_2$.
Therefore $r = 2\sqrt{2}\,s_r$ and $R = 2\sqrt{2}\,s_R$.

Table A9 *Critical values for the Cochran maximum variance test, 1-tail, at the P = 1% level, expressed as a critical variance ratio; and critical values for the Grubbs's test, at the P = 1% level, expressed as the percent reduction in standard deviation caused by removal of the suspect value(s)*

	Cochran test: critical variance ratio *Number of replicates from each laboratory (k)*					*Grubbs's tests*	
L^a	*2*	*3*	*4*	*5*	*6*	*Single value*	*Pair value*
2	—	0.995	0.979	0.959	0.937	—	—
3	0.993	0.942	0.883	0.834	0.793	99.3	—
4	0.968	0.864	0.781	0.721	0.676	91.3	99.7
5	0.928	0.788	0.696	0.633	0.588	80.7	95.4
6	0.883	0.722	0.626	0.564	0.520	71.3	88.3
7	0.838	0.664	0.568	0.508	0.466	63.6	81.4
8	0.794	0.615	0.521	0.463	0.423	57.4	75.0
9	0.754	0.573	0.481	0.423	0.387	52.3	69.4
10	0.718	0.536	0.447	0.393	0.357	48.1	64.6
11	0.684	0.504	0.418	0.366	0.332	44.5	60.5
12	0.653	0.475	0.392	0.343	0.310	41.5	56.8
13	0.624	0.450	0.369	0.322	0.291	38.9	53.6
14	0.599	0.427	0.349	0.304	0.274	36.6	50.8
15	0.575	0.407	0.332	0.288	0.259	34.6	48.3
16	0.553	0.388	0.316	0.274	0.246	32.6	46.0
17	0.352	0.372	0.301	0.261	0.234	31.2	44.0
18	0.514	0.356	0.288	0.249	0.223	29.8	42.1
19	0.496	0.343	0.276	0.238	0.214	28.5	40.4
20	0.480	0.330	0.265	0.229	0.205	27.3	38.9
21	0.465	0.318	0.255	0.220	0.197	26.2	37.4
22	0.450	0.307	0.246	0.212	0.189	25.2	36.1
23	0.437	0.297	0.238	0.204	0.182	24.3	34.9
24	0.425	0.287	0.230	0.197	0.176	23.4	33.7
25	0.413	0.278	0.222	0.190	0.170	22.7	32.7
26	0.402	0.270	0.215	0.184	0.164	21.9	31.7
27	0.391	0.262	0.209	0.179	0.159	21.2	30.8
28	0.382	0.255	0.202	0.173	0.154	20.6	29.9
29	0.372	0.248	0.196	0.168	0.140	20.0	29.1
30	0.363	0.241	0.191	0.164	0.145	19.5	28.3
35	0.325	0.213	0.168	0.144	0.127	17.1	25.0
40	0.294	0.192	0.151	0.128	0.114	15.3	22.5

[a] L = Number of laboratories for the given material.

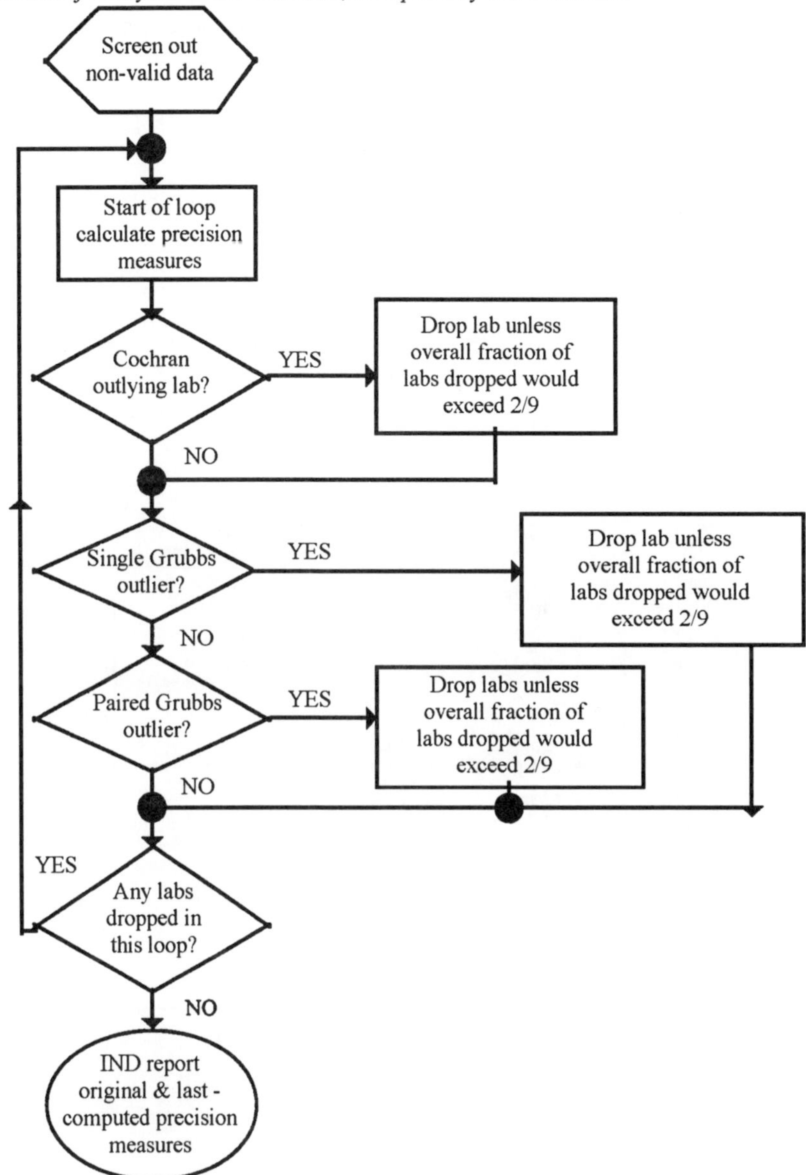

Figure A1 *Outlier rejection scheme*
Reproduced by permission of IUPAC. Copyright IUPAC, 1988.

Table A10 *Critical values for the Cochran maximum variance ratio at the 2.5% (1-tail) rejection level, expressed as the percentage the highest variance is of the total variance*

Number of labs	*Number of replicates (r)* 2	3	4	5	6
4	94.3	81.0	72.5	65.4	62.5
5	88.6	72.6	64.6	58.1	53.9
6	83.2	65.8	58.3	52.2	47.3
7	78.2	60.2	52.2	47.3	42.3
8	73.6	55.6	47.4	43.0	38.5
9	69.3	51.8	43.3	39.3	35.3
10	65.5	48.6	39.9	36.2	32.6
11	62.2	45.8	37.2	33.6	30.3
12	59.2	43.1	35.0	31.3	28.3
13	56.4	40.5	33.2	29.2	26.5
14	53.8	38.3	31.5	27.3	25.0
15	51.5	36.4	29.9	25.7	23.7
16	49.5	34.7	28.4	24.4	22.0
17	47.8	33.2	27.1	23.3	21.2
18	46.0	31.8	25.9	22.4	20.4
19	44.3	30.5	24.8	21.5	19.5
20	42.8	29.3	23.8	20.7	18.7
21	41.5	28.2	22.9	19.9	18.0
22	40.3	27.2	22.0	19.2	17.3
23	39.1	26.3	21.2	18.5	16.6
24	37.9	25.5	20.5	17.8	16.0
25	36.7	24.8	19.9	17.2	15.5
26	35.5	24.1	19.3	16.6	15.0
27	34.5	23.4	18.7	16.1	14.5
28	33.7	22.7	18.1	15.7	14.1
29	33.1	22.1	17.5	15.3	13.7
30	32.5	21.6	16.9	14.9	13.3
35	29.3	19.5	15.3	12.9	11.6
40	26.0	17.0	13.5	11.6	10.2
50	21.6	14.3	11.4	9.7	8.6

Table A11 *Critical values for the Grubbs's extreme deviation outlier tests at the 2.5% (2-tail), 1.25% (1-tail) rejection level, expressed as the percent reduction in standard deviations caused by the removal of the suspect value(s)*

Number of labs	*One highest or lowest*	*Two highest or two lowest*	*One highest and one lowest*
4	86.1	98.9	99.1
5	73.5	90.9	92.7
6	64.0	81.3	84.0
7	57.0	73.1	76.2
8	51.4	66.5	69.6
9	46.8	61.0	64.1
10	42.8	56.4	59.5
11	39.3	52.5	55.5
12	36.3	49.1	52.1
13	33.8	46.1	49.1
14	31.7	43.5	46.5
15	29.9	41.2	44.1
16	28.3	39.2	42.0
17	26.9	37.4	40.1
18	25.7	35.9	38.4
19	24.6	34.5	36.9
20	23.6	33.2	35.4
21	22.7	31.9	34.0
22	21.9	30.7	32.8
23	21.2	29.7	31.8
24	20.5	28.8	30.8
25	19.8	28.0	29.8
26	19.1	27.1	28.9
27	18.4	26.2	28.1
28	17.8	25.4	27.3
29	17.4	24.7	26.6
30	17.1	24.1	26.0
40	13.3	19.1	20.5
50	11.1	16.2	17.3

References

1 Procedural Manual of the Codex Alimentarius Commission, 9th Edition, FAO, Rome, 1995.
2 Council Directive 85/591/EEC Concerning the Introduction of Community Methods of Sampling and Analysis for the Monitoring of Foodstuffs Intended for Human Consumption, *O.J.*, 1985, L372.
3 International Vocabulary for Basic and General Terms in Metrology, ISO, Geneva, 2nd Edition, 1993.
4 Terms and Definitions Used in Connection with Reference Materials, ISO Guide 30, ISO, Geneva, 1992.
5 Protocol for the Design, Conduct and Interpretation of Method Performance Studies, ed. W. Horwitz, *Pure Appl. Chem.,* 1988, **60**, 855.
6 Protocol for the Design, Conduct and Interpretation of Method Performance Studies, ed. W. Horwitz, *Pure Appl. Chem.,* 1995, **67**, 331.
7 Precision of Test Methods, ISO 5725, ISO, Geneva, 1994; previous editions were issued in 1981 and 1986.
8 Guidelines for Referees on Validation of Chemical Analytical Methods: (Utvärdering av kemiska analysmetoder inom NMKL—Handledning för referenter—available in Swedish only), NMKL Secretariat, Finland, 1992, NMKL Report No. 11.
9 The International Harmonised Protocol for the Proficiency Testing of (Chemical) Analytical Laboratories, ed. M. Thompson and R. Wood, *Pure Appl. Chem.,* 1993, **65**, 2123 (also published in *J. AOAC Int.*, 1993, **76**, 926).
10 Guidelines on Internal Quality Control in Analytical Chemistry Laboratories, ed. M. Thompson and R. Wood, *Pure Appl. Chem.,* 1995, **67**, 649.
11 Report of the 21st Session of the Codex Committee on Methods of Analysis and Sampling, FAO, Rome, 1997, ALINORM 97/23A.
12 Council Directive 93/99EEC on the Subject of Additional Measures Concerning the Official Control of Foodstuffs, *O. J.*, 1993, L290.
13 Criteria for Evaluating Acceptable Methods of Analysis for Codex Purposes, Paper CX/MAS 97/3 discussed at the 21st Session of the Codex Committee on Methods of Analysis and Sampling, FAO, Rome, 1997.
14 Council Directive 85/591/EEC Concerning the Introduction of Community Methods of Sampling and Analysis for the Monitoring of Foodstuffs Intended for Human Consumption, *O. J.*, 1985, L372.
15 MAFF Validated Methods for the Analysis of Foodstuffs: VOA—Introduction, General Considerations and Analytical Quality Control, *J. Assoc. Publ. Anal.*, in preparation.
16 MAFF Validated Methods for the Analysis of Foodstuffs: V38—Method for the enumeration of listeria monocytogenes in meat and meat products, *J. Assoc. Publ. Anal.*, in press.
17 Evaluation of Analytical Methods used for Regulation of Foods and Drugs, W. Horwitz, *Anal. Chem.*, 1982, **54**, 67A.
18 Report of the 22nd Session of the Codex Alimentarius Commission, FAO, Rome, 1997, ALINORM 97/37.
19 Establishing Routine Methods to Meet Codex Maximum Residue Limit Requirements, Paper CX/MAS 97/3, App. 1, discussed at the 21st Session of the Codex Committee on Methods of Analysis and Sampling, FAO, Rome, 1997.
20 Validation of Chemical Analytical Methods, NMKL Secretariat, Finland, 1996, NMKL Procedure No. 4.

21 Method Validation—A Laboratory Guide, EURACHEM Secretariat, Laboratory of the Government Chemist, Teddington, UK, 1996.
22 An Interlaboratory Analytical Method Validation Short Course developed by the AOAC International, AOAC International, Gaithersburg, Maryland, USA, 1996.
23 Validation of Methods, Inspectorate for Health Protection, Rijswijk, The Netherlands, Report 95-001.
24 A Protocol for Analytical Quality Assurance in Public Analysts Laboratories, Association of Public Analysts, 342 Coleford Road, Sheffield S9 5PH, UK, 1986.

CHAPTER 4

Use of Recovery Corrections when Reporting Chemical Results

It is recognised that the use of recovery information to correct/adjust analytical results is a contentious one for analytical chemists and particularly for food analysts. Different sectors of analytical chemistry have different practices and, in the case of the food analysis area, different practices for different commodities. Formal legislative requirements with regard to the use of recovery factors also vary sector-to-sector. It is the aim of IUPAC, however, to prepare general guidelines which may be seen to aid the production of the 'best estimate of the true result'. To this end the IUPAC Inter Divisional Working Group on Analytical Quality Assurance is developing Harmonised Guidelines for the Use of Recovery Information in Analytical Measurement. Those guidelines will set out the recommendations to laboratories producing analytical data on the approach to be taken to the use of recovery corrections.

The purpose of the guidelines will be to outline the conceptual framework needed for considering those types of analysis where loss of analyte during the analytical procedure is inevitable. It is accepted that certain questions will not be satisfactorily addressed, and hence remain irreducibly complex, unless such a conceptual framework is established. The questions at issue involve (*a*) the validity of methods for estimating the recovery of the analyte from the matrix of the test material, and (*b*) whether the recovery estimate should be used to correct the raw result. The types of chemical analysis most affected by these considerations are those where an organic analyte is present at very low concentrations in a complex matrix, *i.e.* as occurs in the food analysis sector with increasing frequency.

The draft guidelines[1] are of particular relevance to food analysts, having been discussed in the Codex Alimentarius Commission and now being referred to in various EU Commission Working Group activities (*e.g.* the criteria of methods approach in the aflatoxins area, see Chapter 3). Because of these activities this chapter discusses the use of recovery corrections when reporting analytical results in the food sector. The issue of recovery correction is becoming of increasing importance in that the application, or non-application, of corrections is a major cause of the divergence of analytical data.

This chapter concentrates on the use of recovery corrections in the chemistry sector because reporting microbial analytical data is much more complicated and recovery corrections should generally not be used. The problem of strong or weak, specific or non-specific, binding to matrix surfaces are common within both the chemical and microbial field.

The situation, however, is complicated for the microbiologist as death or growth of micro-organisms can lead to (10 or 100-fold) changes in concentration of micro-organisms within hours. Sometimes the organisms are present in, or transferred into, a viable but non-cultivable state, making them impossible to detect with traditional microbiological techniques.

The type and strength of any surface interaction between micro-organism and matrix as well as growth/death could differ very much from one strain to another within the same species. In addition, the morphology and physiology, and as a consequence the binding characteristics, can change for the one and same strain due to competition from other microbial species, loss of genetic information (*e.g.* plasmids), nutritional stress (*e.g.* starvation or surplus of nutrients), heat, drought and other environmental factors.

Because the field of microbial recovery is still in its early research phase, compensation for different types of microbial losses cannot be recommended when reporting routine analytical results. However, recovery studies do supply information on a microbial method performance and should be used when evaluating them or when a laboratory's competence to use such methods are tested, *e.g.* through accreditation. This chapter therefore only discusses chemical recovery corrections in detail.

1 Introduction

The estimation and use of recovery is an area where practice differs among analytical chemists. The variations in practice are most obvious in the determination of analytes such as veterinary drug residues and pesticide residues in complex matrices, such as foodstuffs and in environmental analysis. Typically, such methods of analysis rely on transferring the analyte from the complex matrix into a much simpler solution that is used to present the analyte for instrumental determination. However, the transfer procedure results in loss of analyte. Quite commonly in such procedures, a substantial proportion of the analyte remains in the matrix after extraction, so that the transfer is incomplete, and the subsequent measurement gives a lower value than the true concentration in the original test material. If no compensation for these losses is made, significantly discrepant results may be obtained by different laboratories. Even greater discrepancies arise if some laboratories compensate for losses and others do not. It is therefore essential that there is a common strategy amongst food analysts to the approach for the use of recovery corrections.

Recovery studies are also an essential component of the validation and use of analytical methods. It is important that all concerned with the production and interpretation of analytical results are aware of the problems and the basis on which the result is being reported, *i.e.* that the food analyst and his customer must discuss and agree on the basis on which results are to be expressed and reported.

At present there is no single well-defined approach to estimating, expressing and applying recovery information. The most important inconsistency in analytical practice concerns the correction of a raw measurement, which can (in principle) eliminate the negative bias due to loss of analyte. The difficulties involved in reliably estimating the correction factor deter analysts in some sectors from applying such corrections.

In the absence of consistent strategies for the estimation and use of recovery, it is difficult to make valid comparisons between results produced in different laboratories or to verify the suitability of data for the intended purpose. This lack of transparency can have important consequences in the interpretation of data. For example in the context of enforcement analysis, the difference between applying or not applying a correction factor to analytical data can mean respectively that a legislative limit is exceeded or that a result is in compliance with the limit. Thus, where an estimate of the *true concentration* is required, there is a compelling case for compensation for losses in the calculation of the reported analytical result.

This approach may be best illustrated by the requirements of the UK Ministry of Agriculture, Fisheries and Food which has stated that for all surveillance data, laboratories adhere to the following instructions:[2]

Correcting for analytical recovery

The Steering Group has promulgated advice on correcting for analytical recovery. It concluded that corrected values give a more accurate reflection of the true concentration of the analyte and facilitate interpretation of data and made the following recommendations:

1. unless there are overriding reasons for not doing so, results should be corrected for recovery;
2. recovery values should always be reported, whether or not results are corrected, so that measured values can be converted to corrected values and *vice versa*;
3. where there is a wide variation around the mean recovery value, correcting for recovery can give misleading results and therefore the standard deviation should also be reported;
4. each Working Party should establish upper and lower limits for acceptable recovery and for variability of recovery. Samples giving recovery values outside the range should be re-analysed or results reported as semi-quantitative; and
5. it is important to report analytical data in a way which ensures that their significance is fully understood. It is particularly important, especially where results are close to legal or advisory limits, to explain those factors, such as sampling and analytical errors and corrections for recovery, which are sources of uncertainties. Both enforcement authorities and

working parties need to consider the most appropriate way to present results.'

In addition the Codex Committee on Methods of Analysis and Sampling, which has historically been concerned with the development and endorsement of methods of analysis and sampling attached to Codex Standards, has recently changed its terms of reference so that it is now concerned not only with the endorsement of appropriate methods of analysis but also with how well its methods are being used in the laboratory. This is evidenced by the recent discussions in CCMAS and the general principles by which laboratories operating under Codex principles should comply (see Chapter 1). These principles are all aimed to enable laboratories to achieve the 'best estimate of the true result'.

Codex has also recognised that this, the use of recovery factors, is an area which of concern to food analysts and which could lead to the production of differing data. The application of recovery factors in analysis is particularly important in Codex work where the difference between a corrected and uncorrected result could mean that a specification (*e.g.* a legislative limit) is exceeded or is in compliance.

It is thus starting to address the problem but there has not yet been consensus on the issue by governments.

2 Procedures for Assessing Recovery

Recovery from Reference Materials

Recoveries can be determined by the analysis of matrix reference materials. The recovery is the ratio of the concentration of analyte found to that certified to be present. Results obtained on test materials of the same matrix could, in principle, be corrected for recovery on the basis of the recovery found for the reference material. However, two problems beset this use of the reference materials, these being: (*a*) the range of certified matrix reference materials available is limited, and (*b*) there may be a matrix mismatch between the reference material and the test material, in which case the recovery value would not be strictly applicable to the test material.

The latter shortfall applies especially in the food analysis sector where reference materials have to be finely powdered and dried to ensure homogeneity and stability. Such treatment is likely to affect the recovery in comparison with that pertaining to fresh foods of the same kind. This matrix mismatch is a problem in the application of recovery information, and is considered below.

Recovery Information from Surrogates

Where (certified) reference materials are unavailable, the recovery of analyte can be estimated by studying the recovery of an added compound or element that is regarded as a surrogate for the native analyte. The degree to which this surrogate is transferred into the measurement phase is estimated separately and this recovery can, if appropriate, also be attributed to the native analyte. This procedure in principle allows the loss of analyte to be corrected, and an unbiased estimate of the concentration of the native analyte in the original matrix to be made. Such a 'correction-for-recovery' methodology is implicit or explicit in several distinct methods of analysis and can be regarded as a valid procedure if it can be shown to be properly executed.

In order for this procedure to be valid the surrogate must behave quantitatively in the same way as analyte that is native in the matrix, especially in regard to its partition between the various phases. In practice that equivalence is often difficult to demonstrate and certain assumptions have to be made. The nature of these assumptions can be seen by considering the various types of surrogate that may be used.

Isotope Dilution

The best type of surrogate is an isotopically-modified version of the analyte which is used in an isotope dilution approach. The chemical properties of the surrogate are identical (or very close) to those of the native analyte and, so long as the added analyte and the native analyte come to effective equilibrium, its recovery will be identical to that of the analyte. In isotope dilution the recovery of the surrogate can be estimated separately by mass spectrometry, or by radiometric measurement if a radioisotope has been used, and validly applied to the native analyte. However, the achievement of effective equilibrium is not always easy. In some chemical systems, for example in the determination of trace metals in organic matter, the native analyte and the surrogate can be readily converted into the same chemical form by the application of vigorous reagents that destroy the matrix. This treatment converts organically bound metal into simple ions that are in effective equilibrium with the surrogate. However, such a simple procedure might not apply to a pesticide residue. In that instance the analyte may in part be chemically bound to the matrix. Vigorous chemical reagents could not be used to release the analyte without the danger of destroying it. The native analyte and surrogate cannot come into effective equilibrium. The recovery of the surrogate is therefore likely to be greater than that of the native analyte. Thus even for this best type of surrogate, a bias in an estimated recovery may arise. Moreover, application of the isotope dilution approach is limited by the availability and cost of isotopically enriched analytes. It is therefore not expected to be a routinely applicable procedure in the food analysis laboratory.

Spiking

A cheaper expedient, and one very commonly applied, is to estimate in a separate experiment the recovery of a portion of the analyte added as a spike. If a matrix blank (a specimen of the matrix containing effectively none of the analyte) is available, the analyte can be spiked into that and its recovery determined after application of the normal analytical procedure. If no field blank is available, the spike can be added to an ordinary test portion that is analysed alongside an unspiked test portion. The difference between these two results is the recovered part of the added analyte, which can be compared with the known amount added. This type of recovery estimate is known as the 'surrogate recovery'. It suffers from the same problem as that encountered with isotopically modified analyte, namely that added analyte may not be in effective equilibrium with the native analyte. If the added analyte is not so firmly bound to the matrix as the native analyte, the surrogate recovery will tend to be high in relation to that of the native analyte. That circumstance would lead to a negative bias in a corrected analytical result.

Internal Standards

A third type of surrogate used for recovery estimation is the internal standard. In internal standardisation, the surrogate is an entity chemically distinct from the analytes, and therefore it will not have identical chemical properties. However, it will normally be selected so as to be as closely related chemically to the analytes as is practicable, thus representing their partition between phases to the greatest degree that is practicable. The internal standard would be used, for example, in recovery estimation where numerous analytes are to be determined in the same matrix and marginal recovery experiments would be impracticable for each of them individually. The question of practicability goes beyond the costs of handling numerous analytes: some analytes (*e.g.*, new veterinary residues, or metabolites) may not be available as pure substances. While it may be the most cost-effective expedient in some circumstances, the internal standard at best is technically less satisfactory than the spike as a surrogate, because its chemical properties are not identical with those of the analytes. Biases in both directions could result from the use of a recovery estimate based on an internal standard. Internal standards may also be used for other purposes.

Matrix Mismatch

Matrix mismatch occurs when a recovery value is estimated for one matrix and applied to another. The effect of matrix mismatch would be manifest as a bias in the recovery in addition to those considered above. The effect is likely to be most serious when the two matrices differ considerably in their chemical nature. However, even when the matrices are reasonably well matched (say two different species of vegetable) or nominally identical (*e.g.*, two different specimens of bovine liver), the analytical chemist may be forced to make the unsubstantiated assumption that the recovery is still appropriate. This would clearly increase the uncertainty in the recovery and in a recovery-corrected result. Matrix mismatch can be avoided in principle by a recovery experiment (*e.g.*, by spiking) for each separate test material analysed. However, such an approach will often be impracticable on a cost–benefit basis so a representative test material in each analytical run is used to determine the recovery.

Concentration of Analyte

The recovery of the surrogate or the native analyte has been treated as if it were independent of its concentration. This is unlikely to be strictly true at low concentrations. For instance a proportion of the analyte may be unrecoverable by virtue of irreversible adsorption on surfaces. However, once the adsorption sites are all occupied, which would occur at a particular concentration of analyte, no further loss on increase in concentration is likely. Hence the recovery would not be proportional to concentration. Circumstances like this should be investigated during the validation of an analytical method, but they may be too time-consuming for *ad hoc* use.

3 Should Recovery Information be Used to Correct Measurements?

It is this question which has caused most controversy and discussion amongst analytical chemists. A strong case can be made either for correcting results for recovery or for leaving them uncorrected. Regardless of these explicit arguments, however, analytical chemists are often obliged to comply with normal practice in their application area.

Arguments for Correction

The following arguments are frequently made for correcting results for recovery:

- The purpose of analytical science is to obtain an estimate of the true concentration of the analyte with an uncertainty that is fit for purpose.
- The true concentration can be estimated only if significantly low recoveries of analyte are corrected.
- An uncorrected bias due to low recovery means that results will not be universally comparable, not transportable and therefore unfit to support mutual recognition.
- Methods of correction advocated are isomorphic with perfectly acceptable analytical techniques such as internal standardisation and isotope dilution and therefore not suspect in principle.
- Although some uncertainty is inevitably associated with correction factors, that uncertainty can be estimated and incorporated into a combined uncertainty for the final result.
- Some legislation imposing maximum limits on contaminants is framed explicitly requiring results to be reported on a corrected basis, *e.g.* aflatoxins and veterinary drug residues.

Arguments against Correction

The following arguments are also frequently made for not correcting results for recovery:

- Estimated recoveries based on a surrogate may be higher than the corresponding value for the native analyte. The resultant corrected result would still have a negative bias.
- Estimated correction factors may be of doubtful applicability because they may vary among different matrices and for different concentrations of analyte.
- Estimated correction factors often have a high relative uncertainty, whereas uncorrected results usually have the smaller relative uncertainty associated with volumetric and instrumental measurement alone. (However, the uncertainty is small only if no contribution from the bias is included.) Therefore corrected results will have a high relative uncertainty, sufficiently high if made explicit to create an unfavourable impression among those unfamiliar with the problems of analysis. This in turn might affect the credibility of science in the enforcement of legislation.
- Relatively small deviations from unity in correction factors could arise largely through random errors rather than a systematic loss of analyte. In that circumstance, correction could make the uncertainty of the result absolutely greater.
- Some legislation imposing maximum limits on contaminants is framed on the understanding that uncorrected results will be used for enforcement purposes, *e.g.* for pesticides in the food sector.

Rational and Empirical (Defining) Methods

Analytical measurements generally strive to estimate the measurand, that is, the true value of the concentration of the analyte, with an uncertainty that is fit for purpose. It is only on that basis that results can be completely comparable. However, it must be recognised that this stance applies equally to 'rational' and 'empirical' methods of analysis.[3] In a rational method the measurand is the total concentration of the analyte in the test material. In an empirical method the measurand is the concentration that can be extracted from the test material by the specific procedure applied, and the result is traceable to the method. Therefore, if the method is regarded as empirical, the concentration extracted is necessarily close to the true value. The concentration extracted will not, however, be identical to the true value because different laboratories may carry out the method protocol slightly differently, thus introducing systematic error; in addition, there will be a repeatability (random) error distribution. In that case the measurand is the concentration of the 'extractable' analyte.

However, regarding methods as empirical does not in itself cause results to comply with the requirement of equivalence. Empirical results will be 'equivalent' throughout a particular analytical sector only where a single method protocol is in use for a particular determination. In some sectors, where methods have stabilised or are specified in regulations, a single empirical method protocol will be used. However, in many sectors the methodology is subject to continuous evolution and single protocols will not be available. In such circumstances there is a strong case for correction of results, as only corrected results will be equivalent.

4 Estimation of Recovery

There is no generally applicable procedure for estimating recovery that is free from shortcomings. It is possible to devise an ideal procedure in which the analyte is determined by a method that gives an unbiased result. However, this would be too resource-intensive for use in routine analysis. Nevertheless there is an alternative procedure which may be used routinely. The recovery obtained in the routine method is tested by using both methods to analyse a large set of typical test materials, a set that covers the required range of matrices and analyte concentrations. This gives the recovery (and its uncertainty) for the routine method for any conceivable situation. In practice there is usually no such definitive method available for reference, so reference materials or surrogate studies have to be used. However, reference materials are few, and lack of resources restricts the range of test materials that can be used to estimate recovery by using surrogates. Additionally, the use of surrogates in itself adds an uncertainty to a recovery estimate because it may not be possible to determine whether some proportion of the native analyte is covalently or otherwise strongly bound to the matrix.

A strategy commonly employed is to estimate recovery during the process of method validation. Recoveries are determined over as wide a range of pertinent matrices and analyte concentration as resources allow. These values are then held to apply during subsequent use of the analytical method. To justify that assumption, all routine runs of the method must contain a reference material (or spiked samples) to act as internal quality control. It is necessary to ensure that the analytical system does not change in any significant way that would invalidate the original estimates of the recovery. The following points are therefore suggested as requiring consideration, even if lack of resources prevents their complete execution in practice.

Representative Recovery Studies

The entire range of matrix types for which the method will be applied should be available for the method validation. Moreover, several examples of each type should be used to estimate the normal range of recoveries (the uncertainty) for that matrix type. If it is likely that the history of the material will affect recovery of the analyte (*e.g.* the technical processing or cooking of foodstuffs), then examples at different stages of the processing should be procured. If this range cannot be encompassed in the validation, there will be an extra uncertainty associated with the matrix mismatch in the use of the recovery. That uncertainty may have to be estimated from experience.

An appropriate range of analyte concentrations should be investigated where that is technically and financially possible, because the recovery of the analyte may be concentration-dependent. Consider adding an analyte to a matrix at several different levels. At very low levels the analyte may be largely chemisorbed at a limited number of sites on the matrix, or irreversibly adsorbed onto surfaces of the analytical vessels. Recovery at this level might be close to zero. At a somewhat higher level, where the analyte is in excess of that so adsorbed, the recovery will be partial. At considerably higher concentrations, where the adsorbed analyte is only a small fraction of the total analyte, the recovery may be effectively complete. The analytical chemist may need to have information about recovery over all of these concentration ranges. In default of complete coverage, it may be suitable to estimate recovery at some critical level of analyte concentration, for example at a regulatory limit. Values at other levels would have to estimated by experience, again with an additional uncertainty.

When spiking is applied to a field blank (*i.e.* a matrix containing effectively zero native analyte) then the whole range of concentration can be conveniently considered. When the concentration of the native analyte is appreciable the spike added should be at least as great, to avoid incurring a relatively large uncertainty in the surrogate recovery.

Internal Quality Control

The principles and application of internal quality control (IQC) are described in Chapter 6, these being partly based on the Harmonised Guidelines.[4] The purpose of IQC is to ensure that the performance of the analytical system remains effectively unchanged during its use. The concept of statistical control is crucial in IQC applied to routine analysis (as opposed to *ad hoc* analysis). When applied to recovery, IQC has some special features that have to be taken into account. This IQC of recovery can be addressed in two distinct ways, depending on the type of control material that is used:

1. A matrix-matched reference material can be used as a control material. The recovery for this material and an initial estimate of its between-run variability is determined at the time of method validation. In subsequent routine runs the material is analysed exactly as if it were a normal test material, and its value plotted on a control chart (or the mathematical equivalent). If the result for a run is in control, then the validation-time estimate of the recovery is taken as valid for the run. If the result is out of control, further investigation is required, which may entail the rejection of the results of the run. It may be necessary to use several control materials, depending on the length of the run, the analyte concentration range, *etc.*

2. Spiked materials can also be used for quality control so long as they are sufficiently stable. Initial estimates of the average recovery and its between-run variability are made during method validation, and are used to set up a control chart. Two variant approaches can be conceived for use in routine analysis: (*a*) a single long-term control material (or several materials) in each run, or (*b*) spiking all or a random selection of the test materials for the run. In either instance the surrogate recovery is plotted on a control chart. While the recovery remains in control it can be deemed to apply to the test materials generally. Of the two alternative methods, the latter (involving the actual test materials) is probably the more representative, but also the more demanding.

There is a tendency for the role of IQC to be confused with the simple estimation of recovery. It is better to regard IQC results solely as a means of checking that the analytical process remains in control. The recovery estimated at the method validation time is usually more accurate for application to subsequent in-control runs, because more time can be spent on studying their typical levels and variability. If real-time spiking is used to correct for recovery, this is more like a type of calibration by standard additions. The same data cannot validly be used both for correction for recovery and for IQC.

5 Uncertainty in Reporting Recovery

Uncertainty is a key concept in formulating an approach to the estimation and use of recovery information. Although there are substantive practical points in the estimation of uncertainty that (at the time of writing) remain to be settled, the principle of uncertainty is an invaluable tool in conceptualising recovery issues. Its application has been presented by Ellison and Williams.[5] This section illustrates the principles there presented. However, it must be appreciated that the estimation of uncertainty in recovery is yet to be studied in detail.

Definition of Uncertainty

Measurement uncertainty is defined by ISO[6,7] as:

'A parameter, associated with the result of a measurement, that characterises the dispersion of the values that could reasonably be attributed to the measurand',

with the note that 'The parameter may be, for example, a standard deviation (or a given multiple of it), or the half width of an interval having a stated level of confidence'. The ISO Guide recommends that this parameter should be reported as either a standard uncertainty, denoted u, defined as the:

'uncertainty of the result of a measurement expressed as standard deviation' or as an expanded uncertainty, denoted U, defined as:

'a quantity defining an interval about the result of a measurement that may be expected to encompass a large fraction of the distribution of values that could be attributed to the measurand'. The expanded uncertainty is obtained by multiplying the standard uncertainty by a coverage factor, which in practice is typically in the range 2 to 3.

To evaluate the uncertainty systematically, it is first necessary to identify the possible sources of uncertainty. The uncertainty arising from each source is then quantified and expressed as a standard deviation, associated with one or more of the intermediate parameters used in calculating the final result. The resulting numerical values, or components, are combined to obtain the overall uncertainty on the basis of their value and the contribution of the parameter affected to the overall result. In practice the task is simplified by the predominance of only a few components; others need not be evaluated in detail.

Some of the sources relevant in chemical measurement are listed in Table 4.1. For discussion of recovery, it is the uncertainties covered resulting from the incomplete definition of the measurand (*e.g.* failing to specify the exact form of the analyte being determined) and the incomplete extraction and/or pre-concentration of the measurand, contamination of the measurement sample, interferences and matrix effects that are of most concern.

Table 4.1 *Sources of uncertainty in analytical chemistry*

Incomplete definition of the measurand (*e.g.* failing to specify the exact form of the analyte being determined)
Sampling: the sample measured may not represent the defined measurand
Incomplete extraction and/or pre-concentration of the measurand, contamination of the measurement sample, interferences and matrix effects
Inadequate knowledge of the effects of environmental conditions on the measurement procedure or imperfect measurement of environmental conditions
Cross-contamination or contamination of reagents or blanks
Personal bias in reading analogue instruments
Uncertainty of weights and volumetric equipment
Instrument resolution or discrimination threshold
Values assigned to measurement standards and reference materials
Values of constants and other parameters obtained from external sources and used in the data-reduction algorithm
Approximations and assumptions incorporated in the measurement method and procedure
Variations in repeated observations of the measurand under apparently identical conditions

Definition of the Measurand

Clear definition of the measurand is crucial to uncertainty estimation and to the relevance or otherwise of recovery values. The most important issue is whether the measurand is the amount of material actually present in the sample matrix (a rational method), or the response to a reproducible, but otherwise essentially arbitrary, procedure established for comparative purposes (an empirical method).

Note that application of a 'standard method' does not render knowledge of the recovery and its uncertainty unnecessary. Both are essential if the results might be compared with other methods, e.g. to establish trends over space or time. In such a case, the measurand is effectively being redefined, leading to (usually) larger uncertainties and relevance of recovery. In routine testing, however, the lesser uncertainty and 'uncorrected' result will normally be cited.

Estimating Uncertainty in a Recovery

The approaches to the estimation of the uncertainty of a recovery provided here are necessarily tentative, and may be expected to be rapidly superseded as detailed studies become available. The important principles are as follows.

1. The recovery and its standard uncertainty may both depend on the concentration of the analyte. This may entail studies at several concentration levels. Subsequent comments in this section apply to a single level of concentration.
2. The main recovery study should involve the whole range of matrices that are included in the category for which the method is being validated. If the category is strict (*e.g.* bovine liver) a number of different specimens of that type should be studied so as to represent variations likely to be encountered in practice (*e.g.* sex, age, breed, time of storage, *etc.*). Probably a minimum of 10 diverse matrices are required for recovery estimation. The standard deviation of the recovery over these matrices is taken as the main part of the standard uncertainty of the recovery.
3. If there are grounds to suspect that a proportion of the native analyte is not extracted, then a recovery estimated by a surrogate will be biased. That bias should be estimated and included in the uncertainty budget.
4. If a method is used outside the matrix scope of its validation, there is a matrix mismatch between the recovery experiments at validation time and the test material at analysis time. This could result in extra uncertainty in the recovery value. There may be problems in estimating this extra uncertainty. It would probably be preferable to estimate the recovery in the new matrix, and its uncertainty, in a separate experiment.

The recovery $R = c_{obs}/c_{ref}$ is the ratio of the observed value c_{obs} obtained in a separate experiment by the application of an analytical procedure to a material containing analyte at a reference value c_{ref}. c_{ref} will be (*a*) a reference material certified value, (*b*) measured by an alternative definitive method, or (*c*) estimated as a spike addition. In a perfect separation R would be exactly unity. In reality, factors such as imperfect extraction often give observations that differ from the ideal. It is therefore good practice in validating an analytical method to estimate a recovery R for the analytical system. In such experiments, the recovery can be tested for significant departure from unity. Such a test considers the question 'is $|R-1|$ greater than (u_R), the uncertainty in the determination of R?', at some level of confidence. Table 4.2 gives some sources of the uncertainty in measured recovery.

Table 4.2 *Sources of uncertainty in recovery estimation*

Repeatability of the recovery experiment
Uncertainties in reference material values
Uncertainties in added spike quantity
Poor representation of native analyte by the added spike
Poor or restricted match between experimental matrix and the full range of sample matrices encountered
Effect of analyte/spike level on recovery and imperfect match of spike or reference material analyte level and analyte level in samples

The analyst then performs a significance test of the form:

$$|R-1|/u_R > t: \quad R \text{ differs significantly from } 1$$

$$|R-1|/u_R < t: \quad R \text{ does not differ significantly from } 1$$

where t is a critical value based either on a 'coverage factor' allowing for practical significance or, where the test is entirely statistical, $t_{(\alpha/2,\, n-1)}$, being the relevant value of Student's t for a level of confidence $1-\alpha$.

Following such an experiment, four cases can be distinguished, chiefly differentiated by the use made of the recovery R:

1. R is not significantly different from 1. No correction is applied.
2. R is significantly different from 1 and a correction for R is applied.
3. R is significantly different from 1 but, for operational reasons, no correction for R is applied
4. An empirical method is in use. R is arbitrarily regarded as unity and u_R as zero. (Although there is obviously some variation in recovery in repeated or reproduced results, that variation is subsumed in the directly estimated precision of the method.)

The uncertainty may be handled in each of these cases as follows:

1 R is not significantly different from 1
The experiment has detected no reason to adjust subsequent results for recovery. It might be thought that the uncertainty in the recovery is unimportant. However, the experiment could not have distinguished a range of recoveries between $1 - ku_R$ and $1 + ku_R$. It follows that there is still uncertainty about the recovery that should be taken into account in calculating the overall uncertainty. u_R is therefore included in the uncertainty budget. (An alternative view is that a correction factor based on $R - 1$ is implicitly applied, but the analyst is uncertain that the value is exactly unity).

2. R differs from 1 and a correction is applied
Since R is explicitly included in the calculation of the corrected result (*i.e.*, $c_{\text{corr}} = c/R$. where c is the raw result with an uncertainty u_c) it is clear that u_R must be included in the uncertainty budget. This leads to a combined uncertainty u_{corr} on the corrected result given by:

$$\frac{u_{\text{corr}}}{c_{\text{corr}}} = \sqrt{\left(\frac{u_c}{c}\right)^2 + \left(\frac{u_R}{R}\right)^2}$$

u_{corr} would be multiplied by k (usually 2) to obtain the expanded uncertainty U.

3. R differs from 1 but no correction is applied
Failure to apply a correction for a known systematic effect is inconsistent with obtaining the best possible estimate of the measurand. It is less straightforward in this case to take recovery into account in calculating the overall uncertainty. If R is substantially different from unity, the dispersion of values of the measurand is not properly represented unless the uncertainty u_R is substantially increased. A simple and pragmatic approach that is sometimes adopted when a correction b for a known systematic effect has not been applied is to increase the expanded uncertainty on the final result to $(U_c + b)$, where U_c is calculated assuming b is zero. For recovery, therefore, $U = U_c + (c/r - c)$. This procedure gives a pessimistic overall uncertainty, and departs from the ISO-recommended principle of treating all uncertainties as standard deviations.

Alternatively, if the correction for recovery is not applied because the analyst's judgement is that the difference is not meaningful in normal use, case 3 may be treated in the same way as case 1 after increasing u_R because the significance test should have used a value larger than u_R. This amounts to estimating u_R as $|1 - R|/t$ where t is the critical value used in the significance test. This amplified uncertainty on the recovery should be included as in case 2. This will normally only be significant where u_R is comparable with or greater than $|1 - R|$.

While either method will provide an estimate of uncertainty, both methods have similar drawbacks arising from the failure to correct the result to give a best estimate of the measurand. Both lead to overstatement of the uncertainty, and the range quoted around the result will include the measurand only near one extreme (usually the upper end), with the remainder of the range unlikely to contain the value with significant probability.

For recoveries of the order of 70%, the additional uncertainty contribution (before applying a coverage factor) will be close to 20% of the result. This is clearly not unreasonable given the size of recovery correction being ignored, but it does point strongly to the consequences for reported uncertainty of neglecting a substantial recovery correction.

There is therefore a clear choice if the customer is not to be misled by a result from a putative rational method. Either the recovery must be corrected or a substantially greater uncertainty must be quoted.

Finally, it should be noted that the foregoing discussion relates to the situation where a result and its uncertainty are obtained on a real scale and reported as such. For the instance where an analyst provides an *interpretation* of a result (*e.g.* by stating that the value is 'not less than . . .' as in the case of food analysis where a comparison is made to a legislative limit) the analyst's professional knowledge of the recovery and overall experimental uncertainty will be taken into account in the interpretation, and accordingly neither the recovery nor an uncertainty need necessarily be reported.

6 Recommendations and Conclusions

Variable practice in handling information recovery is an important cause of the non-equivalence of data. To mitigate its effects the practice of reporting analytical data after the application of an appropriate correction factor is normally encouraged. Where, however, an enforcement limit is based on data which has not had a correction factor applied, the present situation of reporting 'raw' data will continue for the foreseeable future.

Detailed descriptions of recovery experiments and their results should be properly recorded. If it is known or suspected that a proportion of the native analyte in the test material is not extractable by the analytical procedure, the procedure must be qualified as determining only 'available' analyte. Such qualification should be specified on analytical certificates. No valid compensation can be made, or should be attempted, for the 'bound' analyte, which a recovery model does not represent.

It should be recognised that there is a dual role for recovery determinations in analytical measurement, that is, for (*a*) quality control purposes and (*b*) for deriving recovery values. In the latter application, more extensive and detailed data are required.

In the light of the above the following recommendations may be made with regard to the use of recovery factors in the food sector, and which should ideally be followed, or at least considered, by food analysts:

1. Results should be corrected for recovery, unless there are overriding reasons for not doing so. Such reasons would include the situation where a limit (statutory or contractual) has been established using uncorrected data, or where recoveries are close to unity.
2. Recovery values should always be established as part of method validation, whether or not recoveries are reported or results are corrected, so that measured values can be converted to corrected values and *vice versa*.
3. When the use of a recovery factor is justified, the method of calculation should be given in the method.
4. IQC control charts for recovery should be established during method validation and used in all routine analysis. Runs giving recovery values outside the control range should be considered for re-analysis in the context of acceptable variation, or the results reported as semi-quantitative

References

1 Harmonised Guidelines for the Use of Recovery Information in Analytical Measurement, ed. S. Ellison, M. Thompson, P. Willetts and R. Wood, in preparation.
2 Annual Report of the UK Ministry of Agriculture, Fisheries and Food Steering Group on Chemical Aspects of Food Surveillance, HMSO, London, 1996.
3 Sense and Traceability, M. Thompson, *Analyst (Cambridge)*, 1996, **121**, 285.
4 Guidelines on Internal Quality Control in Analytical Chemistry Laboratories, ed. M. Thompson and R. Wood, *Pure Appl. Chem.*, 1995, **67**, 649.
5 S. Ellison and A. Williams in 'Proceedings of the Seventh International Symposium on the Harmonisation of Quality Assurance Systems in Chemical Analysis', ed. M. Parkany, Royal Society of Chemistry, London, 1996.
6 International Vocabulary for Basic and General Terms, ISO, Geneva, 2nd edtion, 1993.
7 Guide to the Expression of Uncertainty in Measurement, ISO, Geneva, 1993.

CHAPTER 5

Measurement Uncertainty/ Measurement Reliability

1 Introduction

In quantitative chemical analysis, many important decisions are based on the results obtained by a laboratory and so it is therefore important that an indication of the quality of the results reported is available. Analytical chemists are now more than ever coming under increased pressure to be able to demonstrate the quality of their results by giving a measure of the confidence placed on a particular result to demonstrate its fitness for purpose. This includes the level that the result would be expected to agree with other results irrespective of the method used. 'Measurement uncertainty' (MU) is a useful parameter which gives this information, and one that is increasingly being discussed in the food analysis community.

In 1993, ISO published the 'Guide to the Expression of Uncertainty in Measurement'[1] in collaboration with other scientific bodies. This guide lays down general rules for the expression and evaluation of measurement uncertainty across a wide range of chemical measurements. Also included in the guide are examples of how the concepts in the guide can be applied in practice. The guide also gives an introduction to the idea of uncertainty and distinguishes between this and error, followed by a description of the steps involved in the evaluation of uncertainty. The EURACHEM Guide to Quantifying Uncertainty in Analytical Measurement[2] provides guidance on the evaluation and expression of uncertainty in quantitative chemical analysis based on the approach laid down in the ISO guide.

There has been some criticism of the practicability of this approach, in which the method is dissected and incremental calculations of uncertainty are made and eventually summed to provide a combined uncertainty. Much of the work to date regarding MU has been theoretical in nature and the amount of supporting analytical data has been limited. This has caused concern to analytical chemists, especially in the food sector where analysts are already required through legislation to have some estimate of the 'variability' of their results, mainly as a result of being required to use methods which have been assessed in a collaborative trial. It has not been established that the two approaches will, or indeed can, provide comparable results. In addition, the term 'measurement uncertainty' has been criticised and alternative terms, *e.g.*

'measurement reliability', are preferred especially by those analysts whose results are liable to form the basis of legal proceedings.

Nevertheless, in view of the importance of the proposed ISO/EURACHEM approach, not least to the accreditation agencies, it is outlined below.

2 ISO/EURACHEM Approach to the Estimation of Measurement Uncertainty in Analytical Chemistry

ISO published the 'Guide to the Expression of Uncertainty in Measurement'[1] in collaboration with BIPM, IEC, IFCC, IUPAC and OIML. The guide lays down general rules for the expression and evaluation of measurement uncertainty across a wide range of chemical measurements. Also included in the guide are examples of how the concepts in it may be applied in practice. The guide also gives an introduction to the idea of uncertainty and distinguishes between this and error followed by a description of the steps involved in the evaluation of uncertainty.

The evaluation of the measurement uncertainty for a method requires the analyst to look closely at all the possible sources of uncertainty within the method, which may take a considerable amount of effort, although the effort involved should not be disproportionate. Usually in practice an initial study will identify the major source of uncertainty associated with the method; this will be the dominating influence on the total uncertainty of the method. It is thus possible to make a reasonable estimate of the uncertainty for the method as a whole by concentrating on the major sources of uncertainty within the method. Following the estimation of the measurement uncertainty for a certain method in a particular laboratory, this estimate can be applied to subsequent results obtained provided that they are carried out in the same laboratory using the same method and equipment; this assumes that the quality control data justify this course of action.

The EURACHEM Guide to Quantifying Uncertainty in Analytical Measurement[2] is a protocol which establishes general rules for the evaluation and expression of uncertainty in quantitative chemical analysis based on the approach laid down in the ISO guide. It is applicable at all levels of accuracy and in all fields including quality control in manufacturing, testing for regulatory compliance, calibration, certification of reference materials and research and development.

Uncertainty

The word uncertainty, when used outside of the science world, portrays doubt. Thus uncertainty of measurement could be understood to mean that the analyst is unsure about the validity and exactness of his result. In the EURACHEM guide the definition attributed to uncertainty is 'a parameter associated with the result of a measurement, that characterises the dispersion of the values that could reasonably be attributed to the measurand'.

The Uncertainty Estimation Process

The estimation process is outlined in the EURACHEM guide and involves the steps given in Figure 5.1. This is best achieved by breaking down a method process into a 'cause-and-effect' diagram, an example of which is attached for the determination of acesulfame-K in a soft drink matrix in Figure 5.2.

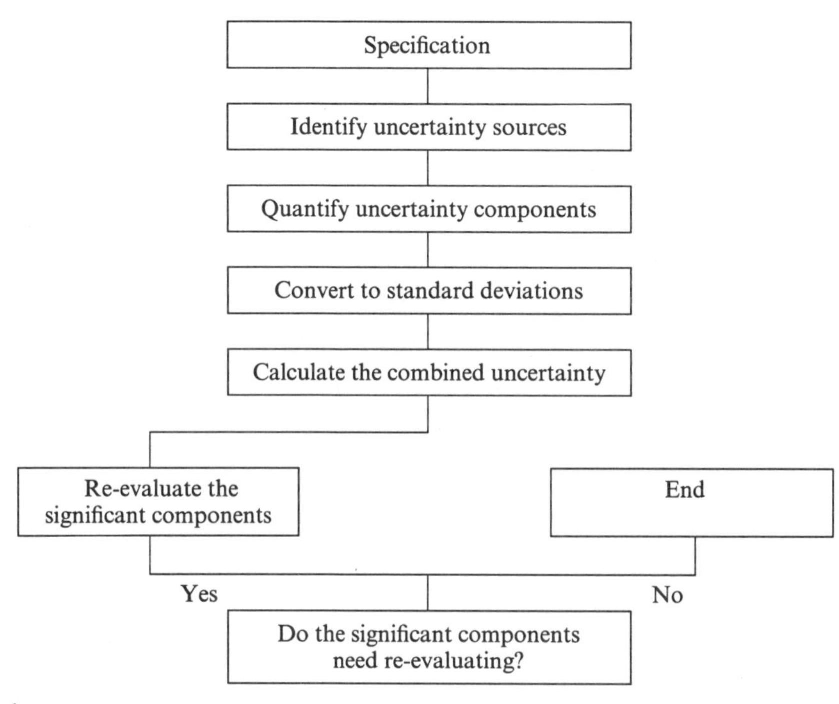

where:

Specification	Write down a clear statement of what is being measured and the relationship between it and the parameters on which it depends.
Identify uncertainty sources	List sources of uncertainty for each part of the process or each parameter.
Quantify uncertainty components	Estimate the size of each uncertainty. At this stage, approximate values suffice; significant values can be refined in subsequent stages.
Convert to standard deviations	Express each component as a standard deviation.
Calculate the combined uncertainty	Combine the uncertainty components, either using a spreadsheet method or algebraically. Identify significant components.

Figure 5.1 *EURACHEM guide to uncertainty estimation*

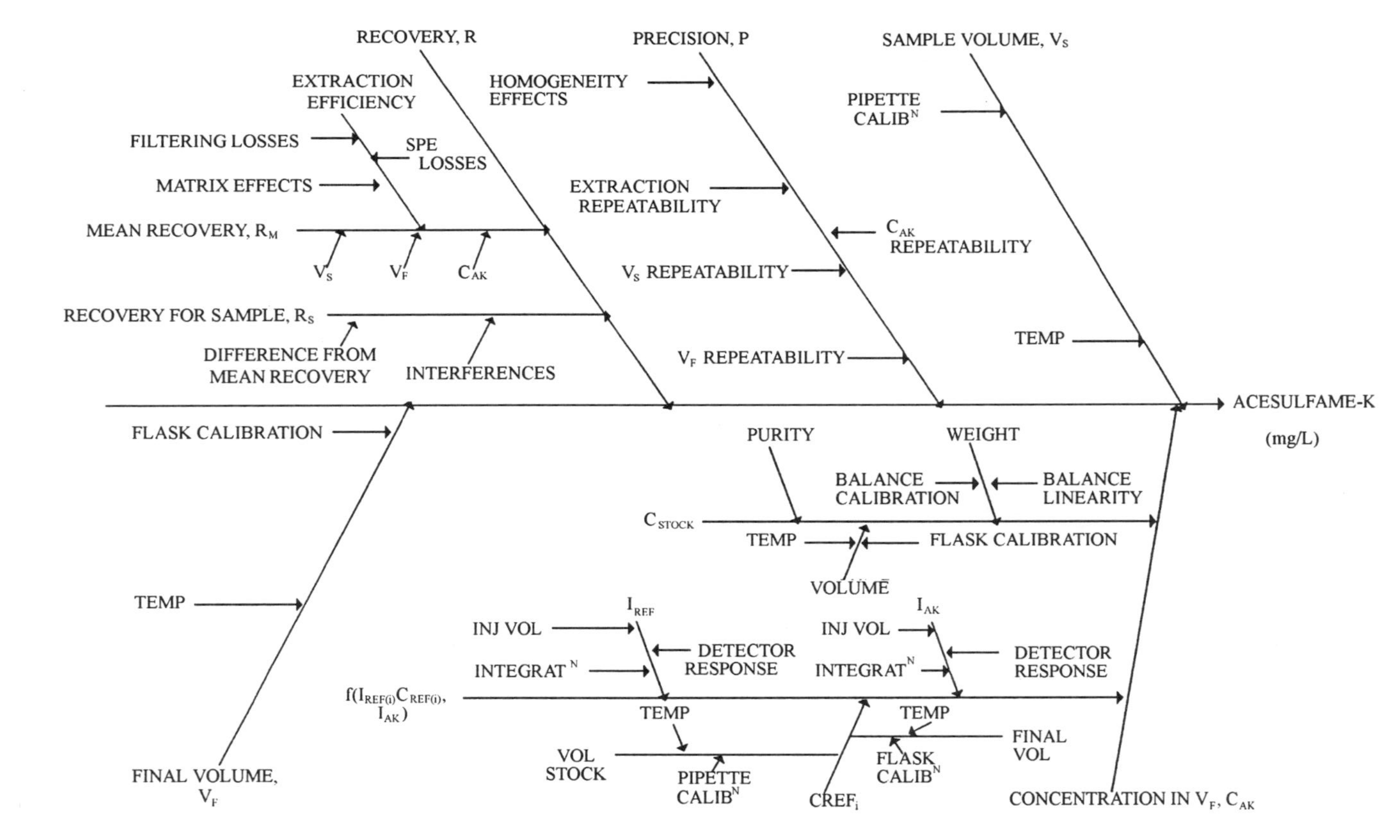

Figure 5.2 *Cause and effect diagram for determination of acesulfame-K in soft drinks*

The final stage is to calculate the expanded uncertainty. This is achieved by multiplying the combined standard uncertainty by a coverage factor, *k*. The coverage factor is chosen after considering a number of issues such as the level of confidence required and any knowledge of underlying distributions. For most purposes a coverage factor of 2 is chosen which gives a level of confidence of approximately 95%.

Reporting Uncertainty

The information required when reporting the result of a measurement ultimately depends on the intended use but should contain enough information that the result could be re-evaluated if new data became available. A complete report should include a description of the methods used to calculate the result and its uncertainty, the values and sources of all corrections and constants used in the result calculations and uncertainty analysis and a list of all the components of uncertainty with full documentation on how each was evaluated. The data and analysis should be given in a way that it can be easily followed and if necessary repeated. Unless it is required otherwise the result should be reported together with the expanded uncertainty, *U*.

3 Alternative Approaches for the Assessment of Measurement Uncertainty

There have been alternative approaches suggested for the assessment of the measurement uncertainty of an analytical result, namely:

1. The Royal Society of Chemistry Analytical Methods Committee[3] is that of the 'top down' approach. Here MU is calculated from the results of previous analyses of reference materials, collaborative trials and/or proficiency tests.
2. The Nordic Committee on Food Analysis, which has prepared a standard entitled 'Procedure For The Estimation And Expression Of Measurement Uncertainty In Chemical Analysis'.[4] This approach is reproduced in full as Appendix I to this Chapter.

4 Accreditation Agencies

Measurement uncertainty has also become an important issue to accreditation agencies (*e.g.* UKAS in the UK). In document M10 of UKAS[5] the instructions to accredited laboratories regarding Uncertainty of Measurement read as follows:

'The laboratory is required to produce an estimate of the uncertainty of its measurements, to include the estimation of uncertainty in its methods and procedures for calibration and testing, and to report the uncertainty of measurement in calibration certificates and in test certificates and test reports, where relevant.

Estimates of uncertainty of measurement shall take into account all significant identified uncertainties in the measurement and testing processes, including those attributable to measuring equipment, reference measurement standards (including material used as a reference standard), staff using or operating equipment, measurement procedures, sampling and environmental conditions.

In estimating uncertainties of measurement, the Laboratory shall take account of data obtained from internal quality control schemes and other relevant sources. The Laboratory shall also ensure that any requirements for the estimation of uncertainty and for the determination of compliance with specified requirements as stated in relevant publications are complied with at all times.'

As UKAS operates to both ISO/IEC Guide 25 and the EN 45 000 series of standards, it is reasonable to anticipate that other accreditation agencies will take the same approach.

Thus formally accredited laboratories have to produce an estimate of the uncertainty of their measurements. It is anticipated that this estimate will have to be developed along the ISO/EURACHEM approach if accreditation agencies maintain their present views.

5 MAFF Project for the Comparison of Results Obtained by Different Procedures for the Estimation of Measurement Uncertainty

The UK Ministry of Agriculture, Fisheries and Food (MAFF) has initiated a project to consider aspects of measurement uncertainty as applied to the food analysis laboratory; the results will be made generally available.[6] Included in this has been a comparison of the two approaches, *i.e.* the top-down (validated method) and bottom-up (ISO) approaches to the estimation of measurement uncertainty. The project is also to assess the practicability of seeking to provide the optimum valid procedures for calculating MU that can be applied in food analysis.

The project has looked at a number of different methods, these being given in the following table.

Each of the methods has been studied both in a collaborative trial conforming to the IUPAC Harmonised Protocol and the measurement uncertainty approach of ISO/EURACHEM outlined above. The results are outlined in Table 5.1.

The results show that in most cases similar uncertainty standard deviations (σ_R and u values) are obtained, so either approach results in similar estimates of the measurement uncertainties for the systems studied.

It could therefore be argued that as food analysis laboratories are required to use collaboratively tested methods, it is unreasonable for accreditation

Table 5.1 *Comparison of uncertainties obtained by collaborative trial data ('top-down') and ISO ('bottom-up') approaches*

Analyte	*Method*	*Matrix*	*Concentration*	*U(x)/x*	RSD_R	Ho_R
Acesulfame-K (mg/l)	HPLC/UV	soft drinks	30	0.328		
		soft drinks	350	0.074		
		cola	352		0.06	0.91
		orange	371		0.06	0.91
Ethanol (l/100kg)	GC	fruit juices	0.5–2.0	0.0717		
		orange juice	0.5		0.034	0.77
		orange juice	1.0		0.017	0.43
		orange juice	1.3		0.038	0.99
		orange juice	1.4		0.018	0.47
		apple juice	0.8		0.018	0.44
		grape juice	1.2		0.021	0.54
		pineapple juice	2.0		0.012	0.33
Heptachlor (mg/kg)	GC-ECD	veg oil	100	0.203		
		veg oil	1.00	0.142		
		olive oil	0.088		0.63	2.73
		olive oil	1.060		0.31	1.96
Total nitrogen (%)	Kjeldahl	cereals & meats		0.0199		
		ground beef			0.0151	
		canned ham			0.0148	
		smoked ham			0.0158	
		pork sausage			0.0180	
		cooked sausage			0.0172	
		dry cured ham			0.0136	
Cholesterol (mg/100g)	GC/FID	oils & fats	95	0.056		
		beef-pork	133		0.047	0.87
		milk fat	273		0.038	0.78
Lead (μg/l)	GFAA	wine	100	0.102		
		wine	26		0.263	0.95
		wine	51		0.298	1.19
		wine	74		0.328	1.39
		wine	147		0.206	0.97
		wine	190		0.288	1.40
		wine	298		0.097	0.51

agencies to increase the burden of work being placed on laboratories which are being accredited for food control work by requiring that a *formal* in-house assessment of measurement uncertainty using the ISO/EURACHEM approach also be carried out if the laboratory is in control, *etc.* Similar concerns have been voiced in the Codex Committee on Methods of Analysis and Sampling at its last session, these being outlined below.

6 Codex Committee on Methods of Analysis and Sampling Approach to Measurement Uncertainty

The Codex Committee on Methods of Analysis and Sampling discussed measurement uncertainty at the 21st Session of the Committee. The subject had been raised because a number of governments were concerned that the ISO/EURACHEM approach presently being advocated by various international organisations was at variance to the approach adopted by Codex, and within the food sector generally, with regard to the expression of uncertainty in methods of analysis and its use by food analysts.

As has been stated previously, for laboratories operating under Codex principles they will have to:

1. use validated methods of analysis;
2. become accredited to ISO/IEC Guide 25;
3. participate in proficiency testing schemes;
4. introduce appropriate internal quality control procedures.

The first of these is incorporated in the Codex Procedural Manual[7] and the last three have been endorsed by the Codex Alimentarius Commission.[8]

Governments were therefore concerned that:

1. The concept of the introduction of measurement uncertainty as developed by the ISO/EURACHEM approach is increasingly being discussed and implemented in other sectors of analytical chemistry and may well therefore have an impact on the work of the Codex Alimentarius Commission. In particular, accreditation agencies are increasingly demanding that measurement uncertainty be quoted on many of the results obtained under accreditation conditions and that such results should have a formal estimate of the measurement uncertainty attached to them. This is amplified below.
2. The term 'measurement uncertainty' is increasing being used. Whereas analytical chemists appreciate exactly what the term measurement uncertainty is, lay people may not do so and that will have unfortunate consequences in legal situations. Some organisations, *e.g.* the UK Royal Society of Chemistry, have recognised this and have recommended the use of an alternative name, *e.g.* 'measurement reliability'.
3. Analysts using Codex approved or endorsed methods of analysis are using methods of analysis which have been fully collaboratively tested. There is an argument to state that if a laboratory has to comply with the four Codex principles outlined above then it is unnecessary for the laboratory to undertake a further estimate of the measurement uncertainty according to the ISO approach.

Nomenclature

As stated above, there is some dispute as to the most appropriate term for 'measurement uncertainty'. There is a some thought that to include the term 'uncertainty' in quoting on analytical measurement is unhelpful and the term 'reliability' or 'confidence' should be substituted.

A suggestion to overcome these difficulties has been to use the terms

1. 'measurement uncertainty' when describing u, the combined standard uncertainty, and
2. 'measurement reliability' when describing U, the expanded uncertainty.

Report from the Codex Committee on Methods of Analysis and Sampling

Governments therefore agreed the following at the 21st Session of the Codex Committee on Methods of Analysis and Sampling regarding 'measurement uncertainty':

1. The Committee will develop for Codex purposes an appropriate alternative term for measurement uncertainty, *e.g.* measurement reliability.
2. The precision of a method may be estimated through a method–performance study, or where this information is not available, through the use of internal quality control and method validation.
3. Consideration should be given as to whether it is necessary to undertake an additional formal evaluation of a method of analysis using the ISO approach in addition to using information obtained through a collaborative trial.
4. Governments should advise accreditation agencies that for national and Codex purposes the measurement uncertainty result need not be calculated using the ISO approach providing the laboratory is complying with the appropriate Codex principles.

The Committee was strongly in support of the above and in addition agreed to request the UK to redraft the paper for consideration by the Committee at its next session.[9]

7 Conclusions

There is concern in the food analysis sector with regard to the current developments in the 'measurement uncertainty' area. This concern is best expressed in the Report of the last Session of the Codex Committee on Methods of Analysis and Sampling. If the recommendations from that Session are followed, laboratories will avoid carrying out duplication of work and the nomenclature will be amended. If an estimate of measurement uncertainty/

reliability has to be carried out, then the 'simple' approach given in the NMKL Procedure should be considered as an alternative to the ISO approach.

References

1 Guide to the Expression of Uncertainty in Measurement, ISO, Geneva, 1993.

2 Quantifying Uncertainty in Analytical Measurement, EURACHEM Secretariat, Laboratory of the Government Chemist, Teddington, UK, 1995, EURACHEM Guide.

3 Uncertainty of Measurement – Implications of its use in Analytical Science, Analytical Methods Committee of the Royal Society of Chemistry, *Analyst (Cambridge)*, 1995, **120**, 2303.

4 Estimation and Expression of Measurement Uncertainty in Chemical Analysis, NMKL Secretariat, Finland, 1997, NMKL Procedure No. 5.

5 General Criteria of Competence for Calibration and Testing Laboratories, UKAS, Queens Road, Teddington, TW11 0NA, NAMAS Accreditation Standard M 10 (1992) and Supplement (1993).

6 An Investigation into the Application of Measurement Uncertainty in Food Analysis: Results from MAFF Project FS2913, CSL, MAFF, Norwich, 1997.

7 Procedural Manual of the Codex Alimentarius Commission, 9th Edition, FAO, Rome, 1995.

8 Report of the 22nd Session of the Codex Alimentarius Commission, FAO, Rome, 1997, ALINORM 97/37.

9 Report of the 21st Session of the Codex Committee on Methods of Analysis and Sampling, FAO, Rome, 1997, ALINORM 97/23A.

APPENDIX I

Procedure for the Estimation and Expression of Measurement Uncertainty in Chemical Analysis Developed by the Nordic Committee on Food Analysis

Contents

1 Foreword
2 Introduction
3 Definitions
4 Estimation of measurement uncertainty
5 Expression of measurement uncertainty
6 Measurement uncertainty and the interpretation of results
7 References in NMKL guide

1 Foreword

Laboratories introducing quality control measures, or seeking accreditation for methods of analysis, face the problem of expressing the measurement uncertainty of results. This procedure has been elaborated by a working group under the Nordic Committee on Food Analysis (NMKL) in order to assist the laboratories. The procedure deals with quantitative analyses only, and is applicable to samples as they arrive in the laboratory. Thus it does not address possible variation resulting from sampling. The procedure is intended to be a practical tool for food chemists. It also aims at de-dramatising the subject of measurement uncertainty. This procedure should be regarded as an attempt to clarify concepts, but since the subject is under vivid discussion the view on measurement uncertainty may change in future. It is therefore possible that this procedure will need to be revised within a few years. NMKL invites all readers and users of this proedure to submit comments and views. Suggestions and comments should be sent to the General Secretariat of NMKL (see below).

The working group under NMKL which elaborated this procedure consisted of the following individuals:

Denmark: Inge Meyland, The National Food Agency
Finland: Esko Niemi, The Customs Laboratory; Harriet Wallin, VTT Biotechnology and Food Research (observer)
Iceland: Arngrímur Thorlacius, Agricultural Research Institute
Norway: Gudmund Braathen, Norwegian Institute for Food and Environmental Analysis
Sweden: Joakim Engman, The National Food Administration (chairman)

The working group extends its thanks to the Nordic accreditation bodies DANAK (Denmark), FINAS (Finland), Norsk akkreditering, NA (Norway) and SWEDAC (Sweden) which forwarded valuable comments on this procedure.

This procedure is available from the General Secretariat of NMKL, c/o VTT Biotechnology and Food Research, PB 1500, FIN-02044 VTT, Finland, tel. +358 9 4565164, fax +358 9 4552103, e-mail: harriet.wallin@vtt.fi

2 Introduction

It is recommended that a chemical analytical result should always be reported as a numerical result, and not simply in the form of a statement of the type 'The concentration of xx is below the maximum limit'. For the result to be complete, the measurement uncertainty of the analysis should be included. In some cases a result is meaingless unless accompanied by its uncertainty.

It is important to understand the difference between measurement error and measurement uncertainty. An error is the difference between a measured and the true value. Uncertainty is the variation resulting from the measurement of the concentration of the analyte in the test material (sample). The error is a difference, while the uncertainty is an interval. In order to quantify the measurement error one must know the true or assigned value, whereas an estimation of measurement uncertainty does not require the true value to be known.

The measurement and its uncertainty are always estimates depending, for example, on the method, the matrix and the analyte concentration. It is therefore important to state how the estimates were made. It should be stated which materials were studied (when estimating the uncertainty). The use of synthetic solutions may result in unrealistically low estimates compared to what may be expected when analysing authentic test materials (samples). However, synthetic solutions may be very useful in the quality control of instruments.

When selecting the materials to be used for estimations of measurement uncertainty, it is important to focus on matrices and analyte levels relevant to routine work. The use of certified reference materials may result in too low estimates, since these materials are often easier to handle (they need less homogenisation, contain no interfering substances occurring in real samples, *etc.*).

The model for the estimation of measurement uncertainty described in

'Quantifying Uncertainty in Analytical Measurement'[1] is based on the principle that the uncertainty is estimated for each step of a method, whereafter the uncertainties are combined in an error budget. This NMKL procedure was elaborated because it was felt that the error budget method is better suited to physical than chemical measurements. The error budget model involves many complex calculations, which leads to a large work-load especially in laboratories using a wide range of analytical methods. It is the view of NMKL that food laboratories need simpler, less time-consuming models for the evaluation of measurement uncertainty. However, even if a simpler model is used to estimate the uncertainty, a qualitative overview of all methods may be justified in order for the laboratory to ascertain the contribution from various steps to the total uncertainty. However, it is important to focus on the analytical process rather than on procedures or statistical calculations.

Compared to the error budget model, the model presented in this procedure is much simpler, and at the same time provides an overall picture of the uncertainty of the entire analytical chain. The approach is based on experimental data generated in the individual laboratory. The measurement uncertainty is estimated on the basis of the internal reproducibility. True reproducibility data (between-laboratory precision) are not used. Performance characteristics estimated on the basis of results of method-performance (collaborative) studies are not used, only data generated by the laboratory itself. Thus the estimated measurement uncertainty applies only to an individual laboratory's own random errors.

3 Definitions

Certified reference material

Reference material, accompanied by a certificate one or more of whose property values are certified by a procedure which establishes its traceability to an accurate realisation of the unit in which the property values are expressed, and for which each certified value is accompanied by an uncertainty at a stated level of confidence. VIM, 6.14 (1993)

Synthetic solution

A solution which may be analysed without the isolation of analytes required by ordinary test samples. It is often a solution prepared in the laboratory by dissolving a known amount of the analyte. The solution often does not contain other substances occurring in an ordinary test solution.

Level of confidence

The probability that the value of the measurand lies within the quoted range of uncertainty. ISO (1 ed. 1993)

Measurement uncertainty

Parameter, associated with the result of a measurement, that characterises the dispersion of the values that could reasonably be attributed to the measurand. VIM, 3.9 (1993)

Repeatability

Closeness of the agreement between the results of successive measurements of the same measurand carried out under the same conditions of measurement. VIM, 3.6 (1993). Measurements are carried out under repeatability conditions, *i.e.* conditions where independent test results are obtained with the same method on identical test items in the same laboratory by the same operator using the same equipment within short intervals of time.

Internal reproducibility

Closeness of the agreement between the results of measurements of the same measurand carried out under internal reproducibility conditions, *i.e.* conditions where independent test results are obtained with the same method on identical test items in the same laboratory by different operators using the same equipment at different times.

Reproducibility

Closeness of the agreement between the results of measurements of the same measurand carried out under changed conditions of measurement. VIM, 3.7 (1993). Measurements are carried out under reproducibility conditions, *i.e.* conditions where independent test results are obtained with the same method on identical test items in different laboratories by different operators using different equipment at different times.

Random error

Result of a measurement minus the mean that would result from an infinite number of measurements of the same measurand carried out under repeatability conditions. VIM, 3.13 (1993)

Systematic error

Mean that would result from an infinite number of measurements of the same measurand carried out under repeatability conditions minus a true value of the measurand. VIM, 3.14 (1993)

Coverage factor

A number that, when multiplied by the combined standard uncertainty, produces an interval (the expanded uncertainty) about the measurement result that may be expected to encompass a large, specified fraction (*e.g.* 95%) of the distribution of values that could reasonably be attributed to the measurand. ISO (1 ed. 1993).

4 Estimation of Measurement Uncertainty

Correction for systematic errors

Systematic errors should not be included in the measurement uncertainty, but should be corrected for. There are several ways to estimate a systematic error. Certified reference materials or materials calibrated in a traceable manner against certified reference materials may be analysed. Participation in recognised proficiency testing schemes may also reveal a systematic error.

Another useful technique is the determination of the recovery of an added analyte. The disadvantage associated with recoveries is that the results must be interpreted with caution. If the added species of the analyte is different from that in the sample, an added analyte may be recovered to 100% without the analyte present in the sample being included in the result. Thus, a poor recovery is an indication of problems, while a good recovery does not necessarily indicate that the result is correct.

The presence of a systematic error may also be investigated by analysing an authentic sample using in parallel the method under study and a well defined, established reference method; this approach is especially useful in the case of 'undefined' parameters such as fat and moisture.

If a systematic error has been identified, the laboratory should preferably eliminate the error. If this is not possible, analytical results obtained using the method should be corrected for the known error unless the error is negligible.

Estimation of measurement uncertainty

In order to make it possible to estimate measurement uncertainty, a parameter describing the uncertainty interval must be identified. The recommended parameter in this connection is the 'internal reproducibility standard deviation'. This can be calculated from analytical data collected on a material analysed under conditions of internal reproducibility. A prerequisite is the availability of a material from which identical test portions can be withdrawn and analysed over time. This model provides a realistic picture of the variation in results that the material may give rise to.

If for some reason it is not possible to analyse a material under internal reproducibility conditions, an alternative model must be used. This may be necessary, for example in cases where no reference materials are available for the purpose and where it is not possible to prepare internal control materials, for example due to an unstable analyte. In such cases (see below) the laboratory is compelled to use analytical data generated from replicate determinations. The repeatability standard deviation is calculated under conditions which do not, however, fully agree with the definition of repeatability. Duplicate determinations are made on materials containing the analyte in similar concentration levels at different times, and if relevant, using different analysts. It should be kept in mind that the repeatability standard deviation estimated in this manner will generally be smaller than the internal reproducibility standard deviation calculated at the same concentration level, and thus will result in an under-estimation of the measurement uncertainty.

Recommended procedure for estimation of the measurement uncertainty interval, based on the internal reproducibility standard deviation

In order to determine the internal relative reproducibility standard deviation (RSD), carry out replicate determinations on an authentic sample, an internal control material or a certified reference material. An internal control material must be stable, and must be available in sufficient amounts so that it can be used over a longer period of time. If a certified reference material is used, data obtained may also be used to investigate the presence of systematic errors. Certified reference materials should as a rule not be used solely to estimate random errors.

Perform at least 10 determinations at different times, if relevant using different analysts. If relevant, RSD should be estimated at different analyte levels, for example at low, medium and high levels. If the determination is associated with a legal norm, the norm level is suitable for the estimation of RSD. All relevant matrices should be included in the experimental work.

Calculate s and RSD from:

$$s = \sqrt{\frac{\sum(x_i - \bar{x})^2}{n-1}}; \mathrm{RSD} = \frac{s}{\bar{x}}$$

where s is the internal reproducibility standard deviation, $\bar{x}$ is the mean value, i is 1, 2, ... n, and, n is the number of determinations.

Below is given an example of the calculation of s from results of single determinations of an arbitrary analyte. A total of ten determinations, 1 to 10, were carried out by different analysts on different days. The results are presented in Table A1.

The values obtained are placed into the formula for the standard deviation:

$$s = \sqrt{\frac{\sum(x_i - \bar{x})^2}{n-1}} = \sqrt{\frac{32.4}{10-1}} = 1.897$$

Table A1 *Raw data for the calculation of s and RSD*

		Result (mg/kg)		
Determination no.	*Date*	x	$x - \bar{x}$	$(x - \bar{x})^2$
1	04.06.96	100.0	−2.4	5.76
2	05.06.96	103.9	+1.5	2.25
3	06.06.96	104.8	+2.4	5.76
4	07.06.96	104.0	+1.6	2.56
5	10.06.96	101.9	−0.5	0.25
6	11.06.96	103.0	+0.6	0.36
7	13.06.96	103.8	+1.4	1.96
8	14.06.96	99.5	−2.9	8.41
9	17.06.96	100.2	−2.2	4.48
10	18.06.96	102.9	+0.5	0.25
Number of analyses $n = 10$		Sum = 1024	Average $\bar{x} = 102.4$	$\sum(x - \bar{x})^2 = 32.4$

The internal relative reproducibility standard deviation is given by:

$$\mathrm{RSD} = \frac{s}{\bar{x}} = \frac{1.897}{102.4} = 0.0185$$

The magnitude of the obtained internal reproducibility standard deviation should be evaluated in relation to the field of application of the method (fitness for purpose).

Correspondingly, the measurement uncertainty may be estimated using results of replicate determinations carried out on reference materials, or from results of studies where the method under study was used in parallel with a well-defined, established reference method. Calculations are carried out as above using the reference value in place of the mean.

If the recommendation cannot be followed: estimation based on repeatability

In order to determine the repeatability, calculate the relative repeatability standard deviation, RSD_r, under conditions which do not, however, fully agree with the definition of repeatability. Perform duplicate determinations on materials with approximately the same analyte level. Carry out at least 10 duplicate determinations at different times, if relevant using different analysts. If relevant, determine RSD_r at different analyte levels, for example at low, medium and high levels. If the determination is associated with a legal norm, the norm level is a suitable level for the estimation of RSD_r. All relevant matrices should be included in the experimental work.

Calculate RSD_r from:

$$\mathrm{RSD}_r = \sqrt{\frac{\sum[(a_i - b_i)/\bar{x}_i]^2}{2d}}$$

where $(a_i - b_i)/\bar{x}_i$ is the relative difference between the duplicate results, i is 1, 2, ... n, and d is the number of duplicate determinations.

In the case of methods having a repeatability standard deviation which is proportional to the concentration of the analyte, the relative standard deviation may be estimated using materials covering a wider concentration range. However, it must be ensured that the same relative error may be associated with the entire range under study. Below is given an example of the calculation of RSD_r using results from duplicate determinations. Materials containing the analyte at different concentration levels are included.

The values obtained are placed into the formula for the relative repeatability standard deviation:

$$\mathrm{RSD}_r = \sqrt{\frac{\sum[(a_i - b_i)/\bar{x}_i]^2}{2d}} = \sqrt{\frac{0.0182}{2 \times 10}} = 0.0301$$

It is recommended that the internal quality control is carried out with the help of control charts based on the differences between results from duplicate determinations. Control charts may reveal whether the initially estimated measurement uncertainty is relevant over a longer period of time. Differences in uncertainties are easily detected in this way, and if they are sufficiently large it may be necessary to estimate a new value for s_r. The construction of a control chart is described in detail in NMKL Procedure No. 3 (1996). Control charts may be based either on absolute or on relative limits. At concentrations above the limit of quantification (10 standard deviations of the blank), many instrumental methods have a relative error only. In such cases a control chart based on relative limits may be used for all samples containing the analyte above the limit of quantification. On the other hand, gravimetric and titrimetric methods often have absolute errors, in which cases control charts with absolute limits are suitable.

The magnitude of the obtained repeatability should be evaluated in relation to the field of application of the method (fitness for purpose).

5 Expression of Measurement Uncertainty

The measurement uncertainty is a parameter which describes in a quantitative manner the variation which the analyte present in the sample gives rise to. In order for an estimated measurement uncertainty to be of value, it must be estimated and expressed in a standardised manner.

Measurement uncertainty can either be expressed as the standard measurement uncertainty, defined as 'measurement uncertainty of the result as one standard deviation', or as an expanded measurement uncertainty, defined as 'a quantity which defines the interval about the result, including a large portion

Table A2 *Raw data for the calculation of RSD_r*

Determination no.	Date	Result (mg/kg)				
		a_i	b_i	$\bar{x}_i$	$(a_i - b_i)/\bar{x}_i$	$[(a_i - b_i)/\bar{x}_i]^2$
1	06.09.96	10.9	11.5	11.2	−0.0536	0.00287
2	07.09.96	109.7	114.1	111.9	−0.0393	0.00155
3	12.09.96	52.3	50.3	51.3	+0.0390	0.00152
4	13.09.96	23.0	23.9	23.45	−0.0384	0.00147
5	17.09.96	11.3	11.9	11.6	−0.0517	0.00268
6	18.09.96	110	105	107.5	+0.0465	0.00216
7	19.09.96	87.9	91.2	89.55	−0.0369	0.00136
8	22.09.96	74.1	77.2	75.65	−0.0410	0.00168
9	24.09.96	13.7	13.2	13.45	+0.0372	0.00138
10	28.09.96	34.4	33.1	33.75	+0.0385	0.00148

Number of duplicate analyses $d = 10$ Sum $\sum[(a_i - b_i)/\bar{x}_i]^2 = 0.0182$

of the variation which would result from the analyte present in the sample, and which is obtained by multiplying the standard measurement uncertainty with a coverage factor'. This coverage factor usually equals 2; in some cases a factor of 3 may be used. Application of a coverage factor of 2 corresponds to a conidence level of 95%, and a coverage factor of 3 to a confidence level of more than 99%.

The expanded measurement uncertainty U is given by:

$$U = k \times \text{RSD} \times c \text{ or } U = k \times \text{RSD}_r \times c$$

where k is the coverage factor, RSD or RSD_r is the relative standard deviation calculated as described above, and c is the concentration of the analyte.

It is recommended that the coverage factor 2 is used.

The calculated expanded measurement uncertainty U represents half of the measurement uncertainty interval. The following format is usually applied to express the entire measurement uncertainty interval: 'measurement result ± U'.

Example 1 (internal reproducibility)

Using the RSD value obtained from the results listed in Table A1 and a measurement result of 99.4 mg/kg, the following expanded measurement uncertainty is obtained:

$$U = k \times \text{RSD} \times c = 2 \times 0.0185 \times 99.4\,\text{mg/kg} = 3.7\ \text{mg/kg}$$

The measured result 99.4 mg/kg accompanied by its expanded measurement uncertainty may be given as 99.4 ± 3.7 mg/kg.

The interval of the expanded measurement uncertainty is thus from 95.7 to 103.1 mg/kg. This interval should include about 95% of the total variation caused by the analyte present in the sample.

Example 2 (repeatability)
Using the RSD_r value obtained from the results listed in Table A2 and a measurement result of 23.4 mg/kg, the following expanded measurement uncertainty is obtained:

$$U = k \times RSD_r \times c = 2 \times 0.0301 \times 23.4\,\mathrm{mg/kg} = 1.4\,\mathrm{mg/kg}$$

The measured result 23.4 mg/kg accompanied by its expanded measurement uncertainty may be given as 23.4 ± 1.4 mg/kg.

Expression of measurement uncertainty in an analytical report

The measurement uncertainty should be reported in a standardised manner. The result, x, should be reported together with the expanded measurement uncertainty, U, calculated using a coverage factor of $k = 2$. The following format is recommended:

(Analyte): $x \pm U$ (units)*
* The reported measurement uncertainty is [an expanded measurement uncertainty according to 'Estimation and expression of measurement uncertainty in chemical analysis', NMKL Procedure No. 5, version 1, 1997] calculated using 2 as the coverage factor [, which gives a confidence level of approximately 95%].

The texts within square brackets [ ] may be omitted or abbreviated in a suitable manner.

Example
Lead: 0.025 ± 0.006 mg/kg
* The reported measurement uncertainty is calculated using 2 as the coverage factor, which gives a confidence level of approximately 95%.

In some cases coverage factors of either $k = 1$ or $k = 3$ may be considered. These correspond to confidence levels of 68% and more than 99%, respectively.

When a coverage factor $k = 1$ is used, *i.e.* the measurement uncertainty is estimated to one standard deviation, the uncertainty is called standard measurement uncertainty, and is desigated u. In such cases the following report format is recommended:

(Analyte): x (units) with a standard measurement uncertainty u (units) [where standard measurement uncertainty is defined according to 'Estimation and expression of measurement uncertainty in chemical analysis', NMKL Procedure No. 5, version 1, 1997, and corresponds to one standard deviation.]

Note that it is not recommended to use the symbol ± when reporting standard measurement uncertainty, since this symbol is usually associated with high confidence intervals.

The text within square brackets [ ] may be omitted or suitably shortened.

Example

Lead: 0.025 mg/kg

Standard measurement uncertainty: 0.003 mg/kg*

* The standard measurement uncertainty corresponds to one standard deviation.

The numerical values of results and measurement uncertainty should not be expressed using an unnecessary number of figures. It is recommended that measurement uncertainty is reported using two significant figures and that the result is rounded off to correspond with the stated measurement uncertainty.

6 Measurement Uncertainty and the Interpretation of Results

The analytical results from a chemical laboratory are usually compared to a norm or a specification. These may be

- a legal norm set in legislation,
- a guideline value given in the legislation, or
- a specification defined by a manufacturer or a client.

Those issuing norms or specifications seldom provide guidance on how the measurement uncertainty should be taken into account in evaluating whether or not a product meets the requirements of the norm or specification. Food legislation, for example, contains numerous norms on contaminants and food additives, but no guidance on how to address measurement uncertainties. Especially in the case of limits concerning low concentration levels, the way in which the measurement uncertainty is taken into account may have a significant influence on the interpretation of results.

In order to harmonise the interpretation of chemical results, the following examples are given on how to take measurement uncertainty into account. The examples follow principles outlined in the NAMAS publication NIS 80.

Examples of how analytical results accompanied by estimates of measurement uncertainties should be interpreted against specification limits (see Figure A1)

Case A

The measured result is within the limits, even when extended by the uncertainty interval. The product therefore complies with the specification.

Case B

The measured result is below the upper limit, but by a margin less than half of the uncertainty interval; it is therefore not possible to state compliance based on the 95% level of confidence. However, the result indicates that compliance is more probable than non-compliance.

Case C

The measured result is above the upper limit, but by a margin less than half of the uncertainty interval; it is therefore not possible to state compliance based on the 95% level of confidence. However, the result indicates that non-compliance is more probable than compliance.

Case D

The measured result is beyond the limits, even when extended downwards by half of the uncertainty interval. The product therefore does not comply with the specification.

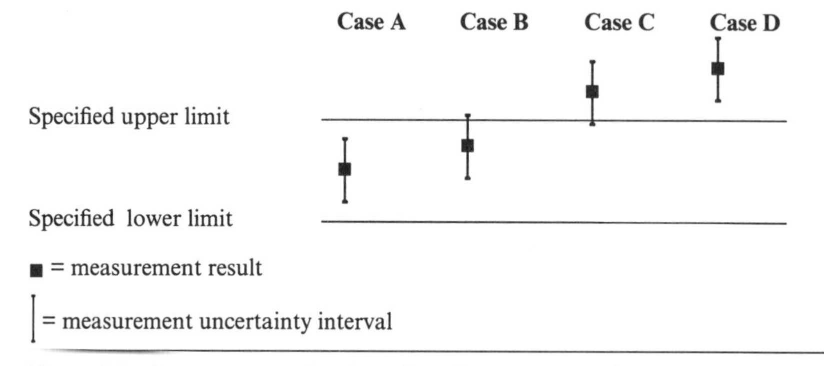

Figure A1 *Interpretation of analytical results against specification limits*

References in NMKL Guide

1 Quantifying Uncertainty in Analytical Measurement, EURACHEM, 1995, ISBN 0-948926-08-2.
2 Control charts and control samples in the internal quality control in chemical food laboratories, NMKL Procedure no. 3, 1996.
3 NAMAS NIS 80: Guide to the Expression of Uncertainties in Testing, 1994, 1st Edition, NAMAS Executive, Teddington, England.
4 Guide to the Expression of Uncertainty in Measurement, ISO, Geneva, 1st Edition, 1993, ISBN 92-67-10188-9.
5 Guidelines for Evaluating and Expressing the Uncertainty of NIST Measurement Results, NIST Technical note 1297, 1994 Edition.
6 Accreditation for Chemical Laboratories: Guidance on the Interpretation of the EN 45000 series of Standards and ISO/IEC Guide 25, EURACHEM Guidance Document No. 1/WELAC Guidance No. WGD 2, 1993.
7 Uncertainty of Measurement: Implications of its Use in Analytical Science, The

Royal Society of Chemistry, Analytical Methods Committee, *Analyst (Cambridge)*, 1995, **120**, 2203.

8 Guidance for Estimating and Expression of Measurement Uncertainty, SWEDAC Doc 95:5, SWEDAC Swedish Board for Technical Accreditation, Borås, 1995 (in Swedish only).

9 International Vocabulary of Basic and General Terms in Metrology, VIM, 2nd Edition, 1993, ISO, Geneva.

CHAPTER 6

Role of Internal Quality Control in Laboratory Quality Assurance Procedures

The IUPAC Inter Divisional Working Group on Analytical Quality Assurance has co-operated with the AOAC International and ISO to produce guidelines on the internal quality control (IQC) of data produced in chemical analytical laboratories; they set out the minimum recommendations to laboratories producing analytical data on the internal quality control procedures to be employed.

The published guidelines[1] are of particular relevance to food analysts, having been discussed in the Codex Alimentarius Commission and are now endorsed for use for Codex purposes.[2] Because of this endorsement for food analysis purposes, this chapter describes IQC as outlined in the guidelines in detail below.

International guidelines for internal quality control, such as those from IUPAC, do not exist for microbiological laboratories. However, the subject has been discussed in different publications and manuals.[3–9]

Although the theoretical basis for internal quality control is the same for both disciplines, the approach to IQC in microbiology has to be slightly different to that in chemistry. As mentioned previously, few microbial reference materials are available for qualitative analyses. In addition, the shelf lives of the current reference samples for quantitative determinations are somewhat limited because of decay in the number of living micro-organisms within a few years. A different approach is necessary for microbiological analysis because it is relatively more dependent on staff competence and manual dexterity, rather than on the performance of analytical instruments. In addition, the fact that very few microbiological methods have been assessed through a collaborative trial causes some difficulties.

1 Introduction

Internal quality control (IQC) is an essential aspect of ensuring that data released from a laboratory are 'fit for purpose'. If properly executed, quality control methods enable the monitoring of the quality of data produced by a laboratory on a run-by-run basis. In runs where performance falls outside acceptable limits, the data produced should be rejected and, after remedial action to the analytical system, the analysis may be repeated and the data then accepted.

Procedures for IQC within analytical chemistry are well known and have been in use for a long period of time. Microbiologists are now developing similar procedures, from experiences in analytical chemistry, to embrace their specific needs.

It must be appreciated that IQC is not foolproof even when properly carried out. It is subject to 'errors of both kinds', *i.e.* runs that are in control will occasionally be rejected and runs that are out of control occasionally accepted. In addition, IQC cannot usually identify sporadic gross errors or short-term disturbances in the analytical system that affect the results for individual test materials. Inferences based on IQC results are applicable only to test materials that fall within the scope of the analytical method validation. Despite these limitations, which experience and diligence can reduce to a degree, internal quality control is the principal recourse available for ensuring that only data of appropriate quality are released from a laboratory. When properly executed it is very successful.

It must be appreciated that a perfunctory execution of any quality system will not guarantee the production of data of adequate quality. The correct procedures for feedback, remedial action and staff motivation must also be documented and acted upon. There must be a genuine commitment to quality within a laboratory for an internal quality control programme to succeed, *i.e.* the IQC must be part of a total quality management system.

Recommendations on the extent that IQC can be usefully incorporated into the work of the laboratory is given in this chapter. It is important that laboratories give serious consideration to the extent of introduction of the recommendations especially as they have now been internationally accepted.

2 Basic Concepts of IQC

IQC is one of a number of concerted measures that analytical chemists and microbiologists can take to ensure that the data produced in the laboratory are fit for their intended purpose. In practice, fitness for purpose is determined by a comparison of the accuracy achieved in a laboratory at a given time with a required level of accuracy. IQC therefore comprises the routine practical procedures that enable the analytical chemist and microbiologist to accept a result or group of results as fit for purpose, or reject the results and repeat the analysis. As such, IQC is an important determinant of the quality of analytical data, and is recognised as such by accreditation agencies.

Internal quality control is undertaken by the inclusion of particular reference materials, reference samples or reference strains, usually known as 'control materials', into the analytical sequence and by duplicate analysis. Chemical control materials should, wherever possible, be representative of the test materials under consideration in respect of matrix composition, the state of physical preparation and the concentration range of the analyte. As the control materials are treated in exactly the same way as the test materials, they are regarded as surrogates that can be used to characterise the performance of the analytical system, both at a specific time and over longer intervals. In microbiology it is rather more difficult to obtain good reference material with natural matrixes. The use of 'spiked samples' could therefore be recommended. Some reference samples with single species of micro-organisms or mixtures of strains are available, however, and may be used for quantitative and qualitative microbiological determinations.

IQC is a final check of the correct execution of all of the procedures (including calibration or the final identification step in a microbiological examination) that are prescribed in the analytical protocol and all of the other quality assurance measures that underlie good analytical practice. IQC is therefore a retrospective procedure. It is also required to be as far as possible independent of the analytical protocol, especially the calibration, that it is designed to test.

Ideally both the control materials, and those used to create the calibration and identification of chemicals and micro-organisms, should be traceable to appropriate certified reference materials or a recognised empirical reference method. When this is not possible, control materials should be traceable at least to a material or strain of guaranteed purity and identity, or other well characterised material. However, the two paths of traceability must not become coincident at too late a stage in the analytical process. For instance, if chemical control materials and calibration standards were prepared from a single stock solution of analyte, IQC would not detect any inaccuracy stemming from the incorrect preparation of the stock solution. In the same way, a strain identified for example with an API-kit would not be useful for functional controls of such kits.

In a typical analytical situation several, or perhaps many, similar test materials will be analysed together in the 'analytical run' and control materials will be included in the run. Often determinations will be duplicated by the analysis of separate test portions of the same material. Runs are regarded for IQC purposes as being analysed under effectively constant conditions. The batches of reagents, microbiological substrates, the instrument settings, the analyst, and the laboratory environment will, under ideal conditions, be assumed to remain unchanged during analysis of a run. Systematic errors should therefore remain constant during a run, as should the values of the parameters that describe random errors. As the monitoring of these errors is of concern, the 'run' is the basic operational unit of IQC.

A run is regarded as being carried out under repeatability conditions, *i.e.* the random measurement errors are of a magnitude that would be encountered in a 'short' period of time. In practice the analysis of a run may occupy sufficient time for small systematic changes to occur. For example, reagents may degrade, instruments may drift, minor adjustments to instrumental settings may be called for, or the laboratory temperature may rise. However, these systematic effects are, for the purposes of IQC, subsumed into the repeatability variations. Sorting the materials making up a run into a randomised order converts the effects of drift into random errors.

3 The Scope of the Internal Quality Control Guidelines

The guidelines which have been endorsed by the Codex Alimentarius Commission are a harmonisation of IQC procedures that have evolved in various fields of chemical analysis, notably clinical biochemistry, geochemistry and environmental studies, occupational hygiene and food analysis.[10–16] There is much common ground in the procedures from these various fields. Thus, the guidelines provide information that is widely applicable, not only in the food sector.

There are a number of particular areas which are not normally of concern to chemical and microbiological food analysts and examiners or for which international recommendations are not yet available. Thus, in order to provide basic guidance on IQC, some types of analytical activity have been excluded; the following are not, therefore, addressed in this chapter:

Statutory and contractual requirements
Quality *cannot* be obtained in chemical and microbiological analytical results without consideration of statutory and contractual requirements. 'Quality' has been defined as 'conformance to customer requirements'.[17] IQC, however, traditionally does not take this point into consideration. Nevertheless, laboratories have to emphasise this part of their quality work as the customer is the ultimate judge in defining quality.

Quality control of sampling
Although it should be appreciated that the quality of the analytical result can be no better than that of the sample, quality control of sampling is a separate subject and in many areas is not fully developed. Moreover, in many instances analytical laboratories have no control over sampling practice and quality.

In-line analysis and continuous monitoring
In the case of these analyses, as there is no possibility of repeating the measurement, the concept of IQC as described in this chapter is not applicable.

Biochemical analysis such as DNA sequencing, ELISA and PCR
Techniques such as DNA sequencing, ELISA and PCR (polymerase chain reaction) are useful both within food chemistry and food microbiology. Indeed, these techniques make the borderline between the two disciplines blurred and so new specialists have to evolve who are competent in both fields. The 'new' biochemical techniques are developing fast but are not yet as common in use as routine techniques as traditional chemical and microbiological procedures. As a consequence, new IQC procedures have to be developed to enable DNA sequencing, ELISA and PCR, *etc.* to be useful for routine analysis.

Multivariate IQC
Multivariate methods in IQC are still the subject of research and are not sufficiently established for discussion in this chapter. The current informa-

tion/recommendations may be considered to regard multianalyte data as requiring a series of univariate IQC tests. Caution is necessary in the interpretation of this type of data to avoid inappropriately frequent rejection of data.

Quality assurance measures
Measures such as checks on instrumental stability before and during analysis, wavelength calibration, balance calibration, tests on resolution of chromatography columns, and problem diagnostics are not discussed in this chapter. They may be regarded as part of the analytical protocol, and IQC tests their effectiveness together with the other aspects of the methodology.

4 Internal Quality Control and Uncertainty

A prerequisite of analytical chemistry is the recognition of 'fitness for purpose', the standard of accuracy that is required for an effective use of the analytical data. This standard is arrived at by consideration of the intended uses of the data, although it is seldom possible to foresee all of the potential future applications of analytical results. For this reason, in order to prevent inappropriate interpretation it is important that the chemical laboratory should be able to give its customers a statement of the uncertainty of its analytical results.

In traditional microbiology the standard accuracy most often is set by the choice of analytical method. The accuracy is seldom known as few methods are collaboratively evaluated or have their method performance characteristics determined in other ways. The analysis for the presence of pathogenic organisms is often carried out as single analysis, whereas microbiological food examinations involve multi-method evaluations for the presence and concentration of different micro-organisms. It is therefore not appropriate to consider uncertainty of measurement for microbiological analyses in the same way as for chemical analysis.

Strictly speaking, a chemical analytical result cannot be interpreted unless it is accompanied by knowledge of its associated uncertainty at a stated level of confidence. A simple example demonstrates this principle. There may be a statutory requirement that a foodstuff must not contain more than 2 mg lead per kg. A manufacturer may analyse a batch and obtain a result of 1.8 mg lead per kg. If the uncertainty of the result expressed as a half range (assuming no sampling error) is 0.2 mg lead per kg (*i.e.* the true result falls, with a high probability, within the range 1.8–2.2 mg) then it may be assumed that the legal limit is not exceeded. If, in contrast, the uncertainty is 0.5 mg lead per kg then there is no such assurance. The interpretation and use that may be made of the measurement thus depends on the uncertainty associated with it.

Analytical results should therefore have an associated uncertainty if any definite meaning is to be attached to them or an informed interpretation made. If this requirement cannot be fulfilled, the use to which the data can be put is limited. Moreover, the achievement of the required measurement uncertainty must be tested as a routine procedure, because the quality of data can vary, both in time within a single laboratory and between different laboratories. IQC comprises the process of checking that the required uncertainty is achieved in a run. However, laboratories have to consider the complications of *always* presenting uncertainties of measurements in analytical reports as discussed previously in this book.

5 Quality Assurance Practices and Internal Quality Control

Quality Assurance

Quality assurance is the essential organisational infrastructure that underlies all reliable chemical and microbiological analytical measurements. It is concerned with achieving appropriate levels in matters such as co-operation with clients, staff training and management, adequacy of the laboratory environment, safety, the storage, integrity and identity of samples, record keeping, the maintenance and calibration of instruments, and the use of technically validated and properly documented methods. Failure in any of these areas might undermine vigorous efforts elsewhere to achieve the desired quality of data. In recent years these practices have been codified and formally recognised as essential. However, the prevalence of these favourable circumstances by no means ensures the attainment of appropriate data quality unless IQC is conducted.

Choice of Analytical Method

It is important that laboratories restrict their choice of methods to those that have been characterised as suitable for the matrix and analyte (*i.e.* the chemical or microbiological specification) of interest. The laboratory must possess documentation describing the performance characteristics of the method, estimated under appropriate conditions. Although this may be particularly difficult for present day microbiological methods, it is something that will have to be developed for procedures in the future.

The use of a method does not in itself guarantee the achievement of its established performance characteristics. There is, for a given method, only the potential to achieve a certain standard of reliability when the method is applied under a particular set of circumstances. It is this collection of circumstances, known as the 'analytical system', that is therefore responsible for the accuracy of analytical data. Hence it is important to monitor the analytical system (including when appropriate the final evaluation step of analytical data) in order to achieve fitness for purpose. This is the aim of the IQC measures undertaken in a laboratory.

Internal Quality Control and Proficiency Tests

Proficiency testing is a periodic assessment of the performance of individual laboratories and groups of laboratories that is achieved by the distribution by an independent testing body of typical materials for unsupervised analysis by the participants.[18] Although important, it must be emphasised that participation in proficiency testing schemes is not a substitute for IQC measures, or *vice versa*.

Proficiency testing schemes can be regarded as a routine, but relatively infrequent, check on analytical errors. Without the support of a well-developed IQC system, the value of participation in a proficiency test is negligible. Probably the main beneficial effect of proficiency tests is that of encouraging participants to install effective quality control systems. It has already been shown that laboratories with effective IQC systems performed better in a proficiency testing scheme.[19]

6 Internal Quality Control Procedures

Introduction

Internal quality control involves the practical steps undertaken to ensure that errors in analytical data are of a magnitude appropriate for the use to which the data will be put. The practice of IQC depends on the use of two strategies, *i.e.* the analysis of reference materials or reference samples to monitor trueness and statistical control, and duplication to monitor precision.

The basic approach to IQC involves the analysis of control materials alongside the test materials under examination. The outcome of the control analyses forms the basis of a decision regarding the acceptability of the test data. Two key points are worth noting in this context.

1. The interpretation of control data must be based on documented, objective criteria, and on statistical principles wherever possible.
2. The results of control analyses should be viewed primarily as indicators of the performance of the analytical system, and only secondarily as a guide to the errors associated with individual test results. Substantial changes in the apparent accuracy of control determinations can sometimes be taken to imply similar changes to data for contemporary test materials, but correction of analytical data on the basis of this premise is unacceptable.

To get an effective and functional IQC, one also has to consider problems such as the evaluation of IQC data being performed long after the analysis has been performed, *etc.* and as discussed later in this book (Chapter 12, Part C).

General Approach—Statistical Control

The interpretation of the results of IQC analyses depends to a large extent on the concept of statistical control, which corresponds with stability of operation. Statistical control implies that an IQC result x can be interpreted as arising independently and at random from a normal population with mean μ and variance σ.[2]

Under these constraints, only about 0.3% of results (x) would fall outside the bounds of $\mu \pm 3\sigma$. When such extreme results are encountered they are regarded as being 'out-of-control' and interpreted to mean that the analytical system has started to behave differently. Loss of control therefore implies that the data produced by the system are of unknown accuracy and hence cannot be relied upon and reported. The analytical system therefore requires investigation and remedial action before further analysis is undertaken. Compliance with statistical control can be monitored graphically with Shewhart control charts which are described below.

Internal Quality Control and Fitness for Purpose

For the most part, the process of IQC is based on a description in terms of the statistical parameters of an ongoing analytical system in normal operation. Control limits are therefore based on the estimated values of these parameters rather than measures derived from considerations of fitness for purpose. Control limits must be narrower than the requirements of fitness for purpose or IQC analyses would not be worthwhile undertaking.

Ad Hoc Analysis

The concept of statistical control is inappropriate, however, when *ad hoc* analyses are undertaken. In *ad hoc* analysis the test materials may be unfamiliar or rarely encountered, and runs are often made up of only a few such test materials. Under these circumstances there is no statistical basis for the construction of control charts. In such an instance the analytical chemist or microbiologist has to use fitness for purpose criteria, historical data or consistency with the visual properties of the test material for judging the acceptability of the results obtained.

Whichever procedure is used, agreed methods of establishing quantitative criteria to characterise fitness for purpose are desirable. However, this is one of the less-developed aspects of IQC. In food analysis the Horwitz curve[20] is frequently used as a fitness for purpose criterion, this particularly in the light of the 'criteria approach' to methods of analysis described in Chapter 3.

The Nature of Errors

Two main categories of analytical error are recognised, namely random errors and systematic errors, which give rise to imprecision and bias respectively. The importance of categorising errors in this way lies in the fact that they have different sources, remedies and consequences for the interpretation of data. They are described below:

Random errors determine the precision of measurement. They cause random positive and negative deviations of results about the underlying mean value.

Systematic errors comprise displacement of the mean of many determinations from the true value. For the purposes of IQC, two levels of systematic error should be considered, these being:

1. *Persistent bias*, which affects the analytical system (for a given type of test material) over a long period and affects all data. Such bias, if small in relation to random error, may be identifiable only after the analytical system has been in operation for a long time. It might be regarded as tolerable, provided it is kept within prescribed bounds.
2. *The run effect*, exemplified by a deviation of the analytical system during a particular run. This effect, where it is sufficiently large, will be identified by IQC at the time of occurrence as an out-of-control condition.

The conventional division of errors between the random and the systematic depends on the timescale over which the system is viewed. Run effects of unknown source can be regarded in the long term as the manifestation of a random process. If a shorter-term view is taken, the same variation could be seen as a bias-like change affecting a particular run.

The statistical model used for IQC used in this chapter is as follows. The value of a measurement (x) in a particular run is given by:

$$x = \text{true value} + \text{persistent bias} + \text{run effect} + \text{random error } (+ \text{ gross error}).$$

The model can be extended if necessary to include other features of the analytical system.

The variance of $x(\sigma_x^2)$ in the absence of gross errors is given by:

$$\sigma_x^2 = \sigma_0^2 + \sigma_1^2$$

where σ_0^2 is the variance of the random error (within run), and σ_1^2 is the variance of the run effect.

The variances of the true value and the persistent bias are both zero. An analytical system in control is fully described by σ_0^2, σ_1^2 and the value of the persistent bias. Gross errors are implied when the analytical system does not comply with such a description.

Statistical evaluation of results from quantitative microbiological examinations have been discussed in greater detail elsewhere.[21,22]

7 IQC and Within-run Precision

Precision and Duplication

A limited control of within-run precision is achieved by the duplication within a run of measurements made on test materials, an approach that is readily applicable to analytical chemistry. Routinely duplicating microbiological analysis, however, is less appropriate, as the variations in cfu-counts are normally large, a consideration that makes the evaluation much more difficult for the microbiologist. One reason to this variation is the low number of microorganisms that are counted and the inhomogeneity of microbial samples.

The objective is to ensure that the differences between paired results are consistent with or better than the level implied by the value of σ_0 used by a laboratory for IQC purposes. The standard deviation of repeatability σ_r should not be estimated from the IQC data and nor should estimates be compared; there are usually too few results for this to be valid. Where such an estimate is needed the formula

$$s_r = \sqrt{\sum d^2/2n}$$

can be used.

The test indicates to the user the possibility of poor within-run precision and provides additional information to help in interpreting control charts. The method is especially useful in *ad hoc* analysis, where attention is centred on a single run and information obtained from control materials is unlikely to be completely satisfactory.

As a general approach all of the test materials, or a random selection from them, are analysed in duplicate. The absolute differences $|d| = |x_1 - x_2|$ between duplicated analytical results x_1 and x_2 are tested against an upper control limit based on an appropriate value of σ_0. However, if the test materials in the run have a wide range of concentration of analyte, no single value of σ_0 can be assumed.

The duplicate material included in the analytical run for IQC purposes must reflect as far as possible the full range of variation present in the run. Duplicates must not be analysed as adjacent members of the run, otherwise they will reveal only the smallest possible measure of analytical variability. The best placing of duplicates is at random within each run. Moreover, the duplication required for IQC requires the complete and independent analysis (preferably blind, even if rarely achieved in practice) of separate test portions of the test material. Duplication of the instrumental measurement of a single test solution is ineffective because the variations introduced by the preliminary chemical treatment of the test material would be absent.

Interpretation of Duplicate Data

The following are the situations in which duplicate data must be interpreted:

Narrow Concentration Range

In the simplest situation the test materials comprising the run have a small range of analyte concentrations so that a common within-run standard deviation σ_0 can be applied. A value of this parameter must be estimated to provide a control limit. The upper 95% bound of $|d|$ is $2\sqrt{2}\sigma_0$ and on average only about three in a thousand results should exceed $3\sqrt{2}\sigma_0$

A group of n duplicated results can be interpreted in several ways.

For example, the standardised difference

$$z_d = d/\sqrt{2}\sigma_0$$

should have a normal distribution with zero mean and unit standard deviation. The sum of a group of n such results would have a standard deviation of $\sqrt{n}$, so only about three runs in a thousand would produce a value of $|\sum z_d| > 3\sqrt{n}$

Alternatively a group of n values of z_d from a run can be combined to form $\sum z_d^2$ and the result interpreted as a sample from a chi-squared distribution with n degrees of freedom (χ_n^2). Caution is needed in the use of this statistic as it is sensitive to outlying results.

Wide Concentration Range

If the test materials comprising a run have a wide range of analyte concentrations, no common standard of precision (σ_0) can be assumed. In such an instance, σ_0 must be expressed as a functional relationship with concentration. The value of concentration for a particular material is taken to be $(x_1 + x_2)/2$, and an appropriate value of σ_0 obtained from the functional relationship, the parameters of which have to be estimated in advance.

8 Suitable Materials that can be Used for IQC Purposes

Control materials are characterised substances that are inserted into the run alongside the test materials and subjected to exactly the same treatment. A control material must contain an appropriate concentration of the analyte, and a value of that concentration must be assigned to the material. Control materials act as surrogates for the test materials and must therefore be representative, *i.e.* they should be subject to the same potential sources of error. To be fully representative, a control material must have the same matrix in terms of bulk composition, including minor constituents that may have a bearing on accuracy. It should also be in a similar physical form, *i.e.* state of comminution, as the test materials. There are other essential characteristics of a control material:

1. It must be adequately stable over the period of interest.
2. It must be possible to divide the control material into effectively identical portions for analysis.
3. It is often required in large amounts to allow its use over an extended period.

Reference materials in IQC are used in combination with control charts that allow both persistent bias and run effects to be addressed. Persistent bias is evident as a significant deviation of the centre line from the assigned value. The variation in the run effect is predictable in terms of a standard deviation when the system is under statistical control, and that standard deviation is used to define action limits and warning limits at appropriate distances from the true value.

Certified Reference Materials

Certified reference materials (CRMs) as defined in Chapter 14 (*i.e.* with a statement of uncertainty and traceability), when available and of suitable composition, are ideal control materials in that they can be regarded for traceability purposes as ultimate standards of trueness.[23] In the past, CRMs were regarded as being for reference purposes only and not for routine use. A more modern approach is to treat CRMs as consumable and therefore suitable for IQC.

The use of CRMs in this way is, however, subject to a number of constraints:

1. Despite the constantly increasing range of CRMs available, for the majority of analyses there is no closely-matching CRM available. This is especially true in the food sector where a very wide range of analyte/matrix combinations may be undertaken by the laboratory.
2. Although the cost of CRMs is not prohibitive in relation to the total costs of analysis, it may not be possible for a laboratory with a wide range of activities to stock every relevant kind of reference material, again a consideration particularly applicable to the food sector.
3. The concept of the reference material is not applicable to materials where either the matrix or the analyte is unstable.
4. CRMs are not necessarily available in sufficient amounts to provide for IQC use over extended periods.
5. Not all apparently certified reference materials are of equal quality. Caution should be exercised when the information on the certificate is inadequate.

6. The use of CRMs is sometimes hampered by the lack of information prepared by CRM producers on how they should be used and by not giving clear guidelines on how to evaluate results from such use. Whereas certified levels often are reported as 95% confidence intervals, many guidelines only supply information on how to evaluate laboratory results compared to CRMs in terms of standard deviations; this has been commented on elsewhere.[24]

If for any of the above reasons the use of a CRM is not appropriate, it falls on to an individual laboratory or to a group of laboratories to prepare their own control materials and assign traceable values of analyte concentration to them. Where a CRM is not available, traceability only to a reference method or to a batch of a reagent supplied by a manufacturer may be necessary. Such materials are referred to as 'house reference materials' (HRM). Suggestions for preparing HRMs are given below. Not all of the methods described are applicable to all situations.

Microbial Reference Organisms; Certified Reference Organisms and Control Strains

Qualitative microbiological analysis (*e.g.* presence/absence or identification) is based on morphological and biochemical characters. However, as the biological variation within species is large, it will never be appropriate and practical for a laboratory to have certified reference organisms available for all types of identification. This is especially the case where organisms slowly dying and being re-cultivated more than a few times leads to a loss of important characteristics. Microbial (certified) reference organisms are available from international culture collections such as the American-type Culture Collection (ATCC), the National Collection of Industrial & Marine Bacteria Ltd (Aberdeen, UK), the National Collection of Type Cultures (NCTC) and the Culture Collection, University of Gothenburg (CCUG).

Reference samples, some of which are 'certified', are available for the qualitative analysis of a few species of micro-organisms. Examples of such are the Community Bureau of Reference (BCR) artificially contaminated dry-milk powder with *Salmonella typhimorium* or *Listeria monocytogenes* present. Freeze-dried mixtures of food or drinking water micro-organisms with known concentration levels, to be used for quantitative and qualitative analysis, can be supplied from the Swedish National Food Administration in Uppsala.

Preparation of Control Materials

Discussions below result from experiences from preparation of chemical reference and control materials. Many of the comments, however, are equally valid for the production of microbial materials.

Assigning a True Value by Analysis

In principle a working value can be assigned to a stable reference material simply by careful analysis. However, precautions are necessary to avoid biases in the assigned value. This requires some form of independent check such as may be provided by analysis of the materials in a number of laboratories and, where possible, the use of methods based on different physico-chemical principles. Lack of attention to independent validation of control materials has been shown to be a weakness in IQC systems.[19]

One way of establishing a traceable assigned value in a chemical control material is to analyse a run comprising the candidate material and a selection of matching CRMs, with replication and randomisation. This course of action would be appropriate if limited amounts of CRMs were available. The CRMs must be appropriate in both matrix composition and analyte concentration. The CRMs are used directly to calibrate the analytical procedure for the analysis of the control material. An appropriate analytical method is a prerequisite for this approach. It would be a dangerous approach if, say, a minor and variable fraction of the analyte were extracted for measurement. The uncertainty introduced into the assigned value must also be considered. However, the constraints on the uses of CRMs described previously must be appreciated.

Materials Validated in Proficiency Testing

These comprise a valuable source of control materials. Such materials would have been analysed by many laboratories using a variety of methods (occasionally a single method has been used). In the absence of counter-indications, such as an obvious bias or unusual frequency distribution of results, the consensus of the laboratories could be regarded as a validated assigned value to which a meaningful uncertainty could be attached. (There is a possibility that the consensus could suffer from a bias of consequence, but this potential is always present in reference values.) There would be a theoretical problem of establishing the traceability of such a value, but that does not detract from the validity of the proposed procedure. The range of such materials available would be limited, but organisers of proficiency tests could ensure a copious supply by preparing batches of material in excess of the immediate requirements of the round. The normal requirements of stability would have to be demonstrable.

The availability of such materials is relatively good in the food sector as the requirement for laboratories to participate in proficiency testing schemes is being encouraged and so there are well defined and organised proficiency testing schemes in the food analysis sector.

Assigning a True Value by Formulation

In favourable instances a control material can be prepared simply by mixing constituents of known purity in predetermined amounts. For example, this approach would often be satisfactory in instances where the control material is a solution. Problems are often encountered in formulation in producing solid control materials in a satisfactory physical state or in ensuring that the speciation and physical distribution of the analyte in the matrix is realistic. Moreover, an adequate mixing of the constituents must be demonstrable. This method is less useful when preparing microbiological reference materials.

Spiked Control Materials

'Spiking' is a way of creating a control material in which a value is assigned by a combination of formulation and analysis. Although spiking is used less often in microbial than chemical analysis, it is also a useful procedure in microbiology.

This method is feasible when a test material essentially free of the analyte (chemical or micro-organism) is available. After exhaustive analytical checks to ensure the background level is adequately low, the material is spiked with a known amount of analyte. The reference sample prepared in this way is thus of the same matrix as the test materials to be analysed and of known analyte level; the uncertainty in the assigned concentration is limited only by the possible error in the unspiked determination. However, it may be difficult to ensure that the speciation, binding and physical form of the added analyte are the same as that of the native analyte and that the mixing is adequate. Although a general problem, this is seldomly discussed in the microbiology sector. Many microbial strains can be strongly bound to surfaces, giving only small fractions of total numbers to be counted in the supernatant of a homogenate.

Recovery Checks

If the use of a reference material is not practicable then a limited check on bias is possible by a test of recovery. This is especially useful when analytes (chemical or microbial) or matrices cannot be stabilised or when *ad hoc* analysis is executed. A test portion of the test material is spiked with a known amount of the analyte and analysed alongside the original test material. The recovery of the added analyte (known as the 'marginal recovery') is the difference between the two measurements divided by the amount that is added. The obvious advantages of recovery checks are that the matrix is representative and the approach is widely applicable; most test materials can be spiked by some means. However, the recovery check suffers from the disadvantage previously noted regarding the speciation, binding and physical distribution of the analyte. Furthermore, the assumption of an equivalent recovery of the analyte added as a spike and of the native analyte may not be valid. A microbial recovery can also be reduced by the presence of a competitive microbial flora of other species. Nevertheless, it can normally be assumed that a poor performance in a recovery check is strongly indicative of a similar or worse performance for the native analyte in the test materials.

Spiking and recovery testing as an IQC method must be distinguished from the method of standard additions, which is a measurement procedure; a single spiking addition cannot be used to fulfil the roles of both measurement and IQC.

Blank Determinations

Blank determinations are nearly always an essential part of the analytical process and can conveniently be effected alongside the IQC protocol. The simplest form of blank is the 'reagent blank', where the analytical procedure is executed in all respects apart from the addition of the test portion. A microbial version of reagent blank is to make a simulated dilution and inoculation, a sterility test that encompass all steps in a microbial analysis.

This kind of blank, in fact, tests more than the purity of the reagents. For example, it is capable of detecting contamination of the analytical system originating from any source, *e.g.* glassware and the atmosphere, and is therefore better described as a 'procedural blank'. In some instances, better execution of blank determinations is achieved if a simulated test material is employed. The simulant could be an actual test material known to be virtually analyte-free or a surrogate (*e.g.* ashless filter paper used instead of plant material, a high-pasteurised milk sample instead of fresh cow milk). Where it can be contrived, the best type of blank is the 'field blank', which is a typical matrix with zero concentration of analyte.

An inconsistent set of blanks in a run suggests sporadic contamination and may add weight to IQC evidence suggesting the rejection of the results. When an analytical protocol prescribes the subtraction of a blank value, the blank value must also be subtracted from the results of the control materials before they are used in IQC.

Traceability in Spiking and Recovery Checks

Potential problems of the traceability of reagents used for spikes and recovery checks must be guarded against. Under conditions where CRMs are not available, traceability can often be established only to the batch of analyte provided by a manufacturer. In such cases, confirmation of identity and a check on purity must be made before use. A further precaution is that the calibration standards and spike should not be traceable to the same stock solution of analyte or the same analyst. If such a common traceability existed, then the corresponding sources of error would not be detected by the IQC.

9 The Use of Shewhart Control Charts

Introduction

The theory, construction and interpretation of the Shewhart chart[25] are detailed in numerous texts on process quality control and applied statistics,[26] and in several ISO standards.[27,28,29,30] There is a considerable literature on the use of the control chart in clinical chemistry.[31,32] Westgard and co-workers have formulated multiple rules for the interpretation of such control charts,[33] and the power of these results has been studied in detail.[34,35] In this chapter only simple Shewhart charts are considered.

In IQC a Shewhart control chart is obtained when values of concentration measured on a control material in successive runs are plotted on a vertical axis against the run number on the horizontal axis. If more than one analysis of a particular control material is made in a run, either the individual results x or the mean value $\bar{x}$ can be used to form a control chart. The chart is completed by horizontal lines derived from the normal distribution $N(\mu,\sigma^2)$ that is taken to describe the random variations in the plotted values. The selected lines for control purposes are μ, $\mu \pm 2\sigma$ and $\mu \pm 3\sigma$. Different values of σ are required for charts of individual values and of means. For a system in statistical control, on average about one in 20 values falls outside the $\mu \pm 2\sigma$ lines, called the 'warning limits', and only about three in 1000 fall outside the $\mu \pm 3\sigma$ lines, the 'action limits'. In practice the estimates $\bar{x}$ and s of the parameters μ and σ are used to construct the chart. A persistent bias is indicated by a significant difference between $\bar{x}$ and the assigned value. A typical control chart showing results from a system in statistical control over 40 runs is shown in Figure 6.1.

Estimates of the Parameters μ and σ

An analytical system under control exhibits two sources of random variation, the within-run, characterised by variance σ_0^2 and the between-run with variance σ_1^2. The two variances are typically comparable in magnitude. The standard deviation σ_x used in a chart of individual values is given by

$$\sigma_x = (\sigma_0^2 + \sigma_1^2)^{1/2}$$

whereas for a control chart of mean values the standard deviation is given by

$$\sigma_{\bar{x}} = (\sigma_0^2/n + \sigma_1^2)^{1/2}$$

where n is the number of control measurements in a run from which the mean is calculated. The value of n therefore must be constant from run to run, otherwise control limits would be impossible to define. If a fixed number of repeats of a control material per run cannot be guaranteed (*e.g.* if the run length were variable) then charts of individual values must be used. Furthermore, the equation indicates that σ_x *or* $\sigma_{\bar{x}}$ must be estimated with care. An attempt to base an estimate on repeat values from a single run would result in unduly narrow control limits.

Estimates must therefore include the between-run component of variance. If the use of a particular value of n can be assumed at the outset, then $\sigma_{\bar{x}}$ can be estimated directly from the m means

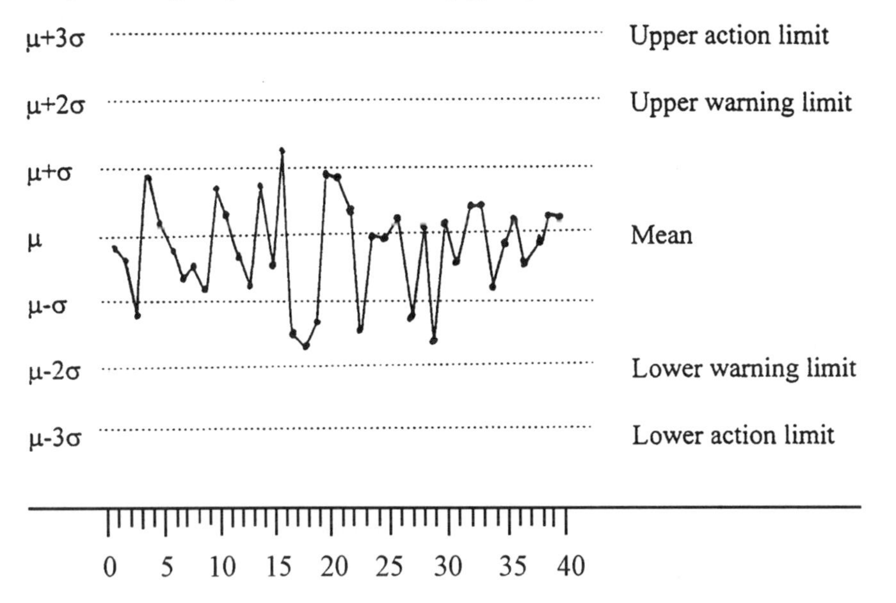

Figure 6.1 *Results from a system in statistical control*

$$\bar{x}_i = \sum_{j=1}^{n} x_{ij}/n$$

where $i = 1, \ldots, m$ of the n repeats in each of m successive runs. Thus the estimate of μ is $\bar{x} = \sum_i \bar{x}_n/m$, and the estimate of $\sigma_{\bar{x}}$ is

$$S_x = \sqrt{\frac{\sum_i (\bar{x}_i - \bar{x})^2}{m-1}}$$

If the value of n is not predetermined, then separate estimates of σ_0 and σ_1 could be obtained by one-way analysis of variance. If the mean squares within- and between-groups are MS_w and MS_b respectively, then

σ_0^2 is estimated by MS_w and

σ_1^2 is estimated by $(MS_b - MS_w)/n$

Often in practice it is necessary to initiate a control chart with data collected from a small number of runs, which may be to a degree unrepresentative, as estimates of standard deviation are very variable unless large numbers of observations are used. Moreover, during the initial period, the occurrence of out-of-control conditions are more than normally likely and will produce outlying values. Such values would bias $\bar{x}$ and inflate s beyond its proper value. It is therefore advisable to recalculate $\bar{x}$ and s after a further 'settling down' period. One method of obviating the effects of outliers in the calculation is to reject them after the application of Dixon's Q or Grubbs's[36] test, and then use the classical statistics given above. Alternatively, the methods of robust statistics could be applied to the data.[37,38]

The Interpretation of Control Charts

The following simple rules can be applied to control charts of individual results or of means.

Single Control Chart

An out-of-control condition in the analytical system is signalled if any of the following occur:

1. The current plotting value falls outside the action limits.
2. The current value and the previous plotting value fall outside the warning limits but within the actions limits.
3. Nine successive plotting values fall on the same side of the mean line.

Two Control Charts

When two different control materials are used in each run, the respective control charts are considered simultaneously. This increases the chance of a type 1 error (rejection of a sound run) but decreases the chance of a type 2 error (acceptance of a flawed run). An out-of-control condition is indicated if any of the following occur:

- at least one of the plotting values falls outside the action limits;
- both of the plotting values are outside the warning limits;
- the current value and the previous plotting value on the same control chart both fall outside the warning limits;
- both control charts simultaneously show that four successive plotting values on the same side of the mean line;
- one of the charts shows nine successive plotting values falling on the same side of the mean line.

A more thorough treatment of the control chart can be obtained by the application of the full Westgard rules, illustrated in Figure 6.2.

The analytical chemist should respond to an out-of-control condition by cessation of analysis pending diagnostic tests and remedial action followed by rejection of the results of the run and re-analysis of the test materials.

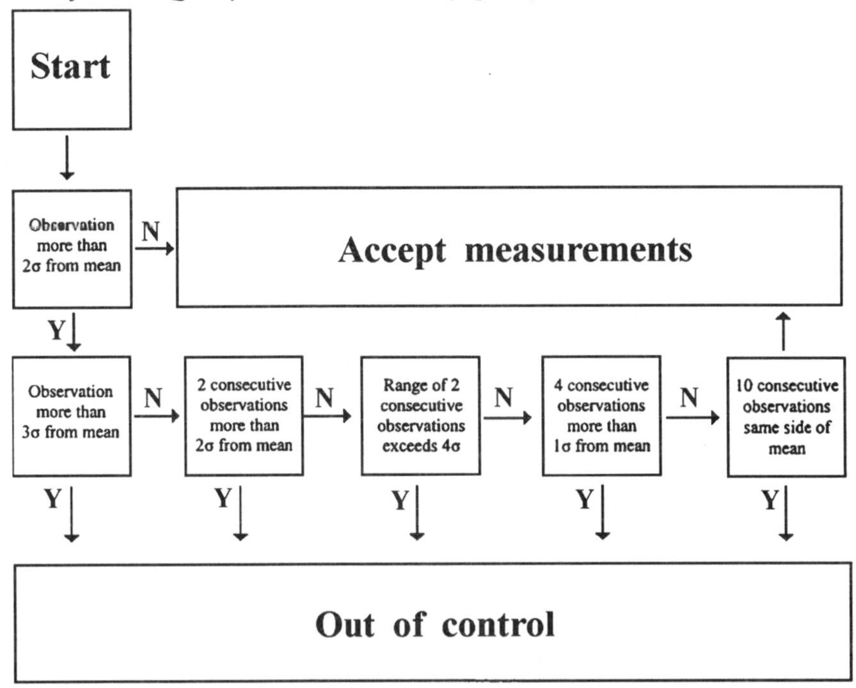

Figure 6.2 *The Westgard rules*

10 Recommendations

The following recommendations represent integrated approaches to IQC that are suitable for many types of analysis and applications areas and have been endorsed for use in food analysis laboratories. Managers of laboratory quality systems will have to adapt the recommendations to the demands of their own particular requirements. Such adaptation could be implemented, for example, by adjusting the number of duplicates and control material inserted into a run, or by the inclusion of any additional measures favoured in the particular application area. The procedure finally chosen and its accompanying decision rules must be codified in an IQC protocol that is separate from the analytical system protocol.

The practical approach to quality control is determined by the frequency with which the measurement is carried out and the size and nature of each run. The following recommendations are therefore made for both chemical and microbiological analysis.

In each of the following the order in the run in which the various materials are analysed should be randomised if possible. A failure to randomise may result in an underestimation of various components of error.

Quantitative Chemical Analysis

Short (e.g. $n < 20$) *Frequent Runs of Similar Materials*

The concentration range of the analyte in the run is relatively small, so a common value of standard deviation can be assumed.

Insert a control material at least once per run. Plot either the individual values obtained, or the mean value, on an appropriate control chart. Analyse in duplicate at least half of the test materials, selected at random. Insert at least one blank determination.

Longer (e.g. $n > 20$) *Frequent Runs of Similar Materials*

Again a common level of standard deviation is assumed.

Insert the control material at an approximate frequency of one per 10 test materials. If the run size is likely to vary from run to run it is easier to standardise on a fixed number of insertions per run and plot the mean value on a control chart of means. Otherwise plot individual values. Analyse in duplicate a minimum of five test materials selected at random. Insert one blank determination per 10 test materials.

Frequent Runs Containing Similar Materials but with a Wide Range of Analyte Concentration

It cannot be assumed that a single value of standard deviation is applicable.

Insert control materials in total numbers approximately as recommended above. However, there should be at least two levels of analyte represented, one close to the median level of typical test materials, and the other approximately at the upper or lower decile as appropriate. Enter values for the two control materials on separate control charts. Duplicate a minimum of five test materials, and insert one procedural blank per ten test materials.

Ad Hoc *Analysis*

The concept of statistical control is not applicable. It is assumed, however, that the materials in the run are of a single type, *i.e.* sufficiently similar for general conclusions on errors to be made.

Carry out duplicate analysis on all of the test materials. Carry out spiking or recovery tests or use a formulated control material, with an appropriate number of insertions (see above), and with different concentrations of analyte if appropriate. Carry out blank determinations. As no control limits are available, compare the bias and precision with fitness for purpose limits or other established criteria.

Quantitative and Qualitative Microbiological Analysis

In principle the same theory can be applied within quantitative and qualitative microbiology as for chemistry. In practice the usage (frequency) of control samples is much lower, because of the lack of good, simple-to-use and readily available control materials. However, it is recommended that, when available, quantitative reference samples are used ideally every week but certainly every second week for each species that is analysed.

In food microbiology, microbial numbers are usually expressed logarithmically, the data not being normally distributed. Results from control samples can therefore be evaluated by plotting them in a control chart, as described previously. There are specific microbial problems with the growth and decay of living micro-organisms in the presently available quantitative reference samples. There is frequently growth as a result of, for example, mistreatment of microbial solutions before inoculation and decay due to bad storage conditions or of 'normal death'. The interpretation and rules of action have to be adopted to accommodate these problems. Furthermore, as one batch of microbial reference samples is replaced with another similar one, it is not always possible to obtain one giving the same average results even though its microbial composition is nominally the same. It is then necessary to use the standard deviation from a single analytical series with one batch as a new reference value is being developed.

In microbiological analyses it is necessary that all critical analytical steps such as incubation temperature are under control. The competence of staff must be assessed through the analysis of natural samples as split duplicates.[39,40] In addition, the regular participation in microbiological proficiency testing schemes are recommended as being even more important within the microbiology analysis sector than within the chemistry analysis sector.

Routine substrate controls by streaking of microbial reference strains cannot replace the use of microbial control or reference samples. Streaking strains is a common practice within medical microbiology. Experience has shown, however, that substrate inhibition on microbiological counts due to 'quality problems' is far more common than substrate problems affecting the morphology of micro-organisms.

References

1 Guidelines on Internal Quality Control in Analytical Chemistry Laboratories, ed. M. Thompson and R. Wood, *Pure Appl. Chem.*, 1995, **67**, 649.
2 Report of the 22nd Session of the Codex Alimentarius Commission, FAO, Rome, 1997, ALINORM 97/37.
3 Handbook for evaluating water bacteriological laboratories, US Environmental Protection Agency, Cincinnati, Ohio, 1975, EPA-670/9-75-006.
4 Quality Assurance in Bacteriological Analysis of Water, A. H. Havelaar, presented at WHO/UNEP training course in Microbiological Methodology for Marine Pollution Monitoring, Rivan, Bilthoven, The Netherlands, 1990.

5 Manual of Food Quality Control; Quality Assurance in the Food Control Microbiological Laboratory, FAO, Rome, 1991, FAO Food and Nutrition Paper 14/12.
6 Quality Assurance and Quality Control in Food Microbiology Laboratories, Particularly when using Freeze-dried Mixed Culture', M. Peterz, thesis, Swedish National Food Administration, 1992.
7 The Microbiology of Water 1994. Part 1—Drinking Water, Report on Public Health and Medical Subjects (PHMS), no. 71, Methods for the Examination of Water and Associated Materials, UK, 1994.
8 Standard Methods for the Examination of Water and Wastewater, ed. A. D. Eaton, L. S. Clesceri and A. E. Greenberg, American Public Health Association, USA, 1995.
9 Quality Assurance Guidelines for Microbiological Laboratories, NMKL Report No. 5, 2nd edition, NMKL Secretariat, Finland, 1995.
10 IFCC Approved Recommendations on Quality Control in Clinical Chemistry. Part 4: Internal Quality Control, *J. Clin. Chem. Clin. Biochem.*, 1980, **18**, 534.
11 Internal Quality Control for Assays of Reproductive Hormones: Guidelines for Laboratories, S. Z. Cekan, S. B. Sufi and E. W. Wilson, WHO, Geneva, 1993.
12 Control Procedures in Geochemical Analysis, M. Thompson, in Statistics and Data Analysis in Geochemical Prospecting, ed. R. J. Howarth, Elsevier, Amsterdam, 1983.
13 Data Quality in Applied Geochemistry: the Requirements and how to Achieve Them, M. Thompson, *J. Geochem. Explor.*, 1992, **44**, 3.
14 Analytical Quality in Workplace Air Monitoring, Health and Safety Executive, London, 1991.
15 A Protocol for Analytical Quality Assurance in Public Analysts' Laboratories, Association of Public Analysts, 342 Coleford Road, Sheffield S9 5PH, UK, 1986.
16 Method Evaluation, Quality Control, Proficiency Testing, (AMIQAS PC Program), National Institute of Occupational Health, Denmark, 1993.
17 Quality Is Free—the Art of Making Quality Certain, Philip Crosby, McGraw-Hill, New York, 1979.
18 The International Harmonised Protocol for the Proficiency Testing of (Chemical) Analytical Laboratories, ed. M. Thompson and R. Wood, *Pure Appl. Chem.*, 1993, **65**, 2123 (also published in *J. AOAC Int.*, 1993, **76**, 926).
19 Effectiveness of Analytical Quality Control is Related to the Subsequent Performance of Laboratories in Proficiency Tests, *Analyst*, 1993, **118**, 1495.
20 Evaluation of Analytical Methods used for Regulation of Foods and Drugs, W. Horwitz, *Anal. Chem.*, 1982, **54**, 67A.
21 Statistical Evaluation of Results from Quantitative Microbiological Examinations, NMKL Report No. 1, NMKL Secretariat, Finland, 2nd Edition, 1983.
22 Statistical Methods Applied in Microbiology and Epidemiology, H. E. Tillett and R. G. Carpenter, 1991.
23 Uses of Certified Reference Materials, ISO Guide 33, ISO, Geneva, 1989.
24 Non-use and misinterpretation of CRMs. Can the situation be improved?, *Fresenius J. Anal. Chem.*, 1998, **360**, 370.
25 Control Charts and Control Materials in Internal Quality Control in Food Chemical Laboratories, NMKL Procedure No. 3, NMKL Secretariat, Finland, 1996.
26 Economic Control of Quality in Manufactured Product, W. A. Shewart, Van Nostrand, New York, 1931.
27 Shewhart Control Charts, ISO 8258, ISO, Geneva, 1991.

28 Control Charts—General Guide and Introduction, ISO 7870, ISO, Geneva, 1993.
29 Control Charts for Arithmetic Means with Warning Limit, ISO 7873, ISO, Geneva, 1993.
30 Acceptance Control Charts, ISO 7966, ISO, Geneva, 1993.
31 S. Levey and I. R. Jennings, *Am. J. Clin. Pathol.*, 1950, **20**, 1059.
32 A. B. J. Nix, R. J. Rowlands, R. W. Kemp, P. W. Wilson and K. Griffiths, *Stat. Med.*, 1987, **6**, 425.
33 Comparing the Power of Quality-Control Rules to Detect Persistent Systematic Error, J. O. Westgard, P. L. Barry and M. R. Hunt, *Clin. Chem.*, 1981, **27**, 493.
34 *Clin. Chem.*, C. A. Parvin, 1992, **38**, 358.
35 Comparison of quality-control rules used in clinical chemistry laboratories, J. Bishop and A. B. J. Nix, *Clin. Chem.*, 1993, **39**, 1638.
36 Protocol for the Design, Conduct and Interpretation of Method Performance Studies, ed. W. Horwitz, *Pure Appl. Chem.*, 1988, **60**, 855.
37 Robust statistics—how not to reject outliers. Part 1. Basic concepts, Analytical Methods Committee, *Analyst*, 1989, **114**, 1693.
38 Robust statistics—how not to reject outliers. Part 2. Interlaboratory trials, Analytical Methods Committee, *Analyst*, 1989, **114**, 1699.
39 Duplicate Split Samples for Internal Quality Control in Routine Water Microbiology, N. F. Lightfoot, H. E. Tillett, P. Boyd and S. Eaton, *J. Appl. Microbiol.*, 1994, **19**, 321.
40 A Semi-emperical Precision Control Criterion for Duplicate Microbial Colony Counts, S. I. Niemelä, *J. Appl. Microbiol.*, 1996, **22**, 315.

CHAPTER 7

Role of Proficiency Testing in the Assessment of Laboratory Quality

The IUPAC Inter Divisional Working Group on Analytical Quality Assurance has co-operated with the AOAC International and ISO to produce a Harmonised Protocol for Proficiency Testing of (Chemical) Analytical Laboratories; this sets out the minimum recommendations on proficiency testing. The published Protocol[1] is of particular relevance to food analysts, having been endorsed for use for Codex purposes by the Codex Alimentarius Commission[2] and also by the EU Council Working Group which developed the Additional Measures Food Control Directive. Because of these endorsements this chapter describes in detail proficiency testing as outlined in the Protocol.

It is to be regretted that proficiency testing was originally introduced into the food analysis laboratory as an aid to help such laboratories enhance the quality of the data they produce through third-party testing, but that it is increasingly being used as a 'qualification'. Thus proficiency testing is increasing in importance in the food analysis laboratory. The laboratory must, therefore, be fully aware of all the steps involved in proficiency testing from both the co-ordinator and participant perspectives and to be able to appreciate in detail the processes and conclusions that may be legitimately drawn from proficiency test results. These are described below.

1 Introduction

What is Proficiency Testing?

A proficiency testing scheme is defined as a system for objectively checking laboratory results by an external agency. It includes comparison of a laboratory's results at intervals with those of other laboratories, the main object being the establishment of trueness.

Proficiency testing schemes are based on the regular circulation of homogeneous samples by a co-ordinator, analysis of samples (normally by the laboratory's method of choice) and an assessment of the results. However, although many organisations carry out such schemes, there has until recently been no international agreement on how this should be done, in contrast to the collaborative trial situation. In order to rectify this, the same international group which drew up the collaborative trial protocols was invited to prepare one for proficiency testing schemes. This led to the development of the IUPAC/AOAC/ISO Harmonised Protocol for Proficiency Testing of (Chemical) Analytical Laboratories. This protocol, although developed for proficiency testing in the chemical analysis sector, has also been used by organisers of microbiology proficiency testing schemes when considering enumeration data.

Participation in proficiency testing schemes provides laboratories with an objective means of assessing and documenting the reliability of the data they are producing. Although there are several types of proficiency testing schemes they all share a common feature: test results obtained by one laboratory are compared to an external standard, frequently the results obtained by one or more other laboratories in the scheme. Laboratories wishing to demonstrate their proficiency should seek and participate in proficiency testing schemes relevant to their area of work. However, proficiency testing is only a snapshot of performance at infrequent intervals; it will not be an effective check on general performance or an inducement to achieve fitness for purpose, unless it is used in the context of a comprehensive quality system in the laboratory.

Other limitations are apparent in proficiency testing. For example, unless the laboratory uses typical analytical conditions to deal with the proficiency testing materials (and this is essentially out of the control of the organiser in most schemes), the result will not enable participants to take remedial action in case of inaccuracy. It is also unfortunate that some participating laboratories for different reasons treat the analysis of proficiency test samples in a non-routine manner. Examples exist where replicate analyses may be carried out whilst reporting only the perceived 'best' result, laboratories wanting to change reported results after submission to the scheme organiser, *etc*. These aspects are discussed later in this chapter. However, conflict between the remedial and the accreditation roles of proficiency testing is now occurring. It is unfortunate that successful participation in proficiency testing schemes has become a 'qualification' (or at least poor performance a 'disqualification') factor in accreditation. Nevertheless, it is recognised by most proficiency testing scheme organisers that their primary objective is to provide help and advice, not to 'qualify' or 'accredit' participants.

It must be appreciated that extrapolation from success in proficiency testing schemes to proficiency in everyday analytical work is an assumption; in most circumstances it would be prohibitively expensive and practically difficult for a proficiency testing scheme organiser to test the proposition experimentally by using undisclosed testing. However, most customers anticipate that performance in a proficiency testing exercise would be the 'best' that is achievable by a laboratory, and that repeated poor performance in a proficiency testing scheme is not acceptable.

History of Proficiency Testing: the Harmonised International Protocol

Proficiency testing emerged from the early generalised interlaboratory testing that was used in different degrees to demonstrate proficiency (or rather lack of it), to characterise analytical methods and to certify reference materials. These functions have now been separated to a large degree although it is still recognised that proficiency testing, in addition to its primary function, can sometimes be used to provide information on the relative performance of different analytical methods for the same analyte, or to provide materials sufficiently well characterised for internal quality control purposes.[3]

The systematic deployment of proficiency testing was pioneered in the US in the 1940s and in the 1960s in the UK by the clinical biochemists, who clearly need reliable results within institutional units and comparability between institutions. However, the use of proficiency testing is now represented in most sectors of analysis where public safety is involved (*e.g.* in the clinical chemistry, food analysis, industrial hygiene and environmental analysis sectors) and increasingly used in the industrial sector. Each of these sectors has developed its own approach to the organisation and interpretation of proficiency testing schemes, with any commonality of approach being adventitious rather than by collaboration.

To reduce differences in approach to the design and interpretation of proficiency testing schemes the three international organisations, ISO, IUPAC and AOAC International, have collaborated to bring together the essential features of proficiency testing in the form of 'The International Harmonised Protocol for the Proficiency Testing of (Chemical) Analytical Laboratories'.[1] This protocol has now gained international acceptance, most notably in the food sector. For the food sector it is now accepted that proficiency testing schemes must conform to the International Harmonised Protocol, and that has been endorsed as official policy by the Codex Alimentarius Commission, AOAC International and the European Union.

Studies on the effectiveness of proficiency testing have not been carried out in a systematic manner in most sectors of analytical chemistry, although recently a major study of proficiency testing under the auspices of the Valid Analytical Measurement (VAM) programme has been undertaken by the Laboratory of the Government Chemist in the UK. However, the results have yet to be published.[4]

Proficiency Testing and Accreditation

Despite the primary self-help objectives of proficiency testing, an acceptable performance in a proficiency testing scheme (where available) is increasingly expected as a condition for accreditation. Indeed, in the latest revision of ISO/IEC Guide 25 it is a requirement that laboratories participate in appropriate proficiency testing schemes whenever these are available.[5] Thus accreditation agencies require laboratories seeking accreditation to participate in an appropriate proficiency testing scheme before accreditation is gained or maintained. Fortunately both the accreditation requirements and the 'self-help intentions' can be fulfilled by the same means at one and the same time. In an evaluation by an accreditation body, not only should the proficiency test results be considered but also the procedures at the laboratory for documentation of such results and any action taken when non-compliances (unsatisfactory results in the proficiency testing scheme) are found.

Proficiency testing schemes themselves will also have to become accredited to the ISO/IEC Guide 43[6] if they are to become internationally accepted and used by participants from many countries. Thus the co-ordinator must document all practices and procedures in a quality manual, the outline of which is given below:

Suggested Headings in a Quality Manual for the Organisation of Proficiency Testing Schemes

1. Quality policy
2. Organisation of agency
3. Staff, including responsibilities
4. Documentation control
5. Audit and review procedures
6. Aims, scope, statistical design and format (including frequency) of proficiency testing programmes
7. Procedures covering
   - sample preparation
   - testing of sample homogeneity
   - equipment
   - suppliers
   - logistics (*e.g.* sample dispatch)
   - analysis of data
8. Preparation and issuing of report
9. Action and feedback by participants when required
10. Documentation of records for each programme
11. Complaints handling procedures
12. Policies on confidentiality and ethical considerations
13. Computing information, including maintenance of hardware and software
14. Safety and other environmental factors
15. Sub-contracting
16. Fees for participation
17. Scope of availability of programme to others

2 Elements of Proficiency Testing

In analytical chemistry, proficiency testing almost invariably takes the form of a simultaneous distribution of effectively identical samples of a characterised material to the participants for unsupervised blind analysis by a deadline. The primary purpose of proficiency testing is to allow participating laboratories to become aware of unsuspected errors in their work and to take remedial action. This it achieves by allowing a participant to make three comparisons of its performance:

- with an externally determined standard of accuracy;
- with that of peer laboratories;
- with its own past performance.

In addition to these general aims, a proficiency testing scheme should specifically address fitness for purpose, the degree to which the quality of the data produced by a participant laboratory can fulfil its intended purpose. This is a critical issue in the design of proficiency testing schemes that will be outlined below.

Economics of Proficiency Testing Schemes: Requirement for Laboratories to Undertake a Range of Determinations Offered Within a Proficiency Testing Scheme

Proficiency testing is in principle adaptable to most kinds of analysis and laboratories and to groups of laboratories of all sizes. However, it is most effectively and economically applied to large groups of laboratories conducting large numbers of routine analyses. Setting up and running a scheme has a number of overhead costs which are best distributed over a large number of participant laboratories. Moreover, if only a small range of activities is to be subject to test, then proficiency testing can address all of them. If in a laboratory there is an extremely wide range of analyses that it may be called upon to carry out (*e.g.* in a food control laboratory), it will not be possible to provide a proficiency test for each of them individually. In such a case it is necessary to apply proficiency testing to a proportion of the analyses that can be regarded as representative. Schemes cannot cover all aspects of some areas of activity, and must be regarded as being *representative* of the particular sector of interest.

It has been suggested that for laboratories undertaking many different analyses, a 'generic' approach should be taken wherever possible. Thus, for general chemical food analysis laboratories, they should participate in, and achieve a satisfactory performance from, series dealing with the testing of

- GC
- HPLC
- trace element and
- proximate analysis procedures

rather than for every analyte that they may determine (always assuming that an appropriate proficiency testing scheme is available). However, the basic participation should be supplemented by participation in specific areas where regulations are in force and where the analytical techniques applied are judged to be sufficiently specialised to require an independent demonstration of competence. In the food sector, examples of such analytes are aflatoxins (and other mycotoxins), pesticides and overall and specific migration from packaging to food products. This is being addressed in the UK interpretation of the requirements of the Additional Measures Food Control Directive.[7]

Limitations of Proficiency Testing Schemes

It is necessary to treat with caution the inference that a laboratory that is successful in a particular proficiency scheme for a particular determination will be proficient for all similar determinations. In a number of instances it has been shown that a laboratory proficient in one type of analysis may not be proficient in a closely related one. Two examples of where the ability of laboratories to determinate similar analytes in the food sector is very variable are given below:

*Total Poly- and (*cis*) Mono-unsaturated and Saturated Fatty Acids in Oils and Fats*

Results from proficiency testing exercises that include such tests indicate that the determinations are of variable quality. In particular, the determination of poly-unsaturated and saturated fatty acids is generally satisfactory but that the determination of mono-unsaturated fatty acids is unduly variable with a bi-modal distribution of results sometimes being obtained. Bi-modality might be expected on the grounds that some participant laboratories were able to separate *cis* from *trans* mono-unsaturated fatty acids. However, examination of the methods of analysis used by participants did not substantiate this; some laboratories reported results as if they were separating *cis* and *trans* fatty acids even though the analytical systems employed were incapable of such a separation. This is clearly demonstrated in reports from the UK Ministry of Agriculture, Fisheries and Food's Food Analysis Performance Assessment Scheme.[8]

Trace Nutritional Elements (Zinc, Iron, Calcium, etc.*)*

Laboratories have been asked to analyse proficiency test materials which contain a number of trace elements of nutritional significance, *e.g.* for zinc, calcium and iron, *etc.* It has been observed that the number of laboratories which achieve 'satisfactory' results for each analyte determined in the same test material differs markedly, thus suggesting that the assumption that the satisfactory determination of one such analyte is indicative that a satisfactory determination would be observed for all similar analytes is not valid. This conclusion is generally assumed even if the elements are determined in a 'difficult' matrix, such as in a foodstuff, where most of the problems may be associated with matrix effects rather than with the end-point determination.

3 Organisation of Proficiency Testing Schemes

Framework

Samples must be distributed regularly to participants who are to return results within a given time. The results will be statistically analysed by the organiser and participants will be notified of their performance. Advice will be available to poor performers and participants will be kept fully informed of the scheme's progress. Participants will be identified by code only, to preserve confidentiality.

The scheme's structure for any one analyte or round in a series should be:

1. co-ordinator organises preparation, homogeneity testing and validation of test material;
2. co-ordinator distributes test samples on a regular schedule;
3. participants analyse test portions and report results centrally;
4. results subjected to statistical analysis; performance of laboratories assessed;
5. participants notified of their performance;
6. advice available for poor performers, on request;
7. co-ordinator reviews performance of scheme;
8. next round commences.

Preparation for the next round of the scheme *may* have to be organised while the current round is taking place; details of the next round may have to be adjusted in the light of experience from the current round.

Organisation

Day-to-day running of the scheme will be the responsibility of the co-ordinator. The co-ordinator must document all practices and procedures in a quality manual (see above). Preparation of test materials will either be contracted-out or undertaken by the co-ordinator. The laboratory preparing the test material should have demonstrable experience in the area of analysis being tested. It is essential for the co-ordinator to retain control over the assessment of performance as this will help to maintain the credibility of the scheme. Overall direction of the scheme should be overseen by a small advisory panel having representatives (who should be practising laboratory scientists) from, for example, the co-ordinator, contract laboratories (if any), appropriate professional bodies, participants and end-users of analytical data.

Test Materials

The test materials to be distributed must, if possible, be generally similar in matrix to the samples that are routinely analysed (in respect of matrix composition and analyte concentration range). This, however, is not always possible with respect to the stability, homogeneity or other reasons when working with micro-organisms and some unstable chemicals. It is essential they are of acceptable homogeneity and stability. The bulk material prepared must be effectively homogeneous so that all laboratories will receive samples that do not differ significantly in analyte concentration. The co-ordinator must clearly state the procedure used to establish the homogeneity of the test material; a suitable procedure is described below. However, as a guide, the between-sample standard deviation should be less than 0.3 times the target value for the standard deviation.

Where possible the co-ordinating laboratory should also provide evidence that the test material is sufficiently stable to ensure that the material will not undergo any significant change throughout the duration of the proficiency test. Prior to distribution of the test samples, the stability of the matrix and the analytes they contain must therefore be determined by carrying out analyses after they have been stored for an appropriate period of time. The storage conditions, most especially of time and temperature, used in the stability trials must represent those conditions likely to be encountered in the entire duration of the proficiency test. Stability trials must therefore take account of the transport of the test samples to participating laboratories as well as the conditions encountered purely in a laboratory environment. The concentrations of the various analytes must show no significant changes during the stability tests, the magnitude of a 'significant change' being assessed from the knowledge of the variance expected for replicate analyses of the bulk material. When unstable analytes are to be assessed it may be necessary for the co-ordinating organisation to prescribe a date by which the analysis must be accomplished.

Ideally the quality checks on the samples referred to above should be performed by a different laboratory from that which prepared the sample, although it is recognised that this may cause difficulties to the co-ordinating organisation.

The number of test materials to be distributed per round will depend mainly on whether there is a requirement to cover a range of compositions. Practical considerations will dictate an upper limit of six to the number of test materials per analyte in the food sector.

Co-ordinators should consider any hazards that the test materials might pose and take appropriate action to advise any party that might be at risk (*e.g.* test material distributors, testing laboratories, *etc.*) of the potential hazard involved.

The recommended procedure to be followed to test for material heterogeneity is:

1. Prepare the whole of the bulk material in a form that is thought to be homogeneous, by an appropriate method.
2. Divide the material into the containers that will be used for dispatch to the participants.

3. Select a minimum (n) of 10 containers strictly at random.
4. Separately homogenise the contents of each of the n selected containers and take two test portions.
5. Analyse the $2n$ test portions in a random order under repeatability conditions by an appropriate method. The analytical method used must be sufficiently precise to allow a satisfactory estimation of s_s.
6. Form an estimate (s_s^2) of the sampling variance and an estimate (s_a^2) of the analytical variance by one-way analysis of variance, without exclusion of outliers.
7. Report values of $\bar{x}$, s_s, s_a, n and the result of the F-test.
8. If σ is the target value for standard deviation for the proficiency test at analyte concentration $\bar{x}$, the value of s_s/σ should be less than 0.3 for sufficient homogeneity.

An example of the application of the above, taken from the Harmonised Protocol, is:

Copper in soya flour ($\mu g\ g^{-1}$)

Sample no.	Copper content	
1	10.5	10.4
2	9.6	9.5
3	10.4	9.9
4	9.5	9.9
5	10.0	9.7
6	9.6	10.1
7	9.8	10.4
8	9.8	10.2
9	10.8	10.7
10	10.2	10.0
11	9.8	9.5
12	10.2	10.0

Grand mean = 10.02

Analysis of Variance:

Source of variation	*df*	*Sum of squares*	*Mean square*	*F*
Between samples	11	2.54458	0.231326	3.78
Analytical	12	0.735000	0.06125	

From statistical tables, the critical value of F ($P = 0.05$, $\nu_1 = 11$, $\nu_2 = 12$) is 2.72 < 3.78. There are significant differences between samples.

$s_a = \sqrt{0.0613} = 0.25 \qquad s_s = [(0.2313 - 0.0613)/2]^{1/2} = 0.29$

σ = 1.1 (This is an example value of a target value for reference standard deviation and is not derived from the data)

$s_s/\sigma = 0.29 / 1.1 = 0.26 < 0.3$

Although there are significant differences between samples (*F*-test), the material is sufficiently homogeneous for the purpose of the proficiency trial, as $s_s/\sigma = 0.26$ is less than the maximum recommended value of 0.3.

Frequency of Sample Distribution

The appropriate frequency for the distribution of test material in any one series depends upon a number of factors of which the most important are:

1. the difficulty of carrying out effective analytical quality control;
2. the laboratory throughput of test samples;
3. the consistency of the results from previous rounds;
4. the cost/benefit of the scheme;
5. the availability of suitable material for proficiency test schemes.

In practice the frequency will probably fall between once every two weeks and once every four months.

A frequency greater than once every two weeks could lead to problems in the turn-round time of test materials and results. It might also encourage the belief that the proficiency testing scheme can be used as a substitute for internal quality control, a concept that laboratories should not accept. If the period between distributions extends beyond four months there will be unacceptable delays in identifying and correcting analytical problems and it then becomes difficult to monitor meaningful trends in a laboratory's performance, making impact of the scheme on the participants small.

There will be circumstances where consideration of the above factors may mean that it is acceptable to have a longer timescale between distribution of test samples. It would be one of the functions of the advisory panel to comment on the frequency of distribution appropriate for a particular scheme.

In addition, this panel would also provide advice on the areas to be covered in any particular sector of analytical chemistry. This is particularly difficult where there are a considerable number of diverse analyses in the sector.

Sample Distribution, etc.

Ideally, samples should be shipped to the participants 'anonymously', thereby stimulating the laboratories to handle them as 'routine' samples. Although this is not usually possible, participating laboratories should try to handle the proficiency test materials in a routine manner with respect to sample receipt, performing registration, storage before and after analysis, *etc.*

However, special procedures for packaging and shipping may have to be considered when handling microbiological samples.

Estimating the Assigned Value (the 'True' Result)

The assigned value is a critical parameter for any proficiency testing scheme. The co-ordinator must, therefore, give details on how the assigned value is to be obtained together, where possible, with a statement of its traceability and its uncertainty.

There are a number of possible approaches to establishing the assigned value for the concentration of analyte and its uncertainty in a test material, but only four are normally considered.

Consensus value from expert laboratories

This value is the consensus of a group of expert laboratories that achieve agreement by the careful execution of recognised reference methods; it is the best procedure in most circumstances for determining the assigned value in representative materials. When such a value is used, the organising body should disclose the identities of the laboratories producing the individual results, the method of calculating the consensus value and, if possible, a statement of the traceability and of its uncertainty. The consensus value will normally be a robust mean from the expert laboratories.

Formulation

This method comprises the addition of a known amount or concentration of analyte to a base material containing none. The method is especially valuable when it is the amount of analyte added to individual test portions that is subject to testing, as there is no requirement for ensuring a sufficiently homogeneous mixture in the bulk test material. In other circumstances problems might arise with the use of formulation, as follows:

1. There is a need to ensure that the base material is effectively free from analyte or that the residual analyte concentration is accurately known.
2. It may be difficult to mix the analyte homogeneously into the base material where this is required.
3. The added analyte may be more loosely bonded than, or in a different chemical form from, that found in the typical materials that the test materials represent.

Unless these problems can be overcome, representative materials (containing the analyte in its normally occurring form in a typical matrix) are usually preferable. Where formulation is used, traceability to certified reference materials or reference methods should be cited if possible.

Direct comparison with certified reference materials

In this method, the test material is analysed along with appropriate certified reference materials by a suitable method under repeatability conditions. In effect the method is calibrated with the CRMs, providing direct traceability and an uncertainty for the value assigned to the test material. The CRMs must have both the appropriate matrix and an analyte concentration range that spans, or is close to, that of the test material. In many areas, and particularly the foodstuffs area, the lack of CRMs will restrict the use of this method.

Consensus of participants

A value often advocated for the assigned value is the consensus (normally the robust mean) of the results of all of the participants in the round of the test. This value is clearly the cheapest and easiest to obtain and is that most appreciated by participants in the scheme. The method usually gives a serviceable value when the analysis is regarded as easy, for instance when a recognised method is applied to a major constituent. In an empirical method (where the method 'defines' the content of the analyte), the consensus of a large number of laboratories can be safely regarded as the true value.

There are a number of drawbacks to using the assigned value derived from the consensus of participants. At a fundamental level it is difficult to establish traceability or an uncertainty to such a value, unless all of the participants were using the same reference method. Other objections that can be levelled against the consensus value are (*a*) there may be no real consensus amongst the participants and (*b*) the consensus may be biased by the general use of faulty methodology. Neither of these conditions is rare in the determination of trace constituents.

Choice between methods

The choice between these methods of evaluating the assigned value depends on circumstances and is the responsibility of the organising agency for the scheme. It is usually advisable to have an estimate additional to the consensus of participants. Any significant deviations observed between the estimates must be carefully considered by the technical panel.

Empirical methods are used when the analyte is ill-defined chemically. In an empirical method, *e.g.* the determination of 'fat', the true result (within the limits of measurement uncertainty) is produced by correct execution of the method. It is clear that in these circumstances the analyte content is defined only if the method is simultaneously specified. Empirical methods can give rise to special problems in proficiency trials when a choice of such methods is available. If the assigned value is obtained from expert laboratories and the participants use a different empirical method, a bias may be apparent in the results even when no fault in execution is present. Likewise, if participants are free to choose between empirical methods, no valid consensus may be evident among them. Several recourses are available to overcome this problem:

1. a separate value of the assigned value is produced for each empirical method used;
2. participants are instructed to use a prescribed method; or
3. participants are warned that a bias may result from using an empirical method different from that used to obtain the consensus.

Notwithstanding the options given above, participants in proficiency testing schemes prefer the assigned value to be estimated from the consensus mean from all participants. Results from proficiency testing schemes in the food area suggest that the difference in assigned values derived from the various methods given above is not significant, and so to use the procedure advocated by the participants has merits in terms of acceptability to them.

Choice of Analytical Method

Participants are normally able to use the analytical method of their choice except when otherwise instructed to adopt a specified method. Methods used should preferably be validated by an appropriate means, *e.g.* collaborative trial, comparison to a reference method, *etc.* before use. As a general principle, procedures used by laboratories participating in proficiency testing schemes should simulate those used in their routine analytical work *including all steps from sample receipt, documentation of analytical results and the evaluation of such results, to their reporting procedures.*

Where an empirical method is used, the assigned value will be calculated from results obtained using that defined procedure, or from results obtained using procedures complying with a common principle. This is a particularly important consideration in the food sector, as demonstrated by the comparison of methods for the determination of 'fat' in the UK Food Analysis Performance Assessment Scheme.[9] If participants use a method which is not equivalent to the defining method, then an automatic bias in result must be expected when their performance is assessed.

Results Reported from Participants

Reports from laboratories to the co-ordinator should be as similar to ordinary reports as possible. The more differences there are in the reporting format, the more likely it is that a participating laboratory uses non-routine procedures in all steps when analysing proficiency test samples.

Performance Criteria

Laboratories will be assessed on the difference between their result and the assigned value. A performance score will be calculated for each laboratory. For each analyte in a round a criterion for the performance score may be set, where appropriate, against which the performance score obtained by a laboratory can be judged. A 'running score' could be calculated to give an assessment of performance spread over a longer period of time; this would be based on results for several rounds.

The performance criterion will be set so as to ensure that the analytical data routinely produced by the laboratory is of a quality that is adequate for its intended purpose. It will not necessarily be appropriate to set the performance criterion at the highest level that the method is capable of providing.

Reporting Results to Participants

Reports issued to participants should be clear and comprehensive and include data on the distribution of results from all laboratories together with the participant's performance score. The test results as used by the co-ordinator should be also displayed, to enable participants to check that their data have been correctly entered. Reports should be made available as quickly as possible after the return of results to the co-ordinating laboratory and, if at all possible, before the next distribution of samples.

Although ideally all results should be reported to participants, it may not be possible to achieve this if the scheme has many participants. In that situation participants should, however, receive at least: (*a*) reports in clear and simple format, and (*b*) results of all laboratories in graphical, *e.g.* histogram, form, even if the codes ascribed to individual laboratories are not included.

It is the laboratories' responsibility to read and take appropriate actions when needed as soon as possible when they have received the report. If the results are discussed among all staff, conclusions on specific microbiological or chemical problems that are highlighted in the report can be generalised.

Liaison with Participants

Participants should be provided with a detailed information pack on joining the scheme. Communication with participants should be *via* a newsletter or annual report together with a periodic open meeting; participants should be advised immediately of any changes in scheme design or operation. Advice should be available to poor performers. Participants who consider that their performance assessment is in error must be able to refer the matter to the co-ordinator.

Feedback from laboratories should be encouraged, so that participants actively contribute to the development of the scheme. Participants should view it as *their* scheme rather than one imposed by a distant bureaucracy. It is therefore important that participants ensure that the scheme to which they subscribe provides this liaison and advice.

Collusion and Falsification of Results

Although proficiency testing schemes are intended primarily to help participants improve their analytical performance, there may be a tendency among some participants to provide a falsely optimistic impression of their capabilities. For example, collusion may take place between laboratories, so that truly independent data are not submitted. Laboratories may also give a false impression of their performance if they routinely carry out single analyses, but report the mean of replicate determinations on the proficiency test samples. A falsely overly-optimistic impression of a laboratory's proficiency will also be given if it fails to distribute the analytical work amongst all analysts routinely involved, but use only their best staff. Proficiency testing schemes should be designed to ensure that there is as little collusion and falsification as possible by, for example, distributing alternative materials within one round, with no identifiable reuse of the materials in succeeding rounds. Also instructions to participants should make it clear that collusion is contrary to professional scientific conduct and serves only to nullify the benefits of proficiency testing to customers, accreditation bodies and analysts alike.

Although all reasonable measures should be taken by the co-ordinators to prevent collusion, it must be appreciated that it is the responsibility of the participating laboratories to avoid it.

Repeatability

Procedures used by laboratories participating in proficiency testing schemes should simulate those used in routine sample analysis. Thus, duplicate determinations on proficiency test materials (*should*) be carried out only if this is the norm for routine work in a laboratory. The result to be reported is in the same form (*e.g.* number of significant figures) as that normally reported to the customer. Some proficiency test co-ordinators like to include duplication in the tests to obtain a measure of repeatability proficiency. This should be allowed as a possibility in proficiency tests, but is not a requirement for participants.

4 Statistical Procedure for the Analysis of Results

The approach described below provides a transparent procedure by using accepted statistics without any arbitrary scaling factors. It is important that food analysts participate in a proficiency testing scheme which complies with the procedures described in order to ensure the acceptability of their data by their 'customers'.

Estimates of Assigned Value

The first stage in producing a score from a result x [a single measurement of analyte concentration (or amount) in a test material] is obtaining an estimate of the bias, which is defined as:

$$\text{bias estimate} = x - X$$

Where X is the true value.

In practice the assigned value, $\hat{X}$, which is the best estimate of X, is used. Several methods are available for obtaining the assigned value; these have been described above. It should be noted that if x is not a concentration measure, a preliminary transformation may be necessary.

Formation of a *z*-Score

Most proficiency testing schemes proceed by comparing the bias estimate (as defined above) with a target value for standard deviation that forms the criterion of performance. A *z*-score is formed from

$$z = (x - \hat{X})/\sigma$$

where σ is the target value for standard deviation.

Although z has the form of a normal standard deviate there is no presumption that this necessarily will be the case. In some circumstances the technical panel may decide to use an estimate of the actual variation ($\tilde{s}$) encountered in a particular round of a trial in place of a target standard deviation. In that case $\tilde{s}$ should be estimated from the laboratories' results after outlier elimination, or by robust methods for each analyte/material/round combination. A value of $\tilde{s}$ will thus vary from round to round. In consequence, the *z*-score for a laboratory could not be compared directly from round to round. However, the bias estimate $(x - \hat{X})$ for a single analyte/material combination could be usefully compared round by round for a laboratory, and the corresponding value of $\tilde{s}$ would indicate general improvement in 'reproducibility' round by round.

A fixed value for σ is preferable and has the advantage that the *z*-scores derived from it can be compared from round to round to demonstrate general trends for a laboratory or a group of laboratories. It is suggested that whatever the value of σ is chosen, it is a practical value and that it is accepted by participants. For some of the tests it is only necessary that the value chosen is sufficient clearly to discriminate in a pass/fail situation.

The value chosen can be arrived at in several ways:

By Perception

The value of σ could be fixed arbitrarily, with a value based on a perception of how laboratories perform. The problem with this criterion is that both perceptions and laboratory performance may change with time. The value of σ therefore may need to be changed occasionally, disturbing the continuity of the scoring scheme. However, there is some evidence that laboratory performance responds favourably to a stepwise increase in performance requirements.

By Prescription

The value of σ could be an estimate of the precision required for a specific task of data interpretation. This is the most satisfactory type of criterion, if it can be formulated, because it relates directly to the required information content of the data. Unless the concentration range is very small, σ should be specified as a function of concentration.

This is frequently used in legislation where method performance characteristics may be specified.

By Reference to Validated Methodology

Where a standard method is prescribed for the analysis, σ could be obtained by interpolation from the standard deviation of reproducibility obtained during appropriate collaborative trials.

By Reference to a Generalised Model

The value of σ could be derived from a general model of precision, such as the 'Horwitz Curve'.[10] However, while this model provides a general picture of reproducibility, substantial deviation from it may be experienced for particular methods. It could be used if no specific information is available.

Interpretation of *z*-Scores

If $\hat{X}$ and σ are good estimates of the population mean and standard deviation, then z will be approximately normally distributed with a mean of zero and unit standard deviation. An analytical result is described as 'well behaved' when it complies with this condition.

An absolute value of z ($|z|$) greater than three suggests poor performance in terms of accuracy. This judgement depends on the assumption of the normal distribution, which, outliers apart, seems to be justified in practice.

As z is standardised, it is comparable for all analytes and methods. Thus values of z can be combined to give a composite score for a laboratory in one round of a proficiency test.

The z-scores can therefore be interpreted as follows:

$|z| < 2$ 'Satisfactory': will occur in 95% of cases produced by 'well behaved results'.

$2 < |z| < 3$ 'Questionable': but will occur in $\approx$ 5% of cases produced by 'well behaved results'.

$|z| > 3$ 'Unsatisfactory': will only occur in $\approx$ 0.1% of cases produced by 'well behaved results'.

It is these 'scores' which will be used to assess the performance of participants in any scheme. It is therefore essential that the advantages and drawbacks of the scores are fully appreciated by participants and their customers. It is important that there is not an 'over-emphasis' placed on individual scores.

Combination of Results Within a Round of the Proficiency Testing Scheme

It is common for several different analyses to be required within each round of a proficiency test. While each individual test furnishes useful information, many participants and their customers may need a single figure of merit that will summarise the overall performance of the laboratory within a round. This approach may be appropriate for the assessment of long-term trends. However, there is a danger that such a combination score will be misinterpreted or abused by non-experts, especially outside the context of the individual scores. Therefore the general use of combination scores is not recommended, but it is recognised that they may have specific applications if based on sound statistical principles and used with due caution.

It is especially emphasised that there are limitations and weaknesses in any scheme that combines z-scores from dissimilar analyses. If a single score out of several produced by a laboratory were outlying, the combined score may well be not outlying. In some respects this is a useful feature, in that a lapse in a single analysis is down-weighted in the combined score. However, there is a danger that a laboratory may be consistently at fault only in a particular analysis, and frequently report an unacceptable value for that analysis in successive rounds of the trial. This factor may well be obscured by the combination of scores.

There are several methods of combining the z-scores produced by a laboratory in one round of the proficiency test described in the Protocol. It must be stressed that all should be used with caution, however. It is the individual z-scores that are the critical consideration when considering the proficiency of a laboratory.

The procedures that may be used are described below:

Procedures for Combining Scores

Several methods of combining independent z-scores produced by a laboratory in one round of the test seem potentially appropriate. For example:

1. the sum of scores, $SZ = \sum z$;
2. the sum of squared scores, $SSZ = \sum z^2$;
3. the sum of absolute values of the scores, $SAZ = \sum |z|$.

These statistics fall into two classes. The first class (containing only SZ) uses information about the signs of the z-scores, while the alternative class (SSZ and SAZ) provides information about only the size of scores, *i.e.* the magnitude of biases. Of the latter, the sum of the squares is more tractable mathematically and is therefore the preferred statistic, although it is rather sensitive to single outliers. SAZ may be especially useful if there are extreme outliers or many outlying laboratories, but its distribution is complicated and its use is not, therefore, recommended.

Sum of scores SZ

The distribution of SZ is zero-centred with variance m, where m is the number of scores being combined. Thus SZ could not be interpreted on the same scale as the z-scores. However, a simple scaling restores the unit variance, giving a rescaled sum of scores RSZ = $\sum z/\sqrt{m}$ which harmonises the scaling. Thus both z and RSZ can be interpreted as standard normal deviates.

SZ and RSZ have the advantage of using the information in the signs of the biases. Thus if a set of z-scores were 1.5, 1.5, 1.5, 1.5, the individual results would be regarded as non-significant positive scores. However, regarded as a group, the joint probability of observing four such deviations together would be small. This is reflected in the RSZ value of 3.0, which indicates a significant event. This information would be useful in detecting a small consistent bias in an analytical system, but would not be useful in combining results from several different systems, where a consistent bias would not be expected, and is unlikely to be meaningful.

Another feature of the RSZ is the tendency for errors of opposite sign to cancel. In a well-behaved situation (*i.e.* when the laboratory is performing without bias according to the designated σ value) this causes no problems. If the laboratory were 'badly behaved', however, the possibility arises of the fortuitous cancellation of significantly large z values. Such an occurrence would be very rare by chance.

These restrictions on the use of RSZ serve to emphasise the problems of using combination scores derived from various analytical tests. When such a score is used, it should be considered simultaneously with the individual scores.

Sum of squared scores SSZ

This combination score has a chi-squared (χ^2) distribution with m degrees of freedom for a well-behaved laboratory. Hence there is no simple possibility for interpreting the score on a common scale with the z-scores. However, the quantiles of the χ^2 distribution can be found in most compilations of statistical tables.

SSZ takes no account of the signs of the z-values, because of the squared terms. Thus, in the example considered previously, where the z-scores are 1.5, 1.5, 1.5, 1.5, SSZ = 9.0, a value that is not significant at the 5% level, and does not draw enough attention to the unusual nature of the results as a group. However, in proficiency tests, there is more concern with the magnitude of deviations than with their direction, so SSZ is appropriate for this use. Moreover, the problem of chance cancellation of significant z-scores of opposite sign is eliminated. Thus the SSZ has advantages as a combination score for diverse analytical tests, and is to an extent complementary to RSZ.

Calculation of Running Scores

Similar considerations apply for running scores as apply to combination scores above.

While the combination scores discussed above give a numerical account of the performance of a laboratory in a single round of the proficiency test, for some purposes it may be useful to have a more general indicator of the performance of a laboratory over time.

While the value of such indicators is questionable, they can be constructed simply and give a smoothed impression of the scores over several rounds of the test. It must be stressed that, as with combination scores, it is difficult to produce running scores that are not prone to misinterpretation, *etc*.

Some procedures that may be used are described below.

The usual procedure for calculating running scores is to form a 'moving window' average. The procedure can be applied to z or to a combination score.

As an example, a running Z score covering the current (n^{th}) round and the previous k rounds could be constructed as follows:

$$RZ = \sum_{j=n-k}^{n} z_j/(k+1)$$

where z_j is the z-score for the material in the jth round.

The running score has the advantage that instances of poor performance restricted to one round are smoothed out somewhat, allowing an overall appraisal of performance. However, an isolated serious deviation will have a 'memory effect' in a simple moving window average that will persist until $(k+1)$ more rounds of the trial have passed. This might have the effect of causing a laboratory persistently to fail a test on the basis of the running score, long after the problem has been rectified.

Two strategies for avoiding undue emphasis on an isolated bad round are available to the co-ordinator. Firstly, individual or combined scores can be restrained within certain limits. For example, a rule such as would be applied is:

if $|z| > 3$ then $z' = \pm 3$, the sign being the same as that of z

where z is the raw value of a z-score, and the modified value z' is limited to the range ± 3.

The actual limit used could be set in such a way that an isolated event does not raise the running score above a critical decision level for otherwise well-behaved conditions.

As a second strategy for avoiding memory effects, the scores could be 'filtered' so that results from rounds further in the past would have a smaller effect on the running score. For example, exponential smoothing uses:

$$\hat{z}_n = \sum_{i=0}^{\infty} \alpha^i z_{n-i}/(1-\alpha)$$

calculated by:

$$\hat{z}_n = (1-\alpha)z_n + \alpha\hat{z}_{n-1}$$

where α is a parameter between zero and one, controlling the degree of smoothing.

Ranking

Laboratories participating in a round of a proficiency trial are sometimes ranked on their combined score for the round or on a running score. Such a ranked list is used for encouraging better performance in poorly ranked laboratories by providing a comparison among the participants. However, ranking is not recommended as it is an inefficient use of the information available and may be open to misinterpretation. A histogram is a more effective method of presenting the same data.

5 An Outline Example of how Assigned Values and Target Values may be Specified and Used in Accordance with the Harmonised Protocol

This is intended to be an example of how assigned values and target values may be calculated and used according to the protocol. Numerical details have been specified for the purposes of illustration only; real schemes will have to take account of factors specific to their area. The example has been taken from the Harmonised Protocol.[1]

Scheme

The example requires that there will be four distributions of materials per year, dispatched by post on the Monday of the first full working week of January, April, July and October. Results must reach the organisers by the last day of the respective month. A statistical analysis of the results will be dispatched to participants within two weeks of the closing dates. This example considers the results from one particular circulation of two test materials for the determination of two analytes.

Testing for Sufficient Homogeneity

In accordance with the procedure described in previously.

Analyses Required

The analyses required in each round will be:

1. hexachlorobenzene in an oil, and
2. Kjeldahl nitrogen in a cereal product.

Methods of Analysis and Reporting of Results

No method is specified, *but* the target values were determined using a standard method, and participants must provide an outline of the method actually used, or give a reference to a documented method.

Participants must report a single result, in the same form as would be provided for a client. Individual reported values are given in Table 7.1.

Assigned Values

Hexachlorobenzene in Oil

Take the estimate of assigned analyte concentration $\hat{X}$ for the batch of material as the robust mean of the results of six expert laboratories.

Reference laboratory	Result (μg/kg)
7	115.0
9	112.0
10	109.0
13	117.0
18	116.2
19	115.0

$\hat{X}$ is 114.23 μg/kg; traceability was obtained using a reference method calibrated using in-house reference standards and the uncertainty on the assigned value was determined to be ± 10 μg/kg from a detailed assessment of this method by the reference laboratories.

Kjeldahl Nitrogen in a Cereal Product

Take the assigned value of analyte concentration $\hat{X}$ for the batch of material as the median of the results from all laboratories.

Target Values for Standard Deviation

Hexachlorobenzene in Oil

In the example used here, the %RSD_R value has been calculated from the Horwitz equation [RSD_R in % = $2^{(1-0.5\log\hat{X})}$]

The target value for the standard deviation (σ) will therefore be:

$$\sigma_1 = 0.222\hat{X}\ \mu\text{g/kg}$$

Table 7.1 *Results used in proficiency testing scheme example*

	Hexachlorobenzene in oil		*Nitrogen in cereal*	
Assigned value of anayte	114.2μg/kg		2.93 g/100g	
Laboratory	*Result*	*z-Score*	*Result*	*z-Score*
001	122.6	0.3	2.97	0.9
002	149.8	1.4	2.95	0.5
003	93.4	−0.8	3.00	1.4
004	89.3	−1.0	2.82	−2.0
005	17.4	−3.8	2.88	−0.9
006	156.0	1.7	3.03	2.0
007	115.0	0.0	2.94	0.3
008	203.8	3.5	3.17	4.7
009	112.0	−0.1	3.00	1.4
010	109.0	−0.2	2.82	−2.0
011	40.0	−2.9	2.99	1.2
012	12.0	−4.0	2.84	−1.6
013	117.0	0.1	2.85	−1.4
014	0.0	4.5	2.93	0.1
015	101.8	−0.5	2.80	−2.4
016	140.0	1.0	2.96	0.7
017	183.5	2.7	2.97	0.9
018	116.2	0.1	2.88	−0.9
019	115.0	0.0	2.92	−0.1
020	42.3	−2.8	2.88	−0.9
021	130.8	0.7	2.78	−2.8
022	150.0	1.4	2.92	−0.1

Kjeldahl Nitrogen in a Cereal Product

In this example, the $\%\text{RSD}_R$ value has been calculated from published collaborative trials.

The target value for the standard deviation (σ) is given by:

$$\sigma_2 = 0.018\hat{X} \text{ g/100g}$$

Statistical Analysis of Results of the Test

Hexachlorobenzene in Oil: Formation of z-Score

Calculate:

$$z = (x - \hat{X}) \,/\, \sigma$$

for each individual result (x) using the values of $\hat{X}$ and σ derived above. These results are shown in Table 7.1.

Kjeldahl Nitrogen in a Cereal Product: Formation of z-Score

Calculate:

$$z = (x - \hat{X}) / \sigma$$

for each individual result (x) using the values of $\hat{X}$ and σ derived above. These results are shown in Table 7.1.

Display of Results

z-Score Tables

The individual results for hexachlorobenzene pesticide in oil and for Kjeldahl nitrogen in a cereal product, together with associated z-scores, are displayed in tabular form in Table 7.1.

Histograms for z-Scores

The z-scores for hexachlorobenzene pesticide in oil and for Kjeldahl nitrogen in a cereal product are also displayed as bar-charts in figures 7.1 and 7.2 respectively. This display is also used by a number of proficiency testing schemes in the food sector, *e.g.* the Food Analysis Performance Assessment Scheme described below.

Decision Limits

Results with an absolute value for z of less than two will be regarded as satisfactory. Remedial action will be recommended when any of the z-scores exceed an absolute value of 3.0.

In this example, such results are: laboratories 005, 008, 012, 014 for hexachlorobenzene pesticide in oil, and laboratory 008 for Kjeldahl nitrogen in a cereal product.

6 Examples of Commercial Proficiency Testing Schemes in the Food Sector

A number of Member States have introduced proficiency testing schemes in the food area. Most operate according to the Harmonised Protocol and thus conform to the requirements internationally accepted for the operation of proficiency testing schemes in the food sector.

The UK Ministry of Agriculture, Fisheries and Food developed a proficiency testing scheme for food analysis laboratories [The Food Analysis Performance Assessment Scheme (FAPAS)]. The scheme complies with the

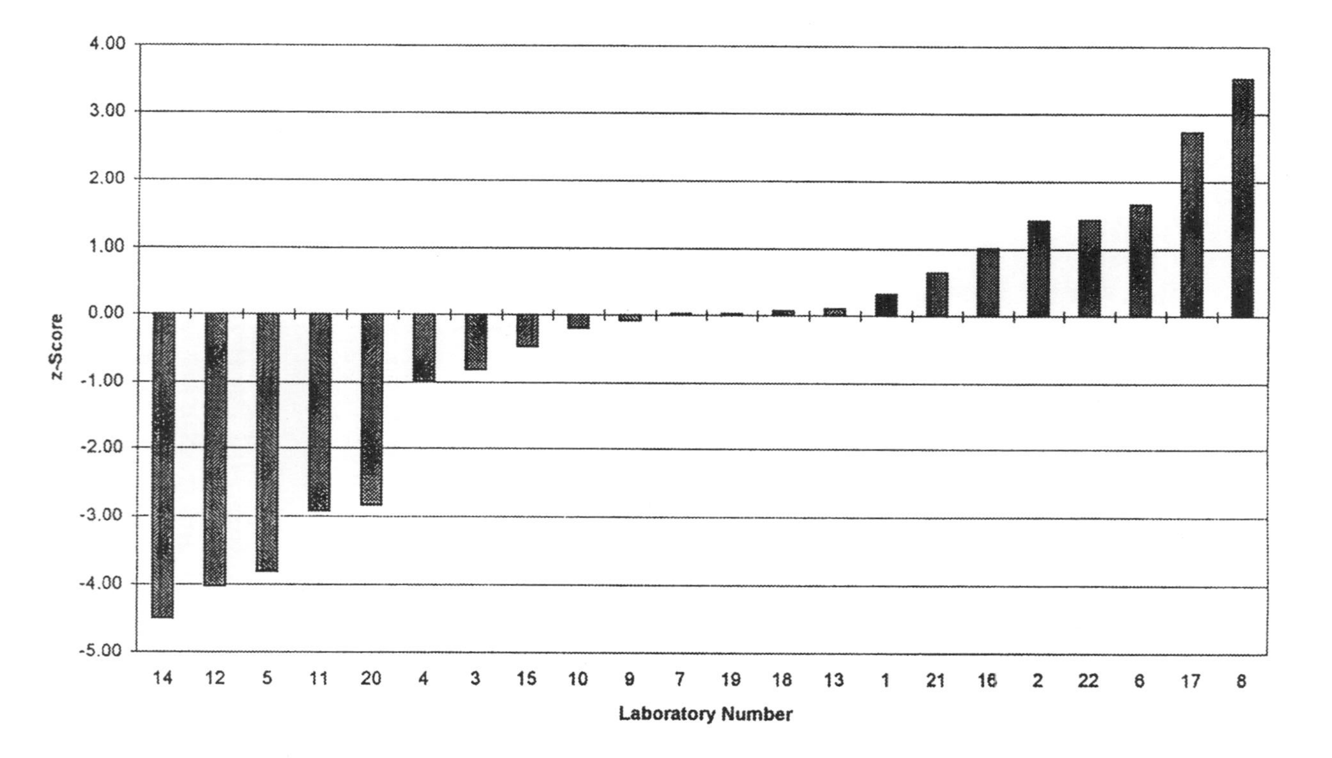

Figure 7.1 *z-Score for hexachlorobenzene in oil (114.2 μg/kg)*

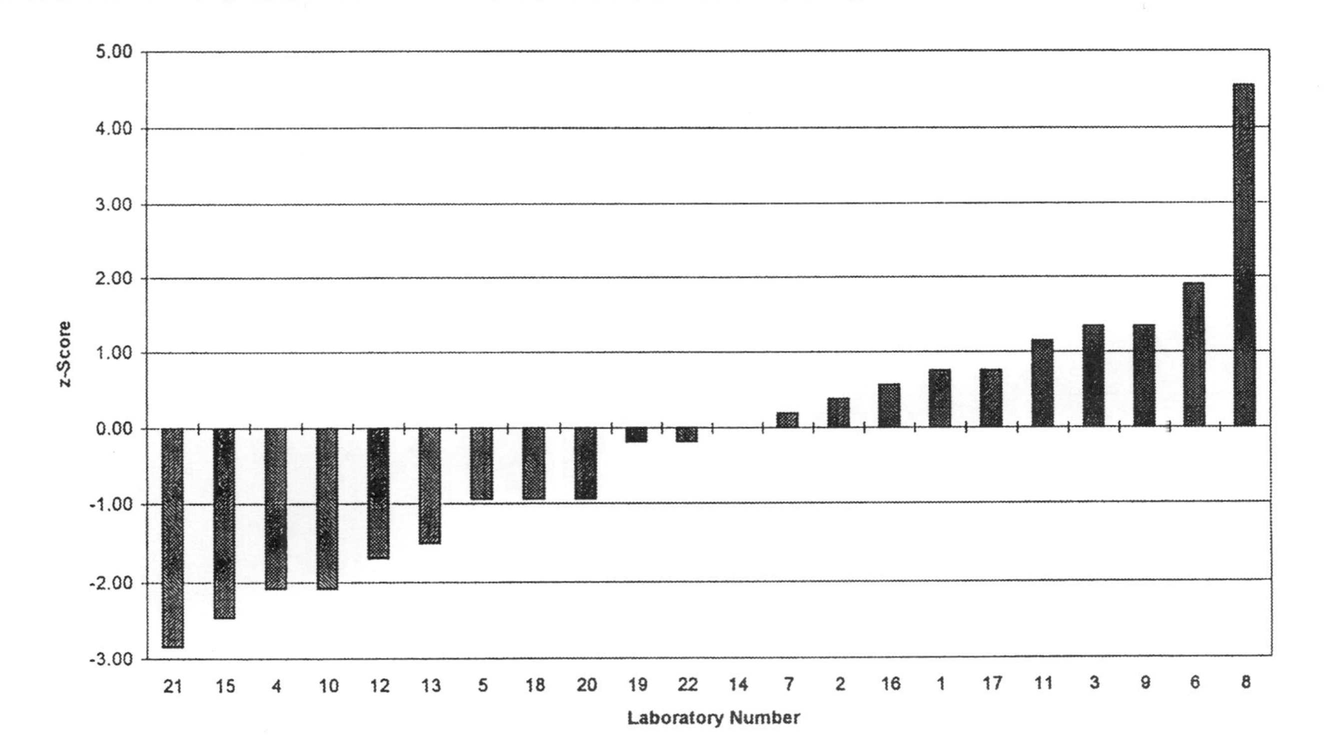

Figure 7.2 *z-Score for Kjeldahl nitrogen in cereal (2.93 g/100g)*

requirements of the Harmonised Protocol referred to above and thus may be regarded as a practical demonstration of the effectiveness of the Protocol.

The Food Analysis Performance Assessment Scheme

The Scheme was developed because it was appreciated that food-related legislation requires that there are data on residues and contaminants, some of which are particularly difficult to analyse. Lack of independent assessment of the data being produced in these consumer safety-related areas would hamper the work of enforcement authorities and would prejudice the recognition of results and certificates which is at the heart of the EC 'Single Market' and 'New Approach' initiatives. It would also limit the scope and reliability of food surveillance work on the UK food supply. FAPAS helps to provide the independent assessment of these data. The Scheme was also developed to assist in the effective implementation and enforcement of regulations made under UK legislation.

The Scheme was developed within the Ministry with help from representatives of all sectors (*i.e.* enforcement, industry/trade and referee analysts) who are concerned with the proficiency of food analysis laboratories. In particular, representatives from the Association of Public Analysts, British Food Manufacturing Industries Research Association, British Retail Consortium, Campden and Chorleywood Food and Drink Research Association, Food and Drink Federation and the Statistical Sub-Committee of the RSC Analytical Methods Committee serve on the Advisory Committee. The Committee has now been joined by user representatives from Denmark and Spain.

The Scheme has a number of series currently in operation, these being for:[11]

Nutritional components
Veterinary drug residues
HPLC procedures
Aflatoxins analysis
OC-pesticides
Trace elements
OP-pesticides (cereals)
Animal feedingstuffs
Overall migration
Specific migration
Alcoholic drinks by GC
Oils and fats by GC
Nitrate analysis
Patulin analysis
Ochratoxin analysis
Nutritional elements
Pesticides (fruit and vegetables)

Participants to the scheme are from all sectors within the UK, with over a half of participants from non-UK laboratories.

Sectors represented include the public sector, enforcement analysts, research associations and other trade bodies, industry, education and consultants.

The results are calculated and displayed in accordance with the Harmonised Protocol recommendations, an example of which has been given previously.

The Nordic Proficiency Testing Scheme for Food Microbiology Laboratories (NFMPT)

The Biology Division at the National Food Administration in Uppsala, Sweden, has, since the beginning of the 1980s, organised a proficiency testing scheme for food microbiology laboratories. Currently, there are about 390 laboratories, mainly from the Nordic countries, participating in the scheme.

An evaluation of the effect of proficiency testing showed that laboratories that have been participating for a long time perform better than laboratories that have recently joined the scheme. The evaluation procedure used in the scheme is available.[12]

Each year, three proficiency test distributions are made and at each occasion 4–5 samples are to be analysed. At present, mixtures of bacteria, but in the future also moulds and yeast, freeze-dried in small glass vials, are used as samples.[13] An indication of the scope of the scheme may be obtained from the microbiological analyses required of the materials distributed in 1996 and 1997, as given in Table 7.2.

Samples will automatically be sent out a few weeks before the required date of analysis. It is not necessary to participate in every determination as the laboratories only have to carry out the microbiological analyses of interest to them. However, if a laboratory does not want to analyse for any organism in a particular distribution, it must inform the National Food Administration before any test material is sent to it from that distribution.

When all the results have been reported to the National Food Administration, a preliminary report is sent out to the participants as soon as possible and within two months a more detailed report, containing all the results and a statistical evaluation of these, is prepared and sent to participants. The final reports are published both in Swedish and English. The participants are often asked to state which media and methods they have used in their examination procedures. This can sometimes give information as to whether some media or methods are to be recommended or not. Such information is also included in the final report.

Other Proficiency Testing Schemes of Interest in the Food Analysis Sector

The proficiency testing schemes given in Table 7.3 are known to be active in the food analysis area and will, therefore, be of interest to food analytical laboratories.

Table 7.2 *Materials distributed under the NFMPT scheme in 1996 and 1997*

Date of distribution	*Microbiological analysis*	*No. of samples*
January, 1996	Enterococci *Clostridium perfringens* Coagulase positive staphylococci *Listeria monocytogenes* Lactic acid bacteria	4
April, 1996	Aerobic plate count Coliform colony count 30 °C Coliform colony count 37 °C Coliform colony count 44 °C Enterobacteriaceae Presumptive *Escherichia coli* *Aeromonas* spp.	4
October, 1996	Coliform colony count 30 °C Coliform colony count 37 °C Coliform colony count 44 °C Enterobacteriaceae Presumptive *Escherichia coli* Salmonella	5
January, 1997	Aerobic plate count Coagulase positive staphylococci Salmonella Campylobacter Moulds and yeast	4
April, 1997	*Clostridium perfringens* Sulfite reducing clostridia *Bacillus cereus* *Listeria monocytogenes,* qualitative and quantitative analysis	4
October 1997	Coliform colony count 30 °C Coliform colony count 37 °C Coliform colony count 44 °C Enterobacteriaceae Presumptive *Escherichia coli* Presumptive *Escherichia coli* O157 Salmonella *Yersinia enterocolitica*	5

Table 7.3 *Proficiency testing schemes in the food analysis area*

Scheme	*Organiser*
AQUACHECK (waters)	WRC, UK
BAPS (analysis of beers)	LGC, UK
CCFRA Microbiology Proficiency Scheme (food examination)	CCFRA, UK
CHEK (food analysis)	Inspectorate for Health Protection, The Netherlands
DAPS (analysis of distilled spirits)	LGC, UK
FAPAS (food analysis)	Ministry of Agriculture, Fisheries and Food, UK
FEPAS (food examination)	Ministry of Agriculture, Fisheries and Food, UK
LEAP (water microbiology and chemistry)	Yorkshire Environmental, UK
NACS (animal feedingstuffs)	Perstorp Analytical, UK
Nordic drinking water microbiology scheme	NFA, Sweden
PHLS food microbiological external quality assurance scheme	PHLS, Colindale, UK
PHLS external quality assessment scheme for water microbiology	PHLS, Colindale, UK
Proficiency testing nutritional components in food	NFA, Sweden,
ProTAS (alcoholic strength of wines, *etc.*)	LGC, UK
Quality in microbiology scheme (food examination)	Quality Management, UK
SMART QA (food examination)	Lab M, UK

References

1 The International Harmonised Protocol for the Proficiency Testing of (Chemical) Analytical Laboratories, ed. M. Thompson and R. Wood, *Pure Appl. Chem.,* 1993, **65**, 2123 (also published in *J. AOAC Int.*, 1993, **76**, 926).

2 Report of the 21st Session of the Codex Alimentarius Commission, FAO, Rome, 1997.

3 Guidelines on Internal Quality Control in Analytical Chemistry Laboratories, ed. M. Thompson and R. Wood, *Pure Appl. Chem.,* 1995, **67,** 649.

4 Personal communication, The Laboratory of the Government Chemist, Queens Road, Teddington, Middlesex TW11 0LW.

5 General Requirements for the Competence of Calibration and Testing Laboratories, ISO/IEC Guide 25, ISO, Geneva, 1990.

6 Development and Operation of Laboratory Proficiency Testing, ISO/IEC Guide 43, ISO, Geneva, 1996.

7 The Additional Measures Concerning the Official Control of Foodstuffs Directive 93/99: Laboratory Quality Standards and Methods of Analysis: Consultation Document, MAFF, London, 1995.

8 Report 0805 of the MAFF Food Analysis Performance Assessment Scheme, FAPAS Secretariat, CSL Food Laboratory, Norwich, 1993.
9 Use of Proficiency Tests to Assess Comparative Method Performance: the Determination of Fat in Foodstuffs, P. Lowthian, M. Thompson and R. Wood, *Analyst*, 1996, **121**, 977.
10 W. Horwitz, Evaluation of Analytical Methods used for Regulation of Foods and Drugs, *Anal. Chem.*, 1982, **54**, 67A.
11 Food Analysis Performance Assessment Scheme, FAPAS Secretariat, MAFF, Norwich, 1997.
12 Evaluation of The Nordic Proficiency Testing Scheme for Food Microbiology Laboratories, *J. Appl. Bacteriol.*, 1992, **72**, 210.
13 Quality Assurance and Quality Control in Food Microbiology Laboratories, Particularly when using Freeze-dried Mixed Cultures, M. Peterz, thesis, Swedish National Food Administration, 1992

CHAPTER 8

Respecting a Limit Value

1 Introduction

After a laboratory obtains and then reports an analytical result, that result is normally compared with either a standard (*i.e.* to a legislative or contractual limit) or with results obtained using another method. There is frequent confusion on the interpretation to be placed on analytical results in such situations. This confusion has been addressed in a number of sectors, one of which has been the EU Working Group in the Milk Market Support Sector. This Group has established general rules for the application of reference and routine methods for the analysis and quality evaluation of milk and milk products covered by common market organisation schemes and for the interpretation of analytical results.[1] Martin *et al.*[2] have also addressed the problem within the fruit juice analysis sector, but their comments can equally be applied to other food analysis sectors.

This chapter aims to bring together a number of considerations in this area to provide guidance to laboratories on the interpretation of results in a number of situations, all of which are of importance to the food analyst. They assume that, in principle, internationally accepted and validated methods should be used as reference or food control methods, *i.e.* the performance characteristics of the methods have been established.

The various situations which are addressed in this chapter:

1. Evaluation of analytical results obtained using validated methods.
2. Results obtained using routine methods and which are found to be close to a specified limit.
3. Procedure for determining the compliance by a laboratory with an established reproducibility limit.
4. Procedure for obtaining a provisional reproducibility limit.
5. Procedures to be adopted when the results of analysis are disputed.

2 Evaluation of Analytical Results Obtained Using Validated Methods

The most common situation where there is need to consider the interpretation of a limit is the simple comparison between a limit value and the analytical result. That limit value may be either legislative or contractual.

If the analytical result indicates that a prescribed or agreed limit may have been exceeded, then the analyst should give consideration to applying the following procedures:

The Analytical Result is Derived from a Single Analysis

When the analytical result is represented by or derived from a single analysis, a second analysis must be carried out under repeatability conditions. If the two analyses cannot be carried out under repeatability conditions, then a further duplicate analysis shall be carried out under repeatability conditions and those results used for the assessment of critical difference compliance.

The arithmetic mean of the two single analyses obtained under repeatability conditions is compared to the (legislative or contractual) limit after calculation of the critical difference as calculated below for the analytical result.

The critical difference for the analytical result is calculated using the formula given below:

$$\mathrm{CrD}_{95}(|\bar{Y} - \bar{m}_0|) = \frac{1.0}{\sqrt{2}}\sqrt{R^2 - \left[r^2 \frac{(n-1)}{n}\right]}$$

where CrD_{95} is the critical difference at the 95% probability value, $\bar{Y}$ is the arithmetic mean of the results obtained, m_0 is the (statutory/contractual, *etc.*) limit, n is the number of analyses per sample, R is the reproducibility of the method at the concentration of interest, and r is the repeatability of the method at the concentration of interest.

If the difference between the (arithmetic mean) analytical result and the limit value is greater than the critical difference as calculated above, then it may be assumed that the sample which has been analysed does not fulfil the statutory or contractual requirements.

The values of r and R may have to be determined by interpolation so as to obtain the values which would apply at the limit concentration/value.

If it is to be expected that most samples comply with the statutory or contractual limit, then the final analytical results may be expected to be less than $[m_0 + \mathrm{CrD}_{95}(|\bar{Y} - m_0|)]$ if the limit is a maximum, or greater than $[m_0 + \mathrm{CrD}_{95}(|\bar{Y} - m_0|)]$ if the limit is a minimum and m_0 is the given limit value.

The Analytical Result is Derived from Indirect Analysis

If the final result x is calculated using a formula of the form $y_1 \pm y_2$ where y_1 and y_2 are the final results of a single type of analysis, then the overall repeatability and reproducibility limits r_x and R_x of the final result x are calculated by:

$$r_x = \sqrt{r_1^2 + r_2^2}$$
$$R_x = \sqrt{R_1^2 + R_2^2}$$

A typical example is the estimation of the fat content of butter by analytically determining the water and fat-free dry matter content of a butter sample to estimate the fat content by difference.

In this example the repeatability limits, r_1 and r_2, and the reproducibility limits, R_1 and R_2, refer to the two different methods for water and fat-free dry matter.

As in the example above, the estimated value, x, is compared to the limit m_0 following the procedures specified.

In this case the critical difference is determined using the formula:

$$\mathrm{CrD}_{95}(|\bar{x} - m_0|) = \frac{1.0}{\sqrt{2}}\sqrt{R_x^2 - \left[r_x^2 \frac{(n-1)}{n}\right]}$$

where $\bar{x}$ is the arithmetic mean of the results x, obtained by calculation.

3 Results Obtained Using Routine Methods and Which are Found to be Close to a Specified Limit

There are frequently occasions when a reference, *i.e.* legislative or contractual, method is specified but a laboratory wishes to use its own routine method with which it is familiar and competent. In such a situation the following procedure can be applied for setting the appropriate decision limits:

The Reproducibility of the Routine Method is Less Than the Reproducibility of the Reference Method

In this situation:

$$R_{\mathrm{Routine}}/R_{\mathrm{Reference}} \leq 1$$

where R_{Routine} is the reproducibility limit of the routine method, and $R_{\mathrm{Reference}}$ is the reproducibility limit of the reference method.

Thus, if m_0 is a limit specified in, for example, legislation or a contract, then the decision limit, s, is equal to the specified limit, *i.e.*

$$s = m_0$$

The Reproducibility of the Routine Method is Greater Than Reproducibility of the Reference Method

Here, if m_0 is an upper limit and if $R_{\text{Routine}}/R_{\text{Reference}} > 1$, where R_{Routine} and $R_{\text{Reference}}$ are as defined above, the decision limit(s) is obtained using the formula:

$$s = m_0 + [(R_{\text{Routine}}/R_{\text{Reference}}) - 1] \times \text{CrD}_{95}$$

If, under the same conditions, m_0 is a lower limit, the decision limit is obtained by:

$$s = m_0 - [(R_{\text{Routine}}/R_{\text{Reference}}) - 1] \times \text{CrD}_{95}$$

where CrD_{95} is the critical difference of the reference method, determined as described previously.

However, where m_0 is an upper limit, a result obtained by a routine method which is greater than the decision limit should be confirmed by repeat analysis using the reference method in either that or in another laboratory before any 'formal' action is taken by the laboratory. The confirmation result must be obtained using at least the same number of analyses/samples as the result obtained using the routine method.

Where m_0 is a lower limit, the same procedure has to be followed for a final result obtained by a routine method which is less than the decision limit.

Applicability of Procedure Where Matrix Effects are Identified

The procedures described above can be applied if there are no detectable matrix effects. Matrix effects can be detected in the following way:

Determine, for each sample used for calibration purposes, the difference (W_i) between the results obtained by the difference and routine methods.

Calculate the standard deviation, s, of the differences by:

$$s = \sqrt{\frac{\sum W_i^2}{2m}}$$

where m is the number of samples used for calibration purposes.

Compare the value of s with the arithmetic mean of the repeatability standard deviation of the reference and the routine methods, *i.e.*:

$$s_{\bar{r}} = \sqrt{\frac{s^2_{r(\text{Reference})} + s^2_{r(\text{Routine})}}{2}}$$

There may be a matrix effect present if:

$$s > s_{\bar{r}}$$

If this occurs, carry out further studies before deciding on a decision limit.

4 Procedure for Determining Compliance by a Laboratory with an Established Reproducibility Limit

A single laboratory frequently wishes to see if it is using a method effectively by determining whether the reproducibility limit is being complied with. In this situation the reproducibility limit is checked by comparing the laboratory's results with the results of an experienced laboratory obtained from an identical test material. Duplicate determinations are carried out in both laboratories and the results are evaluated using the formula:

$$\mathrm{CrD}_{95}(|\bar{Y}_1 - \bar{Y}_2|) = \sqrt{R^2 - \frac{r^2}{2}}$$

where CrD_{95} is the critical difference ($P = 0.95$), $\bar{Y}_1$ is the arithmetic mean of the results obtained in laboratory 1, $\bar{Y}_2$ is the arithmetic mean of the results obtained in laboratory 2, and R is the reproducibility limit and r is the repeatability limit at the concentration of interest. These may have to be determined by interpolation.

If the critical difference is found to have been exceeded, all efforts have to be made to find the cause of this and another comparison exercise should be undertaken. In the case of non-compliance with the reproducibility limit in the second comparison exercise, further investigation must be undertaken.

5 Procedure for Obtaining a Provisional Reproducibility Limit

It is sometimes necessary to 'estimate' the reproducibility limit for a method, *i.e.* to determine a provisional reproducibility limit ($R_{\mathrm{Provisional}}$). This value is calculated from the results of duplicate analyses carried out in two different laboratories using the following equation:

$$R_{\mathrm{Provisional}} = \sqrt{(\bar{Y}_1 - \bar{Y}_2)^2 + \frac{r^2}{2}}$$

where $\bar{Y}_1$ is the mean of two results obtained in laboratory 1, $\bar{Y}_2$ is the mean of two results obtained in laboratory 2, and r is the repeatability limit or provisional repeatability limit.

In some sectors the following recommendations have also been made:

- the $R_{Provisional}$ value can be used to calculate the critical differences required in the first section of this chapter,
- the $R_{Provisional}$ value is fixed at $2r$, if the calculated value for $R_{Provisional}$ is smaller than $2r$,
- if the calculated $R_{Provisional}$ value is larger than $3r$ or is greater than twice the R value predicted from the Horwitz function,[3] then the $R_{Provisional}$ is unacceptably high and cannot be used to calculated critical differences (see also Chapter 3), and
- the $R_{Provisional}$ values should be determined at least once per year on the basis of results obtained in two laboratories.

6 Procedures to be Adopted when the Results of Analysis are Disputed

Some sectors, *e.g.* the EU milk market sector,[1] have laid down procedures to be followed in cases where analytical results are disputed. Although the information applies to discrete lots moving in trade, it is of value to food analytical laboratories and points the way to future analytical result interpretation and is therefore given here. Such interpretation should be agreed by interested parties before any analysis commences.

If Both Laboratories Meet the Repeatability Requirement, and the Reproducibility Requirement is Also Met

The arithmetic mean of the test results obtained by both laboratories is reported as the final result. This final result is evaluated taking the critical difference into consideration, using the following formula:

$$\mathrm{CrD}_{95}(|\bar{y} - m_0|) = \frac{1.0}{\sqrt{2}}\sqrt{R^2 - r^2\left(1 - \frac{1}{2n_1} - \frac{1}{2n_2}\right)}$$

where $\bar{y}$ is the arithmetic mean of all results obtained by both laboratories, m_0 is the statutory or contractual limit, R is the reproducibility value at the limit of interest, r is the repeatability value at the limit of interest, n_1 is the number of results obtained by laboratory 1, and n_2 is the number of results obtained by laboratory 2.

If Both Laboratories Meet the Repeatability Requirement but the Reproducibility Requirement is not Met

The consignment is finally rejected, if the results of both laboratories lead to this conclusion. Otherwise, the consignment is accepted.

If Only One Laboratory Meets the Repeatability Requirement

The final result of the laboratory meeting the repeatability requirement is used for the decision on whether the consignment should be accepted or not.

If Neither Laboratory Meets the Repeatability Requirement, but the Reproducibility Requirement is Met

The arithmetic mean of the final results obtained by both laboratories is used for the decision on acceptability of the consignment, taking the critical difference as described above into consideration.

If Neither Laboratory Meets the Repeatability Requirement, and the Reproducibility Requirement is not Met

The consignment is accepted, if the results obtained by one laboratory lead to this conclusion.

If the Results Have Been Obtained Using Non-validated Methods

The consignment is accepted, if the results obtained by one laboratory lead to this conclusion.

If the Producer Questions the Sampling Procedure

A repeat sampling should be performed if possible, and the producer shall pay the costs of the second analysis if the conclusion drawn by the first laboratory or on the basis of the first sample is confirmed by this second analysis.

References

1 Guidelines for the Interpretation of Analytical Results and the Application of Sensory Evaluation in Relation to Milk and Milk Products under the Common Market Organisation, DG VI, Commission of the European Communities, Brussels, Document VI/2721/95 Rev. 1 of 20.07.1995.
2 Discussion of Some Statistical Concepts Applied to Fruit Juice Analysis, G. G. Martin and Y.-L. Martin, *Fruit Processing*, 1996, **5**, 186.
3 Evaluation of Analytical Methods used for Regulation of Foods and Drugs, W. Horwitz, *Anal. Chem.*, 1982, **54**, 67A.

CHAPTER 9

Experiences in the Implementation of Quality Assurance and Accreditation into the Food Analysis Laboratory: Laboratory Aspects

In this, and the following chapters, specific areas of interest are addressed which, in the experiences of the authors, have been shown to cause some problems when being introduced into the quality assurance procedures of a food analysis laboratory. The information provided may not concur with present UK practice in every respect but does meet the approval of accreditation agencies which are EAL signatories. The areas of concern addressed in this chapter deal with the general structure of and personnel in the laboratory; this chapter is divided into three sections dealing with:

1. Requirements on Management Towards Staff and the Organisation of the Laboratory.
2. Requirements on the Staff in the Laboratory.
3. Requirements on the Laboratory Premises.

1 Requirements on Management Towards Staff and the Organisation of the Laboratory

The laboratory must have a management with the authority to take appropriate decisions to allow the implementation of quality assurance measures and have access to the necessary resources for the laboratory to be able to carry out the work it has agreed to do.

The laboratory should specify and document its organisation so that all members of staff involved in analytical activities are aware of relevant responsibilities and rights of decision. It is important that all members of the staff know where their responsibilities start and where they end. This is aided by a laboratory organisation chart, showing the organisation of the laboratory and the allocation of the various responsibilities within it.

It is the responsibility of the management to ensure that all members of the staff have the necessary education, training and experience to carry out analyses including being able to use any necessary equipment. In some cases it may be necessary to determine objectively their competency in using the equipment.

Thus, it is the responsibility of the management to ensure:

- that the laboratory's quality policy statement is prepared and issued;
- that laboratory work is carried out according to the quality policy and established rules (*e.g.* customer contracts, governmental rules, environmental demands, documentation of analytical work);
- that sufficient staff having the required qualifications and experience are always available;
- that laboratory equipment is suitable and well adapted to its intended use;
- that a quality co-ordinator with well-defined authority and direct access to top management has been appointed to be responsible for the quality system and its documentation;
- that a technical manager with overall responsibility for technical operations in the laboratory has been appointed;
- that the reports issued by the laboratory satisfy established requirements; and
- that the quality system is systematically and periodically reviewed to ensure its continued effectiveness.

The management must ensure that all members of staff receive the necessary additional training, either internally or externally, which is required in order to maintain or add to competence. Training must definitely be provided when staff change responsibilities within the laboratory and when new methods of analysis or techniques are introduced.

It is important that management ensures that all members of staff appreciate that everybody's contribution to the work of the laboratory is important. Work of the required quality is produced only if all staff are motivated to work according to jointly established procedures. It is also important that management encourages the staff to suggest any changes they see as necessary, as well as any improvements to instructions and procedures.

Experience from different quality systems and audits of analytical laboratories in the food area show that, in many laboratories, there is often a long period of time before any substantial positive effects following the introduction of a quality system become obvious. However, it is important that the management actively and visibly support the implementation of the quality system and provides the necessary resources (*e.g.* time, equipment, suitable training) for it during that time.

2 Requirements on the Staff in the Laboratory

Technical Manager

The management should appoint a technical manager having the overall responsibility for the laboratory's technical operations. In a smaller laboratory it is often the laboratory manager who is the technical manager for the entire laboratory. Larger laboratories usually have several technical managers of different seniorities.

The technical manager of a larger laboratory is the manager of a well defined unit, *e.g.* the person responsible for pesticide analyses, microbiological determinations or LC determinations. As a general rule, the technical managers of a larger laboratory should not be so senior within the organisation that they are divorced from practical work in the laboratory. A technical manager should be familiar with the daily work: analytical methods and procedures, the aim of the testing and the way in which the test result is evaluated.

It is the responsibility of the technical manager to ensure that results which leave the laboratory are derived from a well-planned exercise, which was accurately carried out by competent staff using suitable methodology, and that the results are appropriately reported.

Thus, it is the responsibility of the technical manager to ensure that:

- the laboratory uses suitable, documented and appropriately approved methods of analysis, which are kept up to date and which are available to those who are required to carry out the analytical work;
- the laboratory has documented operating instructions for the use of all pieces of equipment used in the handling, preparation and analysis of samples where the absence of such instructions could jeopardise the quality of the analyses;
- the laboratory assistants and technicians clearly understand and master the various steps of analytical work;
- the quality of the analysis is regularly monitored, applying suitable quality control measures;
- all results are documented and recorded (archived);
- analytical results are reported according to established rules;
- staff receive general and specialised training in quality assurance;
- he/she (the technical manager) is completely familiar with and follows the laboratory's quality policy; and
- he/she (the technical manager) follows developments within his or her own analytical field.

There are no formal qualification requirements for the technical manager but in many countries it is usually regarded as appropriate for food analyses to be carried out by or under the supervision of a qualified and experienced analyst with an academic degree. Other qualifications may be acceptable in combination with work experience. A recently graduated analyst should normally have at least two years of relevant work experience in order to be regarded as an 'experienced analyst'.

In some EU Member States, *e.g.* in the UK and Sweden, there are legal requirements that some staff members of official food control laboratories must have particular qualifications, *e.g.* in the UK food analysts and food examiners must possess specified qualifications. However, these analysts or examiners should not be confused with the technical managers; it is not always the case that the technical manager has the specialised knowledge and experience required for official food control.

Quality Co-ordinator

The management of the laboratory must appoint one or more persons to be responsible for the laboratory's quality system and its documentation. It is convenient if only one person is given the overall responsibility for the quality system and that this person, depending on the size of the laboratory, has the appropriate number of co-workers. The person with overall responsibility is sometimes called the quality manager, which may not be the most appropriate designation, since it may give other staff in the laboratory the impression that only that person is responsible for the quality of everybody's work. A title such as quality co-ordinator is to be preferred, since it emphasises that this person co-ordinates quality matters, whilst every member of staff is responsible for the quality of work carried out by the laboratory. Naturally, the technical manager has responsibility for the quality of results whilst, taking the matter even further, the ultimate responsibility for quality lies with the laboratory management.

The quality co-ordinator (however named) is explicitly responsible for keeping the quality system and its documentation up to date. The co-ordinator with overall responsibility should have a named deputy, and all quality co-ordinators (however named) should have direct access to senior management.

The laboratory should document the competence requirements of the quality co-ordinator(s), *e.g.* that the person should have had an appropriate education with at least two years of relevant work experience, be familiar with quality assurance issues and the general principles of quality systems as well as have practical experience of the implementation of quality systems in the laboratory. The laboratory should define the tasks and the responsibility of the quality co-ordinator(s), for example as set out below.

The quality co-ordinator should be charged with reporting to management all quality problems having a direct influence on the quality policy or the principles of the quality manual. The quality co-ordinator should be given the authority, either directly or *via* the management, to intervene in work if there is reason to suspect serious quality problems. The quality co-ordinator may, for example, be authorised to stop a test report being issued or sent out, or, in serious cases, to require the cessation of all analytical activities.

The field of work of the quality co-ordinator may also include keeping the laboratory's records on staff, staff competence, methods and instruments, but others may also be responsible for these tasks.

Thus, it is the responsibility of the quality co-ordinator to:

- be acquainted with, follow and work for the implementation of the quality policy of the laboratory;
- draw up guidelines for the quality assurance in co-operation with (senior) management;
- organise and regularly carry out internal quality audits;
- document the quality system of the laboratory in a quality manual and in standard operating procedures;
- keep records of documents relating to the quality system;
- support efforts to ensure that all members of the staff are appropriately trained in quality matters;
- report on quality matters to the management of the laboratory, and to take part in periodical management quality reviews and report their outcome to the staff; and
- participate, as the representative of the laboratory, in all external audits and report the findings to (senior) management.

Since all activities relating to quality assurance activities should be regarded as joint work aiming to achieve mutual goals, best results are achieved if the entire staff can be motivated, as far as possible, to participate in the work, especially in the elaboration of the laboratory's internal quality procedures and other documentation. It is important that the quality co-ordinator consults other staff early on in the development process and takes into account their experience and wishes. Such consultation and 'listening' is a prerequisite for the final procedures functioning well in practice. To be successful in his or her work, the quality co-ordinator must have active and visible support from the senior management.

Laboratory Staff

The laboratory should have sufficient staff with the necessary education, training, technical knowledge and experience to be able to carry out its contracted analyses. Procedures should be in place to ensure sufficient staff are available when members are absent as a result of illness or for other reasons, *e.g.* leave.

The management of the laboratory should establish minimum requirements for the education and experience of the staff charged with supervising and carrying out the analytical work of the laboratory.

Staff without formal laboratory education or persons undergoing training may carry out analytical work provided it can be proved that they have been sufficiently trained and that their work is appropriately supervised.

When new staff are employed they should be instructed in detail regarding the work for which they will be directly responsible, and also be given information on the operation of the entire laboratory. Newly employed staff should also be given an opportunity to familiarise themselves with the laboratory's quality policy and its quality manual. Analytical work may be carried out only by members of staff who have been deemed competent for the tasks in question. The competence of staff should be monitored in suitable ways, *e.g.* in accordance with documented quality control routines. Sometimes retraining may be necessary, *e.g.* in cases when an analytical method or technique has been out of use for a long time. All members of staff should be given training in quality assurance matters.

Thus, it is the responsibility of the laboratory assistant/technician to:

- closely follow given instructions and analytical methods;
- document all analytical work;
- be acquainted with and follow security and environmental regulations;
- report deviations noted in the course of analytical work; and
- be acquainted with and follow the quality policy of the laboratory.

The laboratory should keep records on the qualifications of the staff, relevant training given (both internal and external courses) and other training given, *e.g.* courses on the use of instruments, on quality assurance, *etc.* The records may also, when suitable, contain information on scientific publications, participation in important conferences as well as data from participation in any proficiency testing schemes.

The laboratory should also document in the records referred to above the technical competence of individual members of the staff to carry out analyses or parts of them. The aim of such documentation is to ensure that the laboratory can, should it become necessary, demonstrate that individual staff members have been suitably trained and that their competence to perform specific analytical tasks has been evaluated and is regarded as appropriate.

It is necessary to specify all methods of analysis or parts of methods which belong in the person's field of competence, *e.g.* it is not sufficient to state 'all methods for the determination of metals'. In some cases it may prove more practical to document competence in relation to specific analytical techniques. The competence should be verified by the person's nearest supervisor.

It is important that competence records are kept up to date by regular review, *e.g.* annually. During such a review the laboratory should add any new competence achieved since the last review and also consider deleting from the field of competence analytical methods or techniques which have not been used over a longer period of time, *e.g.* not used during the previous year.

3 Requirements on the Laboratory Premises

As a general rule the laboratory should have suitable and sufficiently spacious premises. Adequate space reduces the risk of accident and other dangers, and allows staff the necessary freedom of movement. Separate rooms must be available for some specific applications, *e.g.* where sensitive equipment is used and for the washing and sterilisation of glassware.

Working Environment

The working environment (usually the laboratory) must never be such that the analytical results are invalidated or that the required accuracy of measurement is not achieved. The laboratory must be protected from exceptional conditions (heat, dust, humidity, steam, noise, electromagnetic interference, *etc.*) which could affect its analytical work. It is particularly important to evaluate these factors when activities are undertaken at sites other than at the permanent laboratory.

The ventilation must be balanced and constructed in a way which eliminates draught and dust formation. Special attention should be given to equipment such as analytical balances, which, because they are sensitive instruments, should not be placed in an environment where there is a draught. If possible, the laboratories should not form parts of routes of transit.

Excessive heat and light may also adversely affect analytical activities. Balances and incubators for microbiological work should not, for example, be placed so that they are subject to direct sunlight. Efficient air-conditioning of laboratory premises is often necessary.

Strong electromagnetical fields may influence some analytical instruments and irregularities in the supply of electricity may halt analytical systems, especially those using computerised systems.

Where relevant to the work of the laboratory, there should be suitable procedures for monitoring the environment, *e.g.* temperature, humidity, microbiological and chemical contamination, special requirements on the intensity of light, radiation, *etc.* Critical environmental factors must be checked and documented, for example the temperature of refrigeration rooms.

Laboratory Facilities

Ideally, laboratory facilities should not be subject to direct sunlight (*i.e.* not be situated on the sunny side of a building). The laboratory rooms and the work should be arranged so that passage through the laboratory of people, samples, chemicals, *etc.* are minimised. The ceilings and walls of facilities should be constructed of materials which are easy to clean. In certain cases, *e.g.* in microbiological work and in trace element analysis, the ceilings may have to be sealed in order to minimise contamination due to dust. Laboratory benches should be made of resistant and easily cleaned material.

Microbiological and chemical activities must be performed in separate rooms. If the laboratory also performs sensory evaluations, these must take place in rooms separate from chemical and microbiological work, and the laboratory environment in which such tests are performed must meet stringent requirements, as described in Chapter 13.

There may also be reasons to segregate analytical work which is especially sensitive to the influence of other activities, or work where there are particular problems or risks. Trace element analyses, for example, should not be carried out in laboratory rooms where the same elements may be analysed at higher concentrations. Work involving carcinogenic substances or genetically modified organisms should be performed in separate laboratories. Before a 'separate laboratory' is taken into use it must be ensured that it is free of contamination and fulfils other requirements such as legislation requirements.

Separate storage spaces for chemicals, glassware, acids, solvents, rubbish, *etc.* and sufficient refrigeration and freezing for storage of samples, chemicals and reagents must be available. The refrigeration capacity should preferably be such that 'pure' chemicals and solvents can be stored separately from 'impure' samples. Special attention should be paid to the storage of samples to ensure their integrity. The laboratory must pay attention to the possibility of deterioration and contamination of samples and reagents, and to ensure that samples cannot be mixed together.

Rooms intended solely for staff, such as changing rooms and staff rooms, should be available, and there should be separate facilities in the changing rooms for protective clothing and personal clothing. It is not suitable, and indeed it is frequently prohibited, to eat, drink or smoke in laboratory premises. Sometimes, for example in trace analysis or microbiological work, activities such as smoking and eating may compromise the quality of the analytical work. The laboratory management should therefore strictly forbid such activities.

The laboratory should have documentation on the various facilities of the laboratory, the intended use of each room, any restrictions and other important characteristics.

Access to Facilities and Confidentiality

Access to the laboratory premises must be restricted to staff; it is a serious quality risk from the point of view of confidentiality and safety if unauthorised persons are allowed free access to the laboratory. Rules should be established for external visitors. In general, all members of staff have access to all laboratory premises. However, it may be necessary to restrict staff access to some areas (*e.g.* where particularly confidential work or where work with radioactive or carcinogenic substances is carried out or where there is a risk of contamination, *etc.*). Staff should be adequately informed of such restricted areas, what they are used for, what the restrictions are and why access in these special cases is restricted.

Quality models such as the European standard EN 45001 require that laboratories give clients the possibility of monitoring the performance of the testing laboratory in relation to the work to be carried out if so requested. In practice this means that the client or his representative is given access to relevant areas of the testing laboratory, to witness the tests carried out for the client. Laboratories must ensure that such access does not conflict with rules of confidentiality of work for other clients and with safety. Furthermore, it is important that the laboratory has established procedures for the supervision of outside service staff and contractors who may work in the laboratory on a temporary basis. It has, for example, occurred more than once that unsupervised contractors have, without prior warning, switched off electricity supplies, thereby ruining analytical runs.

Housekeeping

The management must provide for good housekeeping in the laboratories and ensure that they are at all times kept clean and tidy.

Good housekeeping may influence the economy and efficiency of work in a positive way, and minimise the number of sources of errors. In general a better working atmosphere and safer conditions result from improvements in tidiness and cleanliness of the laboratory premises. An outside visitor gets a positive impression from a clean and tidy laboratory, and it is always worth remembering that obvious untidiness does not instil confidence in the customer. It is the responsibility of all staff members to contribute to good housekeeping. In general, the storage of items not immediately required (such as superfluous stocks of reagents and glassware, empty cardboard boxes, *etc.*) in the laboratory is to be avoided.

Routine cleaning is usually undertaken by separate cleaning staff. It is important that the cleaning staff are informed about safety and confidentiality matters of the laboratory. Specific, written instructions should be available, especially in cases when external contract cleaning companies are employed. It is recommended that it be the responsibility of the technical laboratory staff to clean hoods, stores of chemicals and media, refrigerators, incubators, *etc.* The laboratory staff should also be responsible for the order in cupboards and drawers. It is recommended that laboratory staff assist the cleaning staff with (spring) cleaning, washing of windows, *etc.*, if equipment and other laboratory utensils have to be moved. This enables cleaning to be thorough, while minimising the risks involved when persons having little laboratory training handle delicate laboratory equipment.

Waste Disposal

Laboratories must have instructions on the handling and disposal of waste. When elaborating the procedures, priority should be given to the safety of the staff, protection of premises and associated constructions (*e.g.* corrosive acids and substances blocking drains) and protecting the external environment. National regulations must be taken into account, especially in the case of disposal of hazardous waste.

When disposing of samples, care should be taken to preserve the confidentiality of work, especially in the case of samples from external customers. If it is known or suspected that a sample is infectious, the sample and cultures prepared from it and any disposable equipment used in handling the sample should be autoclaved before disposal.

CHAPTER 10

Experiences in the Implementation of Quality Assurance and Accreditation into the Food Analysis Laboratory: Sampling, Sample Handling and Sample Preparation

In this chapter, specific areas of interest are addressed which have been shown to cause some problems in their introduction into a food analysis laboratory, these being the sampling, sample handling and sample preparation procedures.

1 Introduction: Quality Throughout the Analytical Chain

The 'handling of samples' is often the first step in the analytical chain encountered by analytical food laboratories and so is often the first aspect included in the quality assurance measures of the laboratory. In order to achieve adequate quality of the entire analytical chain, quality assurance measures must be introduced in all steps of the chain, from planning and sampling to the final evaluation. There are many examples where there have been errors in the planning and/or taking of samples such that the analytical report has been useless even though the laboratory has otherwise ensured the quality of its analytical results.

The introduction of quality assurance measures into the food industry and the official food control laboratory is currently changing the nature of analytical activities in the food sector. Previously, laboratories were often requested only to analyse final products; it is now anticipated that control activities will increasingly be carried out so as to monitor the manufacturing process itself. The number of analytical determinations carried out on individual samples, using defined methods of analysis, may therefore decrease. Sampling for official food control is also changing and is now often carried out in the form of more extensive and selective projects; this means that analytical laboratories may anticipate that they will have to offer new and more extensive services than previously.

The objective of sampling and the way in which it is carried out must be thoroughly described. The objective will often determine the way in which samples are to be transported, the analytical methods to be used and the format of the report of the laboratory. A laboratory, because it has experience in ensuring the quality of its own work and also of accreditation, enables it to assist industry and control authorities with advice in quality assurance matters. It is therefore advisable that laboratories which are not at present involved in planning or conducting of sampling extend their field of competence to include these considerations.

Quality models such as EN 45001 require laboratories to afford clients co-operation to enable them to clarify requests. The ISO 9000 series of standards emphasises the need to establish procedures for contract review and evaluation of the needs of the client. 'Evaluation of the needs of the client' and 'quality as experienced by the client' are, according to the Total Quality Management model described in, for example, 'The European Quality Award', crucial factors contributing to successful quality (see Chapter 2). Experience shows that today's food laboratory needs to develop standard operating procedures for the evaluation of client needs. An analytical result can never be of adequate quality (*i.e.* fit for purpose) until all such needs are met.

2 Sampling as a Part of the Experimental Plan

Sampling should always be carried out on the basis of an established objective. The objective may be the investigation of the causes of food poisoning, the monitoring of a food production line, or the determination of nutritional values of a number of foodstuffs. The objective should be documented in an experimental or sampling plan, which is presented by the client—industry, food authority, consumer, *etc.*—to the laboratory when drawing up a contract for the analytical work. This plan forms the basis of, but is not identical with, the analytical request. Usually the plan is more extensive, also including a description of the client's own quality activities.

It is recommended that experimental plans are established and applied, even if such plans have, up until now, been relatively unusual in the food sector. A prerequisite to obtaining quality in the final product, *i.e.* the analytical result, is that all work is carried out systematically and according to predetermined plans.

Experimental and Sampling Plans

The information on the following should be included in an experimental or a sampling plan:

- the objective of the investigation including information on the components or organisms to be determined;
- the parties involved (client, person taking the sample, laboratory, *etc.*);
- the nature of the sample, sampling location and time of sampling;
- the number of samples, the way in which they are to be taken, packed and transported (requirements on sterility, containers and equipment, sampling model, *etc.*);
- any requirements regarding the pre-treatment of samples and the selection of analytical method(s);
- the time and cost requirements (for the entire investigation, for the person taking the sample, for the laboratory);
- any possible legal requirements and international agreements which have to be observed;
- the requirements on documentation (on sampling, reports from the laboratory, the client's own summary); and
- the quality assurance aspects of the investigation (the client's own activities, any requirements on the person taking the sample, the laboratory and any others involved).

In cases where several parties are involved, the plan needs to be more extensive than, for example, in the case where one laboratory performs the entire investigation from planning to reporting. Plans covering a longer period of time, or including several stages, may need to be continuously revised in co-operation with all involved.

Additional information on the contents of an experimental plan can be found in OECD Environmental Monograph No. 45 on GLP (see Chapter 2). Information is also included in the report 'Orientation to Quality Assurance Management' of the US Environmental Protection Agency (EPA), covering the quality assurance of the chain of environmental investigations from planning to final reporting. It may be used as the basis for ensuring the quality of larger investigations in the food sector.

3 Sampling

Representative Samples

As far as possible, sampling should be carried out aiming at ensuring that the sample is representative of the consignment to be investigated. It is frequently difficult to take samples from a food consignment in a statistically correct manner. The results of the investigation will probably be erroneous if the sampling was not adequately carried out.

Established information which will help in the consideration of the development of sampling plans and procedures has been given in the following:

1. The Instructions on Sampling for the Codex Commodity Committees on

the Application of the General Principles of Codex Sampling Procedures.[1] This document was developed by the Codex Committee on Methods of Analysis and Sampling and is currently undergoing revision in the light of the new role and importance of the Codex Alimentarius Commission.
2. The Code of Practice on Sampling for Analysis or Examination prepared under the UK Food Safety Act 1990.[2]
3. International Commission on Microbiological Specifications for Foods.[3]

Sufficient Amount of Sample

A prerequisite for the analytical result to have adequate quality is that a sufficient amount of sample is taken. Factors such as the homogeneity of the sample, whether the sample is prepacked or not, the required amount of sample needed for its pre-treatment, the characteristics of the analytical method, economical and legal aspects, *etc.* all have an impact on the amount of sample to be taken.

As many samples are heterogeneous in nature, the amount of material taken will influence the sensitivity of the analytical method. If it is initially established that a certain amount of sample is to be taken and analysed for the presence of, for example, pesticides or *Salmonella*, this amount cannot be changed without also changing the meaning/significance of the analytical result.

Equipment and Packaging

Samples of prepacked foods should, if possible, be delivered to the laboratory in unbroken packages. Suitable equipment and containers must be used to ensure that they do not affect the sample. The laboratory will sometimes be required to provide the person taking the sample with equipment, *etc.*

All parts of the equipment coming into contact with the food must be sterile when sampling for microbiological examinations. If equipment has not been pre-sterilised, sterilisation may be achieved by immersing in or rinsing with ethanol followed by flaming off the residual ethanol, or by direct flaming over an open flame. Equipment should be protected against contamination during cooling, and then be used immediately.

Labelling of Samples and Documentation

In order that samples can be positively identified, they must be unambiguously labelled using permanent ink. The time and temperature of the sampling and other information needed by the laboratory should be documented on a submission form which is attached to the sample. The extent of the documentation will naturally vary according to the objective of the sampling. Sampling for production control purposes will often require very little documentation, whereas sampling for official food control purposes requires more extensive documentation. Experience shows that documentation often is incomplete and that food laboratories thus receive insufficient information for the analytical work to be carried out as competently as it could have been.

Sampling for Official Control of Foods

Sampling for official control of foods is the responsibility of the relevant food authorities and is often carried out by employees of the authorities. However, it is becoming more common that laboratories carrying out analytical work for official food control are also required to carry out the initial sampling. Such arrangements require that sampling plans are available. Below are given examples on the requirements on sampling for official food control used in Sweden and which may serve as a general model:

- A minimum of a 200 g sample should be taken from unpacked foods or from larger packages. Unpacked refrigerated or frozen meat products for chemical analysis may, however, require that larger samples are taken, and that samples are taken from two or more sampling sites.
- A minimum of a 200 g sample should be taken from prepacked refrigerated or frozen meat products. Samples of other foods should not be smaller than 100 g. The smallest available package sold to the consumer may be taken as the sample, if the amount of the product is sufficient.
- At least a 500 g sample should be taken for the examination of the presence of mould. This is particularly important due to heterogeneity if, for example, cereals are to be investigated.
- Samples should, whenever possible, be taken and labelled in the presence of the owner of the food, or his representative. If the owner so requires, or for other special reasons, an extra sample should be taken, sealed and given to the owner. One reason for taking an extra sample is if it is already clear, even at the sampling stage, that the analytical result will be used for corrective actions or other binding decisions, or in legal proceedings. In such cases it is advantageous if the analytical work can be repeated. The taking of an extra sample increases the legal safety of the party subjected to the control.

It is interesting to note that other countries have different procedures, *e.g.* the UK requires three equivalent samples to be taken at the time of sampling for distribution to the control authorities, the food retailer or manufacturer and a portion for the Court to be held should there be cases of dispute. Interestingly, there is now a requirement in some Regulations, *e.g.* the UK Aflatoxin regulations, for sample treatment preparation to take place before sample subdivision.[4]

4 Packing and Transportation

All food samples should be packed in such a way that they are protected from external influence, and that there is no leakage into or out of the package.

Requirements on the transportation temperature of samples vary depending on the nature of the sample, the type of analytical work to be carried out, and on the condition of the sample. In some cases it is sufficient for the sample to be transported in boxes intended for refrigerated or frozen material. The maximum permitted time between sampling to initiation of the analytical work depends on the nature of the sample and the analyte (substance or micro-organism) to be considered. However, no matter what the material, it is important that any legislative requirements are observed. Some foods, for example fish or shellfish, cannot be transported for long periods of time since they perish quickly. In order to ensure that such samples arrive in good condition in the laboratory, special requirements and precautions may have to be taken regarding refrigeration (*e.g.* crushed ice) and transportation.

The packaging should be labelled 'food sample' or in a similar way. Laboratories must be familiar with any special requirements which the postal service place on the packaging; usually a description of and information about the sample is given on the packaging.

Easily perishable samples must be transported in a way which ensures that they arrive in the laboratory within a specified period of time. Experience shows that such samples are not to be transported by normal postal services.

Laboratories subjected to severe competition may be tempted to extend their catchment area for sampling to such an extent that it becomes difficult to meet transportation times and temperature requirements. This practice is not to be encouraged if it would lead to analytical results being questioned later.

5 Receipt of Samples in the Laboratory

It is the responsibility of the laboratory to ensure that samples are handled in an adequate manner from the moment they reach the laboratory. A refrigerated sample left unattended will reach room temperature very quickly! The receipt of samples must be recorded, *e.g.* in a sample log. In addition the laboratory should take the actions listed below:

Checking of the Sample Description

When samples arrive in the laboratory, they should be checked to ensure that they are accompanied by a sample submission form and that this agrees with the nature and the labelling of the sample. Any manifest accompanying a sample for official food control purposes should contain the following information (many of the pieces of information are included in the experimental plan if such a plan exists):

- the objective of the sampling;
- the name, address and telephone number of the client;
- the name and address of the laboratory;
- the place of sampling, the nature, name/code and labelling of the sample;
- the time of sampling (date and, if applicable, the exact hour);
- the requested analytical work (including any special requirements regarding pre-treatment);
- the party to be invoiced, and the party to which the analytical result should be sent (if not identical with the client); and
- any special requirements (*e.g.* if the sample is to be stored at a specific temperature).

The laboratory should, if possible, contact the client or the person who took the sample if important information is missing in order to complete the information on the sample submission form. Any such contacts and the information so gained should be documented. The laboratory should consider not carrying out any analytical work if requests as to the required analytical work are unclear or if the requests are not legally binding (signed and provided with a name and address). Disputes between laboratories and clients are known to have resulted from incomplete analytical requirements being given. The client may, for example, refuse to pay for the work, claiming that the laboratory selected and used too-expensive analytical methods.

Time of Arrival of the Sample and its Condition

When receiving samples, laboratories should document the date, time and give a unique identification to the sample (*e.g.* a code number), together with the identity of the staff member who took care of the sample. Permanent ink should be used. The package and the condition of the sample must be checked, and, if applicable, it must also be checked that the sample has been handled in the prescribed manner. In some cases, the laboratory should document the temperature of the sample on arrival.

Possible errors or deficiencies should be noted. When the condition of the sample does not lead to any comments, this may be documented by the person receiving the sample in the form of a signature on the protocol without any further annotations. In cases where samples on receipt have clearly not been transported in the prescribed manner, or where there are other circumstances which may have an effect on the analytical result, the laboratory should seek the advice of the client as to whether the analytical work is to be carried out.

6 Storage of Samples Prior to and after Analysis

The laboratory should have access to sufficient storage room for its samples. Samples must be protected against any chemical, physical and mechanical influences which may cause changes in the samples. Easily perishable samples to be stored for longer periods of time should, if possible, be frozen.

Refrigerators for storing samples should have a temperature between 0 °C and +4 °C, freezers (or cold rooms) a temperature of not higher than −18 °C. The temperature of the storage areas should be regularly checked and documented. Freezers (or cold rooms) for storage of samples should be equipped with alarms, and there should be written instructions on actions to be taken should an alarm be given. Samples to be stored for a specific time at a specific temperature must be stored at the required temperature but the margins of error for temperature adjustment and reading must also be taken into account. The storage is a part of the microbiological examination and the temperature of the incubator/refrigerator should therefore be checked and documented at least every 24 h. Samples of dried or preserved foods can be stored at room temperature unless this affects the particular analyte to be estimated.

The laboratory should also have written instructions for the storage of samples after analysis. In the case of chemical analysis, samples should, if possible, be subdivided and subsamples be stored for possible re-analysis.

Samples Which may be the Subject of Legal Action

If it is expected that an analytical result will be involved in legal disputes, or has particular economic significance, the sample should be handled with a higher degree of care than usual. In such situations it is advisable that the packaging employed when transporting the sample to the laboratory is equipped with a proper seal or with sealing tape.

Handling of 'Anonymous' Samples

In some cases, laboratories need to establish procedures for the handling of 'anonymous' samples. Such situations arise where the client may wish the reason for the investigation or for his identity to be kept confidential. Anonymous handling may also be called for in cases of blind duplicates of samples from external sources, or internally as a part of a laboratory's own quality control measures.

7 Pre-treatment of Samples

The pre-treatment of samples is carried out differently for samples intended for chemical analysis, microbiological examination or for sensory analysis. Pre-treatment normally has the objective of preparing a homogeneous mixture from which test portions can be removed for analysis. Pre-treatment may also aim at solubilising, extracting or concentrating the analyte (chemicals or micro-organisms) prior to analytical work. It is important that the chosen method of pre-treatment does not adversely effect the sample or the analyte.

Analytical methods often include procedures for pre-treatment of samples. However, some methods do not contain any such instructions. If in such cases the laboratory fails to document the procedures it has used, it may be difficult afterwards to evaluate or repeat this analytical step of the analytical chain (which is a basic requirement of any quality assurance measures).

Prior to starting analytical work, laboratories must ensure that the selection of the method of pre-treatment is documented. It must be ensured that special pre-treatment requirements are identified and documented. Examples of special pre-treatment conditions include prevention of exposure to air in order to avoid oxidation of the analyte, prevention of activity of proteolytic enzymes when homogenising meat, and homogenising in a way which prevents liberated heat killing or damaging micro-organisms present in the sample.

If a sample is to be both microbiologically examined and chemically analysed, the microbiologist should be given the opportunity to work on his test portion first. It is important that the laboratory pays attention to labelling, transportation and storage, as well as confidentiality aspects when dividing up a sample between various departments of the laboratory. It has happened more than once that a well-controlled transportation and refrigeration chain, *i.e.* controlled from the site of sampling to the laboratory, has been broken in the laboratory, for example because the chemical analyst has been unaware of the needs of the microbiological analyst as regards storage conditions or sterility, or because the original label has disappeared at the time of subdivision of the sample.

Pre-treatment of Samples for Chemical Analysis

The pre-treatment of 'wet food samples' normally starts in the same way as when foods are consumed. Bones and intestines are removed. Vegetables, fruit and berries should be rinsed and stalks removed. The pre-treatment of samples for pesticide analysis is an exception to this rule, since maximum limits of pesticides refer to the concentration found in the entire vegetable, *etc.*

Prior to chemical analysis, dry samples are often quartered, *i.e.* the sample is mixed and shaped into a symmetrical cone, which is then divided into four parts. Diagonal parts are removed, after which the remaining parts of the sample are mixed and again shaped into a symmetrical cone, and divided into four parts. This procedure is repeated until a suitable sample size is obtained.

Sometimes samples need to be ground in a mill. Suitable mills and homogenisers for chemical analysis include the Omini-Mixer, Waring Blender, Ultra-Turrax and Moulinex. Pre-treatment equipment should be easy to clean and it should not transfer damaging amounts of extra heat to the sample. In some cases, *e.g.* in trace elemental analysis, particular requirements are to be placed on homogenisers and mixers in order to ensure that they do not contaminate samples.

Pre-treatment of Samples for Microbiological Examination

Frozen samples should, as a rule, be thawed at a maximum temperature of + 4 °C, and the thawing should be completed within 18 h. Provided the sample thaws within 15 min, small samples may be thawed in an incubator at a maximum temperature of +17 °C.

The odour, appearance and other observations of samples for microbiological examination should be inspected immediately after opening the sample package and taking a subsample. Deviations that occur that were not anticipated should be documented.

Microbiological examinations usually require test portions of 10 g. In the case of mould examination, a test portion of 40 g is suitable. In bacteriological investigations, peptone water (0.85% NaCl, 0.1% peptone) is normally used for homogenising and diluting. Mould and yeast investigations require the use of peptone water without the addition of salt. Dilutions are often made in steps of 1:10, unless otherwise stated in the method. The use of other types of homogenisation and dilution liquids may become necessary in future as it is recognised that not all micro-organisms of foods are liberated on homogenisation with the traditionally used peptone water.

In microbiological examinations, the Stomacher type of homogeniser should preferably be used, normally for 1 min. Samples should be homogenised for 2 min in the case of mould examinations. The Stomacher type homogeniser is, however, not suitable for hard samples, which may damage the plastic bags. In such cases, it is recommended to that a Waring Blender or equivalent is used and that homogenisation is continued for 30–60 s.

Pre-treatment of Samples for Sensory Evaluation

Foods intended for sensory evaluation should be pre-treated in such a way that all members of the sensory panel evaluate test portions of the sample which are as similar to each other as possible. Prior to evaluations, samples must be stored in a way that does not cause any changes in the samples or to their appearance. Samples should be labelled in a neutral manner so as not to give the assessors any indication as to their origin or type, or to reveal any other information which may prevent the evaluation from being impartial. Prior to evaluation, the samples should be equilibrated to the same temperature for all assessors.

8 The Future

Although sampling and sample preparation has been hidden until now, it is to be anticipated that food control laboratories will be required to document their procedures in a form suitable for accreditation purposes, to enable a generic approach to be developed (see Chapter 2).

References

1 Instructions on Sampling for the Codex Commodity Committees on the Application of the General Principles of Codex Sampling Procedures, FAO, Rome, CX/MAS 1–1987.
2 Food Safety Act 1990: Code of Practice No. 7: Sampling for Analysis or Examination, HMSO, London, 1991.
3 Microorganisms in Foods 2. Sampling for Microbiological Analysis: Principles and Specific Applications, 2nd Edition, ICMSF (International Commission on Microbiological Specifications for Foods), Blackwell Scientific Publications, Oxford, 1986.
4 Aflatoxins in Nuts, Nut Products, Dried Figs and Dried Fig Products Regulation 1992, HMSO, London, SI 1992 No. 3236.

CHAPTER 11

Experiences in the Implementation of Quality Assurance and Accreditation into the Food Analysis Laboratory: Equipment, Calibration and Computers

In this chapter, specific areas of interest are addressed which have been shown to cause some problems in their introduction into a food analysis laboratory. The specific areas are concerned with the use of equipment in the laboratory. The chapter is divided into three sections dealing with:

1. Apparatus and equipment.
2. Calibration.
3. Computers.

1 Apparatus and Equipment

Introduction

The quality of analytical work depends to a large extent on the quality of the equipment used. It is therefore important that before acquiring equipment the laboratory investigates and defines the requirements the equipment must meet. It is also important that equipment in use is properly maintained and checked.

The purchasing of equipment should be carefully planned. Instruments must be able to operate to the standard required for their intended use. Other important factors to consider are installation and operation costs, space requirements and, not least, availability of service and spare parts. If possible, it is sensible to consult another laboratory using the same instrument in order to discuss their experience of the instrument.

When purchasing equipment, the laboratory should pay special attention to the selection of the supplier. If possible, the laboratory should select suppliers operating to a recognised quality system. In the case of more expensive equipment, the laboratory should consider carrying out a second party audit at the supplier and also consider to what extent it requires the quality system of the supplier to be certified. A certification against, for example, ISO 9001 provides confidence that the supplier not only operates a quality system in its production but also that its service department is included in the quality system. Many laboratories require, as a result of demands from their customers, that certain equipment, *e.g.* microbiological media, are purchased only from suppliers having been certified against ISO 9000. The laboratory should keep records of all suppliers of equipment and services.

Generally, equipment must be suitable for its intended use and be capable of achieving the accuracy required. Equipment should be suitably labelled for unequivocal identification. Some laboratories also label instruments with their calibration status.

Before being put into service, new instruments should be subjected to checking and verification or calibration. Equipment which has been moved or is taken out of store after not being used for some time needs to be re-verified before use. It is, for example, necessary to check an analytical balance when it is moved internally from one room to another. All such checks should be documented.

Some equipment should only be used by named, authorised staff. The laboratory must have instructions on the use and operation of all relevant equipment where the absence of such instructions could jeopardise the technical operations.

A rule-of-thumb is that if something could to go wrong without instructions, then instructions are needed! Such instructions are sometimes called standard operating procedures and can be of at least three different kinds, for example:

- instructions on how to use the equipment;
- instructions on the routine check to ensure that the instrument is functioning correctly, and
- instructions on service and/or calibration procedures.

All instructions must be included in the document control system of the laboratory.

Laboratories must themselves decide when operating instructions are needed; most laboratories do not have instructions on the use of laboratory analytical balances, but perceive a need for instructions, for example for pH meters. Such operating instructions must be approved by authorised persons and be readily available to the staff. Operating instructions accompanying equipment are usually in English, German or French. In countries where these languages are not spoken, it must be arranged for the operators of the equipment to understand the operating instructions to a sufficient degree and for each relevant operating instructions to be translated into a language understood by the operator. It is usually not necessary, though, to translate lengthy manuals, *etc.* since such documents are primarily used by scientist and service staff with a sufficient command of the major languages.

Instruments and other pieces of equipment should be serviced and maintained in order that they function in a satisfactory manner at all times. Sometimes it is desirable to establish service contracts with the supplier which will facilitate the better functioning of the instrument. A service contract does not, however, guarantee proper operation of the equipment and there must therefore always be a suitable programme for routine maintenance, calibration and checks of instruments and their components established by the laboratory.

All operating instructions, manuals, records of checks, calibrations and services, as well as records of malfunctions and how they have been remedied, must be readily available to the staff who operate the instrument.

Instruments normally in use in a food analytical food laboratory fall into the following categories:

1. *General equipment* not normally influencing measurements or having very little effect on the analytical result (*e.g.* hot plates, stirrers, non-volumetric glassware or glassware used only to estimate volumes) and the heating and ventilation equipment of the laboratory.
2. *Volumetric equipment* such as volumetric flasks, pipettes, pycnometers, burettes, *etc.*
3. *Measuring equipment* such as balances, chromatographs, electrochemical equipment, hydrometers, microbiological incubators, microscopes, spectrophotometers, thermometers, viscometers, balances, *etc.*
4. *Reference standards of measurement* such as 'certified' reference materials, microbiological reference strains, standard weights and reference thermometers.
5. *Computers* and other similar instruments.

These are described below.

General Requirements

The laboratory must keep records of all its major instruments. It is for the laboratory to decide which instruments are 'major'. All instruments which have an influence on analytical measurements, *i.e.* measuring equipment (see 3 above) fall into this category, including incubators, heating cabinets, refrigerators and freezers.

The records should contain the following information and be kept either centrally or in close proximity to each instrument:

- name of the equipment;
- make, model and serial number of the instrument and its unique identification;
- supplier, date of receipt, date placed in service;

- current location in the laboratory;
- reference to an instrument logbook as described below; and
- if relevant, name of person responsible for the instrument.

Instruments should also have a logbook which can either be an integral part of the records of the instrument or, possibly a more convenient solution, be kept in connection with each instrument.

However, it is sometimes practical to have a single logbook for an instrumental set-up, for example an LC pump supplemented with an auto-injector, UV detector and an integrator.

An instrument logbook should contain:

- information on the identity of the instrument and, if applicable, its various components;
- full information of damage and repairs including dates when the defect was identified, thorough description of defects, date of repair, nature of repairs, including possible changes of components, name of the repairer;
- when necessary, procedures for the performance check of the instrument, their nature and frequency, acceptance limits and corrective actions to be taken in cases where the limits have been exceeded;
- procedures for the calibration of the instrument together with acceptance limits of calibration results;
- references to operating instructions and manuals and their location in the laboratory; and
- dated and signed results of performed performance checks and calibrations together with an indication of the acceptability of the check or calibration.

In order that the laboratory is able to design a meaningful programme for the performance check and calibration of the instrument, the laboratory must be aware of the critical control points of the instrument and those of the analytical systems in which it will be used. Only points having an influence on the analytical result should be included in checks. Laboratories may, when possible, find it more cost effective to check and calibrate the entire analytical system rather than taking an instrument-oriented approach.

After identifying points or stages to be checked, the laboratory must establish numerical limits which must be reached (or not exceeded, as applicable) in order that the instrument or the analytical system passes the check. As the third step the laboratory must clearly define and document what steps the personnel should take in cases where action limits are exceeded. Finally, the operation of a check or calibration programme must be continuously monitored in order to detect any needs for changes. Experience has shown that many laboratories tend to forget the last two aspects.

It is often helpful that one single person is appointed to be responsible for an instrument or for a group of similar instruments. This person naturally must have the necessary competence for the job, and management must ensure that this person is given the time and opportunity to maintain and develop any necessary specialist skills. If an instrument has several operators it is an advantage if the laboratory keeps records of all users of the instrument.

Equipment which gives suspect results, or which has been shown to be defective, shall be taken out of service, clearly labelled as such and, if possible, stored separately from other similar equipment.

Such equipment should not be used until it has been repaired and shown by calibration, check or test to function satisfactorily.

The laboratory may establish specific security rules for certain instruments. If such rules are needed they should appear at the start of the operating instructions.

General Equipment

General equipment in the laboratory, such as hot plates and stirrers, should be kept clean and be suitably maintained. Periodical performance checks should performed in the case of instruments that can be adjusted if the adjustments influence the analytical results (*e.g.* the temperature in a muffle furnace or in a water bath).

Documented instruction procedures must be prepared for the purchase, reception and storage of consumable materials used for the technical operations of the laboratory.

Volumetric Equipment

The use of volumetric equipment such as volumetric flasks and pipettes has a large influence on the analytical results and these must therefore be correctly maintained and calibrated. There are various quality classes of volumetric glassware, and it must be decided which quality class is required to be used for each analytical determination. The need for checks before taking the glassware into use and the need for periodical checks afterwards must be evaluated. Experience shows that some suppliers do not always deliver, for example, class A glassware conforming to specifications.

The accuracy of certain types of volumetric equipment, *e.g.* automatic pipettes, is affected, for example, by the way in which they are cleaned. It may be necessary to check and calibrate such equipment more frequently and care should be taken that the instructions for their operation and maintenance are sufficiently detailed.

In the case of analytical work requiring the highest degree of accuracy, it is recommended that reagents are weighed instead of measured volumetrically.

Care should be taken to avoid contamination from the volumetric equipment itself or as a result of earlier use of the equipment. Where necessary, the laboratory should have written instructions on what volumetric equipment should be used, how it should be stored and cleaned as well as any instructions on how to keep such equipment separate from other similar equipment. Examples exist where volumetric glassware has become damaged, with great effects on the precision, as a result of being stored in a contaminated atmosphere. These aspects are of particular importance in the case of the analysis of trace elements.

Measuring Equipment

In the case of measuring equipment such as spectrophotometers, chromatographs, electrochemical equipment, balances, *etc.*, their correct operation, regular maintenance, cleaning and calibration is not always sufficient to ensure the correct functioning of the instrument. Measuring equipment should therefore be subjected to regular performance checks in a systematic way.

The laboratory must decide with what frequency performance checks should be carried out and there must be written procedures for the performance checks. It is recommended that laboratories, when first introducing regular performance checks of an instrument, decide on frequent checks. With experience and if it can be shown that instruments are generally stable, the time intervals between checks can be made longer. Performance checks are discussed later in this chapter.

Reference Standards and Reference Materials

For calibration purposes, the laboratory should have available at all times relevant reference standards of measurement such as standard weights and reference thermometers and, if appropriate, reference materials and microbiological reference strains (certified or non-certified) to be used in the control of the entire analytical chain.

Reference standards of measurement shall be used for calibration purposes only, and shall be duly protected and stored separately from other similar equipment. If the laboratory wishes or is required to demonstrate traceability of calibration to international or national standards of measurement, then the reference standards held by the laboratory must be accompanied by current certificates issued by an organisation that has been accredited for the calibration in question.

The laboratory must also hold certificates for its certified reference materials if these are used to demonstrate traceability to a certified concentration or characteristic of the material. Most certified reference materials are certified to a stated date, and therefore laboratories should make sure that the materials they use to demonstrate traceability are not out of date. It should also be appreciated that manufacturers who carry out the certification of reference materials usually have not themselves been certified to ISO 9000, *i.e.* the quality system of those preparing reference materials have generally not been subjected to a third party assessment. Therefore the quality of certified reference materials may vary widely.

The use of reference standards and certified reference materials are discussed further in Section 2 of this chapter.

Glass- and Plasticware

Items of glass and plastic must be suitable with respect to their quality and design for their intended uses. The laboratory should take into account any special requirements regarding particular resistance to acids or bases, protection of reagents against light, resistance to large differences in heat or pressure or possible special contamination requirements. In general, bases should not be stored in glassware and organic solvents not stored in plastic containers. Plastics should not be used as weighing vessels because their static electricity properties may interfere with the electronics of balances. There are specific, strict requirements on glass- and plasticware to be used in the sensory analysis of foods.

Water, Chemicals and Gases

Chemicals, gases, solvents, *etc.* are commercially available in various quality classes.

Water

The analytical task will determine which purity class is to be used. A 'pro analysis' purity or 'puris' is sufficient in many analytical situations. It is recommended that reagents of these purity classes are used where methods do not specify the purity of reagents. However, it is in the laboratory's interest from a cost point of view not to use reagents of higher purity than is required in any specific determination. Analysts should avoid using pro analysis reagents for 'safety's sake' if a technical quality is acceptable.

It must be appreciated that distilled, RO (reverse osmosis) purified or de-ionised water may contain impurities. These are, however, as a rule so small that distilled or de-ionised water may be used for many determinations without further purification. However, it must be ensured that the purification system and its pipe system is selected, installed and maintained sufficiently thoroughly in order to ensure a sufficient degree of purity of the water. The purity of the water should be checked regularly, *e.g.* by measuring its microbiological quality, its conductivity, heavy metal content or the presence of other compounds which may have an effect on the analytical result. For ultra-pure water, the measurement of conductivity is not recommended, because the carbon dioxide which will be absorbed from the air will influence the measurement and make it worthless.

If the laboratory's pipe system for purified water is a very large one, then the laboratory should consider checking the water both at the production site and at the ends of the system. If purified water is stored in containers before use, these should be cleaned regularly.

The laboratory has to decide on action limits for the purity measurements of purified water. It is difficult to give any general recommendations.

In the analysis of trace elements, the water used should be further purified, *e.g.* using a specific procedure with dispensable cartridges for prefiltration, organic adsorption, de-ionisation and final filtration.

Chemicals

Chemicals and prepared reagents and solutions should be properly stored. The temperature, moisture and dust formation in storage spaces may need to be monitored.

It is an advantage if the laboratory has storage space for refrigerators to be separated for the storage of 'pure' reagents and for 'impure' samples. It is recommended that chemicals are labelled both with the date of receipt of the chemical and with the date of opening the container. If the chemical has a finite shelf-life, this should be marked on the container. A date printed by the manufacturer on a container often refers to the shelf-life of the unopened package; this should be taken into account if chemicals are stored for a long time after the package has been opened.

Unless there are over-riding reasons, laboratories should use older chemicals first. The laboratory should have written instructions for the preparation of reagents and solutions. It is recommended that containers or bottles containing dry chemicals which have been stored cold are brought to room temperature before they are opened.

All containers with prepared reagents and solutions should be clearly marked with their contents, dated and signed by the person who prepared the reagent or solution. If the reagent or solution has a finite shelf life, the expiry date should be marked on the container.

Gases

Gases used should be of a purity required for the specific analytical situation. If the laboratory is equipped with a central gas supply, measures should be taken to ensure that switches between the supply of gas to different pipes cannot occur. The laboratory should also ensure that warnings that a gas supply is running low are given in good time in order to avoid a stoppage of gas occurring during analytical work.

2 Calibration

Introduction

There is much confusion about the concepts of calibration, standardisation and adjustment both in the laboratory and in the analytical literature.

According to international definitions:

Calibration means establishing and recording the measurement uncertainty of measuring equipment. A true calibration does not involve any adjustment of an instrument, but may demonstrate the need for adjustment.

An *adjustment* of an instrument is carried out when an estimated systematic error is corrected for by changing a setting of an instrument.

Standardisation of an instrument means establishing a relationship (*e.g.* numerical or graphical) between the reading of an instrument and known properties of a material. This material is often called a standard and may, for example, be a weight or a solution with a known analyte concentration. An example is the standardisation of a spectrophotometric determination by measuring absorbance values of a series of standard solutions and establishing a relationship between these and the known concentrations of the solutions in the form of a standard curve or a calculated mathematical function. The use of terms such as calibration curve or calibration function in this connection is unfortunate and should be abandoned.

Often there is confusion between calibration and the combination of standardisation and adjustment. For example, a pH meter is incorrectly said to be calibrated when it is standardised with one or more buffer solutions of known pH value(s) and the reading is then adjusted to show the correct value. Modern electronic balances include a 'calibration function'. When in use the balance is loaded with internal or external weights and the balance is automatically adjusted so that its reading corresponds to the known mass of the weight(s). In both these cases, the difference between the measured value and the 'true' value remains disconnected and therefore in neither case has a calibration been performed.

Different Degrees of Calibration

The definition of the concept of calibration covers actions which vary considerably with respect to the amount of work and degree of complexity. A simple performance check consisting, for example, of weighing one single weight, recording the reading and the comparison of the reading to the true mass of the weight, can formally be called a calibration. Such a check may be carried out in the course of a few minutes, whereas the amount of work for the full traceable calibration of an analytical balance will be measured in hours. Between these extremes there are calibrations with varying degrees of traceability.

Common to all calibrations is the requirement that the laboratory must have a written procedure for the performance of the calibration and must have set clear limits for accepted deviations. The laboratory must have documented rules for actions to be taken if the check or the calibration reveals deviations which exceed the limits set. Such actions include both decisions on the future use of the instrument and decisions on the evaluation of the results that the instrument has produced in the time period from the previous check or calibration in which acceptable performance was noted up to the time the unacceptable performance was detected.

If, as a result of a check or calibration, an unacceptably large deviation is detected, then the instrument should either be taken out of service or, in some cases, all future results should be corrected with a value corresponding to the observed deviation. The laboratory must be able to justify (and document) such corrections and they must be made until the instrument is repaired/ adjusted and has been shown by calibration to operate acceptably. In addition, all results achieved using the instrument in the time period from the previous successful check must be repeated unless it can be shown that the detected deviation has not significantly affected the trueness of the results received. In order to be able to prove this, control charts or results from analysing certified reference materials may be used.

When establishing acceptable deviations for an instrument, it is important to clarify the extent of the influence of the deviations on the analytical determina- [illegible] the instrument will be used. This evaluation should be carried [illegible] ysts responsible for the various analytical determinations. The [illegible] ptable deviation of the instrument must be set taking into [illegible] hod having the strictest requirements on the instrument.

Calibration Programme

It is necessary for the laboratory to develop and operate a suitable calibration programme including all instruments that have a bearing on the final analytical result. In the case of an accredited laboratory, reference standards of measurement such as the weights used in the calibration of balances must themselves be calibrated by a body that can provide traceability to a national standard of measurement, *e.g.* an accredited calibration laboratory. Reference standards of measurement held by the laboratory shall be used for calibration only and for no other purpose. It is important to appreciate that an annual calibration is not sufficient; equipment must be subjected to periodic intermediate checks, *i.e.* to verification between calibrations. Frequent checking will detect possible malfunctions early and therefore minimise efforts needed for examining the effect of a defect on previous tests.

In Table 11.1 some typical equipment used in an analytical food laboratory and examples of frequently used calibration and verification intervals are given. Intervals must be adjusted to the needs of individual laboratories depending on the frequency of use of the instrument. It is recommended that calibration and verification intervals are selected to be initially shorter, but with experience, if calibrations and verifications show that the instrument is stable, intervals may be extended.

The intention of such a programme for calibration and check of equipment is to provide confidence that measurements made by the laboratory are, whenever relevant and possible, traceable to national or international standards, or in some cases to certified reference materials, such as certified solutions used in the calibration of the absorbance of a spectrometer. Traceability of, for example, balances and thermometers can be demonstrated by calibration certificates of weights and reference thermometers, respectively, issued by competent (accredited) calibration laboratories that show that there is an unbroken link in the calibration chain to the national standard. When carrying out calibration, the uncertainty of the instrument should also be calculated. If it is known that the uncertainty of a particular instrument contributes very little to the total measurement uncertainty, then that instrument need not be rigorously calibrated; a simple performance check is usually enough.

Table 11.1 *Examples of frequently used calibration and check intervals for typical equipment of an analytical food laboratory*

Equipment	*Calibration interval*	*Check interval*
Reference weights	3 years	1 year
Reference thermometer mercury type	5 years	1 year (ice-point control) Daily[1] if the mercury column is damaged
Reference thermometers electronic type	1 year	
Thermometer (mercury)	1 year	
Thermometer (electronic)	1 year	3 months
Analytical balances	1 year	1 week
Automatic pipettes	3 months	Daily
pH meter	–	Daily
Spectrometer	6 months	Daily[2]
Chromatograph	6 months	Daily[2]
Certified reference materials		1 year
Microbiological reference strains		After two sub-culturings

[1] 'Daily' means each day the instrument is used. [2] Spectrometers and chromatographs should be subjected to some sort of verification every working day, *e.g.* the measurement of standard solutions. Spectrometers and chromatographs in frequent use should be calibrated more often than every six months.

Frequency of Checks

The frequency with which measuring equipment should be subjected to performance checks depends on the nature of the instrument, on how much it is used and to what extent its measurement uncertainty affects the final analytical result. Thus, an instrument used on a daily basis should be checked more frequently than an instrument used once a month. The frequency must be balanced against the amount of extra work that would need to be carried out if analytical work is required to be repeated as a result of the detection of unacceptable instrumental deviations. Where control charts constructed on the basis of results from determinations made on a suitable control material are not appropriate, or when there are no certified reference materials available, it may become necessary to repeat all determinations in the entire period back to the previous successful check/calibration. In such cases, the advantages of frequent checks are obvious; frequent checks will take much less time than a possible exercise in which all determinations are to be repeated.

It is difficult to recommend what interval checks should be made. Many laboratories check the following equipment on a daily basis: automatic pipettes, balances, the temperature in incubators, pH meters. Refrigerators and freezers are often checked on a weekly basis, except where the analytical work specifically requires more frequent checks.

It is usual that a check, but not a full performance check, is carried out each day on spectrophotometers and chromatographs in use. More thorough checks of such instruments are often performed on a weekly or monthly basis, depending on the scope and the frequency of use of the instrument (see Table 11.1 above).

Calibration of Balances

Larger laboratories using many analytical balances in analytical work which requires high accuracy of balances may find it convenient and more cost-effective to carry out internal traceable calibration of all balances. If traceable calibration is needed for one or only a few balances, as is frequently the situation in smaller laboratories or laboratories undertaking work where weighings do not significantly influence analytical results, it may be more appropriate to use externally purchased calibration services. Some suppliers of balances have gained accreditation for the calibration of balances and offer calibration services in connection with an annual maintenance and adjustment of balances.

A laboratory wishing to carry out internal calibrations of its balances must hold its own reference standard weights covering the relevant masses to be weighed. These weights must only be used for calibration purposes and should be handled with great care in order to protect them. The laboratory must have a detailed written procedure for calibration work, and this must be carried out only by an appointed member of the staff having sufficient training and experience in the field. If the laboratory is accredited or plans to seek accreditation, then calibration of balances must be performed in a traceable manner, which in practice means that the weights held by the laboratory must be calibrated traceable to international or national standard weights by an organisation competent to do so, *i.e.* by an organisation accredited for the calibrations of weights. This calibration of the weights used in internal calibration of balances must be demonstrated by a certificate signed by the organisation that carried out the calibration.

Analytical balances in routine use should be calibrated annually and their performance should be checked at least once a week. Calibrations are performed using standard weights, which preferably have themselves been calibrated and are traceable. In performance checks, other weights or stable objects may be used. In order to establish the 'true' mass of the weight or object which is used in routine performance checks, it is recommended that the weight or object is weighed on a balance at the same time as the annual calibration of the balance, because at that specific time the measurement uncertainty of the balance is known.

Practical guidance on the traceable calibration of analytical balances is given in NMKL Procedure No. 1.[1]

Calibration of Thermometers

Most food laboratories have many thermometers in use which need to be calibrated, *e.g.* those placed in refrigerators, incubators, water baths, heating cabinets, *etc.* The extent of the calibrations and their frequency depend very much on the type of the thermometer being considered.

Generally, glass–liquid thermometers are stable and so need not be calibrated very frequently. Some laboratories calibrate such thermometers every five years, but then subject them to an annual performance check.

Electronic thermometers, however, are much more unstable and are often calibrated annually and checked every three months. Different types of temperature probes require different calibration intervals, *e.g.* thermocouples are more unstable than PT 100 resistors. Probes equipped with a protective coating are more stable than probes without.

Calibrations and performance checks should be carried out using the laboratory's reference thermometers. These must cover the temperature interval in which measurements are to be made, and the thermometers must have a suitable scale division. As with reference standard weights, these reference thermometers must only be used for calibration and check purposes and must be stored separately from the laboratory's working thermometers. If the laboratory needs to demonstrate traceability in its temperature measurements, the reference thermometers must be traceably calibrated by an organisation accredited for the calibration of temperature and hold a certificate for each thermometer. As with balances, the laboratory should set acceptance limits and decide on actions to be taken if these limits are exceeded.

There is a document available giving practical guidance on the traceable calibration of thermometers, the NMKL Procedure No. 2.[2]

Calibration of Pipettes

Automatic pipettes have become among the most frequently used instruments in the modern laboratory. The rules for the calibration and performance check of such pipettes will largely depend on how often they are used, by whom and for what sort of solution. It is relatively easy to calibrate and check the performance of such pipettes; this need not take too much time. The very frequent use of these pipettes justifies frequent checks. Many laboratories check pipettes every day. It is important that acceptance limits are set, and that these limits are set taking into account the requirements on accuracy of the method in which the pipette is used. If it is to be used in many methods, it is the most strict requirement for accuracy required by any of the methods that is to be selected.

The calibration of an automatic pipette requires an analytical balance, a thermometer and a liquid which, as far as possible, corresponds to the solutions which are pipetted in the analytical work which is to be undertaken by the laboratory. In order to avoid changes in temperature during the calibration, the liquid used should be at ambient temperature. First the temperature of the liquid is measured and then the density of the liquid is obtained from reference tables. The automatic pipette is calibrated by delivering a portion of the pipette to a tared container and weighing the container. If the analytical balance and the thermometer are traceably calibrated, the traceability will be transferred to the automatic pipette.

The calibration is usually carried out using distilled water. If the pipette is used for measuring liquids having a density considerably different from that of water, another 'typical' liquid should be selected. Many pipettes have simple devices for the adjustment of volumes which means that most members of the staff, after suitable training, can carry out calibrations/checks with subsequent adjustments. For the performance checks to be meaningful, acceptance limits have to set for the various volumes of interest. In calibration, the work commences with a simple check and a possible adjustment if needed. After adjustment follows a more complete investigation using all volumes of interest (at least three levels if the pipette is used for a variety of volumes), and the pipette is adjusted when necessary.

Check and Calibration of an Entire Analytical Procedure

Some of the sources of analytical errors cannot be eliminated even using the most thorough calibration procedures. Examples of such sources are homogenisation errors, matrix effects, poor quality chemicals, human mistakes and differences between analysts, large environmental temperature differences and electrical interferences from other instruments. The most difficult problem with these types of errors is that they are often difficult to detect. If the source of the error is always present and causes constant errors, replicate results will agree well and lead to the error not being detected. If the error appears only intermittently, it may be detected when it has an effect on replicate determinations, for example on the results of control materials, and then there is the risk that the result is rejected as a statistically deviating result (outlier). To some extent it is possible to reduce the occurrence of such errors by describing the entire analytical procedure more accurately, and especially the manual steps, in order to minimise differences caused by human error. The description should, as far as possible, cover all problems connected with different types of samples. This is the main idea behind standard or reference methods which usually describe the analytical procedure very thoroughly and have been tested in interlaboratory method–performance tests using many different types of samples. The disadvantage of such methods is that they are often based on out-dated analytical principles which have been replaced by more modern techniques achieving better trueness and accuracy.

If the laboratory suspects that a certain determination is not fit for its purpose, it should try to find another laboratory experienced in carrying out the same analytical task and send samples to that laboratory in order to make a comparison. Recovery tests are also useful in the identification of difficult sources of errors. Poor recoveries indicate problems, whereas good recoveries are not necessarily a guarantee of good results, as matrix effects may have a different impact on the analyte of the sample and the added form of analyte.

Over time, the most efficient way to avoid unexpected and unknown errors is, as far as possible, regularly to analyse both certified reference materials and internal control samples which have been calibrated against certified reference materials, and to participate in well-organised proficiency tests. In the accreditation process, the use of analytical methods which have been validated is emphasised. Methods should have been validated using certified reference materials if materials with a suitable matrix and a suitable concentration level of the analyte are available. Laboratories are also encouraged to participate regularly in proficiency tests using as many as possible of the methods in use in the laboratory. Reference materials and sample materials used in proficiency tests do normally not require homogenisation prior to analysis. Therefore the analysis of such materials will not aid the determination of sources of errors caused by or associated with inhomogeneity or a possible contamination during homogenisation.

The Use of Certified Reference Materials

Possibly the best way of tackling unknown errors and errors difficult to foresee is regularly to analyse certified reference materials. The use of these materials is, however, associated with some disadvantages. Such determinations will not detect homogenisation errors since reference materials often are particularly homogenous. Reference materials are also relatively expensive, which limits their usefulness especially in analytical work requiring large amounts of the sample. Finally it must be noted that certified reference materials sufficiently similar to the analytical task at hand may not always be available.

If a certified reference is available from a reputable supplier, the concentrations given for the analyte in the certificate can be accepted. In such cases the analysis of the material represents a traceable calibration of the entire analytical procedure, except for the homogenisation procedure. It is probably the best way of ensuring traceability in analytical work. The traceability then holds as long as it is accepted that the matrix and the concentration level of the analyte of the reference material is representative of the samples analysed. If in the same analytical run both an internal control sample and a certified reference material are included and good agreement with the certificate value is achieved, then to some extent the traceability of the reference material is transferred to the internal control material, but with reservation regarding matrix similarity. A certified reference material should be included in an analytical run every time a new internal control material is used. This applies even if the similarity of the matrix and the concentration levels of the analyte in the best matched available certified reference material and those in the new internal control material are not ideal.

3 Computers in the Laboratory

The performance of computers and computerised equipment used in the laboratory should be ensured using the same principles as for all quality assurance activities, *i.e.* the quality assurance measures should focus on the critical points of the use of the equipment in such a way that possible functional errors are prevented. The quality assurance measures should be documented so that they can later be reconstructed and reported if required.

In practice it may be difficult to implement these principles, since many of the operations carried out by computers are not readily available to the user. A computer suffers to a considerable extent from the so-called 'black box' syndrome, which further contributes to the user's perception of computers as inaccessible and uncontrollable. However, computers are not more 'uncontrollable' than many other complicated tools and structures used on a daily basis in the laboratory. In the case of computers, the user must define the level of control and documentation, which assures the intended level of security and which must be possible in practice.

Guidelines for the quality assurance of computerised equipment and work practice in accreditation circumstances been described elsewhere,[3,4] as have more detailed descriptions of quality assurance measures of computerised equipment.[5,6]

Critical Points

The critical points in the use of computers depend both on the nature of the equipment (hardware) and on the nature of the programs (software). The use of computers in the laboratory ranges from the use of PCs for storing and processing data to computerised measurement equipment with or without data transmission to totally integrated laboratory information management systems (LIMS).

LIMS is the term for software which electronically gathers, calculates and distributes analytical data, which in most cases is gathered directly from measurement equipment. LIMS includes word processing, databases, spreadsheets and calculation modules. LIMS can usually carry out a wide range of functions, typically registration and traceability of samples, computing, quality control, economical control and reporting. Special validation requirements include the control of access to the various functions, traceability of amendments to registered data and control of files.

The quality assurance measures to be introduced depend on the type of equipment in the laboratory as well as the type of programmes and their use. The user should critically examine his work procedures to establish the critical points and elaborate suitable control procedures in order to prevent errors. A balance will always have to be made between that which is practically feasible and economically possible, to that which is seen to be ideal. As is generally the case with quality assurance, the measures should focus on conditions of real significance to the quality of work. Thus there is no need establish procedures for the control of systems if the control does not contribute to the quality assurance of the entire system or parts of it in the form of prevention or isolation of errors in the final products, which in the laboratory is the measurement result, either in the form of raw data or calculated data. When assessing the critical points, a good rule is to evaluate, separately for each point, the seriousness of an error at that particular point, and the probability for an error to occur. On this basis it should be possible to decide on the necessary level of assurance and control. Examples are given below of conditions which need to be considered.

It may be practical to have as a starting point the following three main stages of quality assurance of computers and computerised equipment:

- initial validation of the equipment *and of the software (programs)*;
- continuous control of the equipment *and the software*; and
- safe-keeping of data.

Common to all three stages is the requirement for documentation.

Measurement System and Validation of the Equipment

A measurement system in the analytical laboratory will typically consist of an analogue sensor, an amplifier and an analogue-to-digital converter together with a personal computer (PC). The computer collects and stores the data, and may also further process the raw data to production data.

The quality assurance of such a system may vary according to the size and the use of the system but should as a minimum include:

- validation of data; and
- identification of data.

Validation of data must ensure that the input from the sensor at all times results in the expected output of raw and production data, provided that the procedures for the collection and processing of data are identical, *i.e.* measurements should always be reproducible.

Identification of data is accepted to mean that it is possible to assign measurement results to a given measurement in an unambiguous way so that at all times it is possible to retrieve results from earlier measurements.

The individual components of the system may be integrated or isolated and the user is therefore not always able to evaluate their individual performance. In such cases the data output of the system must be validated as output from a 'black box', *i.e.* it is the final output which is evaluated.

Before the Equipment and Software is Taken into Use

A given measurement system will always have a specific field of application. It must therefore be ensured that the measurement system is suitable for the particular purpose. Depending on the complexity of the system and of its purpose the system should, before being taken into use, be validated against the laboratory's own or the manufacturer's specifications. As a minimum, the laboratory must carry out a series of test measurements in which a relevant measurement range is studied and the correct functioning of the entire system is verified.

Such work should focus on the most important points and the laboratory should distinguish between internally developed, specially developed (hardware/software typically commercially developed for laboratories, *e.g.* data logs with software) and standard hardware/software (PCs and accompanying peripheral equipment, operating systems, and shelf software such as spread sheets, *etc.*).

Developments in the area of computers are proceeding very quickly both with respect to hardware and software. As a result, new revisions of software are available, outdating previous versions within a few years.

When investing in new computers and programs, laboratories have to consider not only the present needs but also the compatibility of the new acquisitions with previously generated data using older equipment. Laboratories should also consider the need to use data many years after it has been generated; this is a general problem in laboratories working to the GLP principles of OECD, as there it is required that data must be retrievable up to periods of 10–25 years after its production.

The Programs

Internally developed equipment may and should be thoroughly specified regarding laboratory needs. A plan should be formulated in advance of development to establish which validation measures the system must successfully meet before the equipment can be taken into use.

Specially developed equipment must meet its purchase specifications. It will generally be sufficient to use reputable suppliers, ideally suppliers which are certified to a quality control standard. However, the equipment must be systematically checked (see above) before being taken into use.

A laboratory is not normally able to influence the design of standard commercial equipment. However, PCs are now generally accepted to be equivalent and thus possible deviations do not normally occur sufficiently to influence data integrity.

When using statistical programs or spreadsheets, the formulas used within the programs must be investigated and confirmed by recalculating examples to ensure that the program actually uses the formulas claimed. When using commercial spreadsheet programs, special attention should be given to procedures by which formulas can be locked/frozen. There is a risk that formulas are inadvertently changed by operators and steps should be taken to reduce and avoid this possibility. Examples exist where such changes in spreadsheets have been introduced, resulting in incorrect data and extensive economic losses for the client of the laboratory. In the case of more complicated calculations, such a control procedure may not be possible and the laboratory, in such cases, will have to accept the documentation and guarantee provided by the producer. In some cases it may be sensible not to use the most recent versions of software programs; the immediately previous version of software is frequently prone to fewer errors than the latest version. The use of internally developed applications or macros, based on standard programs, should be validated in line with other internally developed equipment.

Validation procedures and their documentation should be elaborated and kept on file for later control and possible renewal.

Continuous Control

After equipment has been found suitable to be taken into use, it must be continuously assessed as remaining stable over its entire life. A plan for its systematic control should be established. This plan may contain parts of the elements of the system or the entire system, depending on what is necessary and most practical.

Where do Errors Occur?

It is frequently more effective and necessary to control the analogue components of a system (the sensor, amplifier and analogue-to-digital converter) than to control the PC's digital components (the PC and associated programs). This is because it is much less probable that the function of the digital components change with time. It may be generally assumed that a computer will not wear and therefore it is expected to produce the same output for a given input during its entire lifetime. Nevertheless laboratories must always be aware of, for example, the risk of formulas being inadvertently changed, as discussed above, as well as of other risks inherent with the use of computers such as the introduction of viruses, *etc.*

The laboratory should concentrate its efforts on aspects where the sources of errors are greatest, *i.e.* on the control of adjustment and calibration of the analogue components and the verification and maintenance of procedures to ensure the safekeeping of data. In many cases the analogue components are not available for separate control, and the equipment must then be checked as a whole as a 'black box'.

Control Program, Instrument Lists and Software Inventory

The laboratory should elaborate principles for the continuous control and establish a control program for its equipment. An appropriate starting point for the control program is the instrument register in which all computers associated with analytical work should be listed. Control requirements and intervals should be defined separately for each item. The register should also include the software used by the laboratory together with its installation information. Such information should include, for example, the program version, copy number of disk containing the original program, licence number and ID. When computer software and computerised equipment is replaced, the equipment should be checked for correct operation before it is taken into use again.

In the case of more extensive systems (*e.g.* LIMS) there must be sufficient and up-to-date system documentation which, for a given time, describes the configuration of the computer system, how components depend on each other and which procedures are in place to ensure correct, continuous running and development of the system.

Quality Assurance

The Equipment

As with all laboratory equipment, computers should be placed in a suitable environment in which attention has been paid to temperature, humidity, chemicals, *etc.* In the case of computers, special attention should be paid to electromagnetic radiation and strong magnetic fields. It is known that problems may occur if computers are placed next to thermostatically regulated equipment, *e.g.* ovens. Care must be taken to ensure that staff are sufficiently educated and trained in the use of computer equipment. Guidelines and instruction manuals must always be available.

Data

During the generation of data, several factors may influence and affect the 'correctness' of the measurement data that have been collected. As a general rule, it may be assumed that missing data are better than faulty data.

When ensuring correctness of data, it is understood that data which are to be stored must be kept safe against errors or changes made by individuals or by other means. These include deliberate efforts to change data, virus attacks, computer errors, *etc.*

Viruses

In a laboratory environment, computer viruses will only rarely influence the correctness of measurements. It is more probable that the presence of a virus will cause the entire system to be unusable and thus prevent any measurements being made or destroy any measurements already made by deleting data.

It is clearly sensible to guard laboratory computers against viruses by installing all computers in laboratory use with anti-virus software, and ensuring that the software is continuously updated.

Identification of Data

During the generation of data, two identification aspects must be observed: (*a*) who carried out the measurements in question and was this person authorised to do so, and (*b*) the identification of the stored data in relation to the carried out measurement or series of measurements.

The highest degree of safety in identifying data will be obtained by using a network operating system. With such a system, access to instruments, programs, data, *etc.* can be controlled and checked.

If automatic network control is not available, the laboratory must create a manual system such that users initial or sign data input to allow the identity of the person who carried out the measurements to be determined and who registered the data if these are subsequently found to be necessary.

Many laboratories use computerised equipment for the production of analytical data, which then are registered/transcribed manually into an electronic storage medium, typically into a larger database application. The person transcribing is not necessarily the same person who produced the data. It is important to create suitable control practices and data structures in order to ensure that registered data are correct and that information on the identity of the analyst, apparatus and measurement series is saved so that it is always possible to trace and identify raw data.

Archiving

Raw data should be stored in a retrievable way such that the data can be identified as being associated with any particular measurement or series of measurements. In order to ensure this, several different procedures can be employed. A few possibilities are listed below:

- to develop a specific format under which data from a given application are stored;
- to design a suitable structure for any databases in which raw data are stored; and
- manually to label storage media containing raw data.

In most cases, measurement programs will be able to handle and identify several measurements simultaneously. Cross-references to other data stored in other systems (*e.g.* laboratory reports, measurement reports, log books, *etc.*) will often be necessary.

By establishing standard procedures, it must be demonstrated before measurements are started that there exists an unambiguous method which will identify measurement data from various measurements.

The long-time storage of computerised data requires special procedures. Previous versions of software programs may need to be available, and sometimes also older, out-dated and long-since replaced equipment will need to be available.

Storage Premises

Data are often stored locally or on a network server. Usually it is easiest to achieve the highest degree of security of stored data by storing on a network server. Contrary to most stand-alone PC equipment, a network server can be secured against:

- theft, fire and other environmental hazards present in a laboratory, *e.g.* by placing the server in a locked and secured room;
- voltage disturbances and hardware faults, by equipping the server with an emergency current supplier and specific error-tolerant and correcting disk systems as well as controlling the server using network management systems; and
- unauthorised access *via* encrypting, security procedures and control functions in the operating system.

Network operating system are often transaction based and therefore such a transaction log will offer an opportunity to identify events and re-establish earlier states if that is needed. This provides a greater possibility of control and traceability. In addition, such a system has better control at the file level, and it is possible to identify data owners to a much higher degree than on local stand-alone PCs.

Safety of data will typically be reduced if for some reason data cannot be stored on a network server but have to be stored locally. However, special rules/facilities may increase safety, such as encrypting of data, controlling the access to the individual PCs and other procedures which, when established locally, will require more administration and offer a lesser degree of safety of data.

Storage Media

Usually data are stored on standard hard disks. When the data to be stored are very extensive, or when there are special requirements that stored data cannot later be modified, data may be stored on optical media, *e.g.* on WORM (write-once-read-many) drives or on CD-ROMs.

Back-up Procedures

In the laboratory, as in any other environment using computers, it is essential to introduce and maintain procedures for backing-up of data irrespective of whether the data are stored locally or on a network server. Backing-up of locally stored data can be done manually or automatically *via* a network to a back-up server or to a local back-up medium. It is important to ensure that local back-up media (*e.g.* diskette, tape or CD-ROM) are not stored in the same place as the computer from which the data was taken. In cases of theft, fire or similar occurrences, there is a greater chance that back-up data are kept safe if they are physically stored elsewhere.

Storage of data on a local netserver is the administratively and functionally best and safest solution. Advanced automatic back-up will keep safe data for all users irrespective of whether the individual PCs in the laboratories are available or not at the time of backing-up. At the same time, such a system ensures that back-ups are actually made, since this duty can then become the responsibility of a computer division or a computer staff member. In addition, procedures for re-reading backed-up data must be developed, and these should be checked periodically.

Thus one of the most important tasks of the staff responsible for the computers in the laboratory is to develop a back-up procedure which gives the required degree of data security and offers a possibility for retrieval of data when computer failure has occurred. The procedure should also include the storage of back-up data, verification of correct backing-up and the re-establishment of data.

References

1 Performance check and in-house calibration of analytical balances, NMKL Procedure No. 1 (1995), NMKL Secretariat, Finland.
2 Performance check and in-house calibration of thermometers, NMKL Secretariat, Finland, 1995, NMKL Procedure No. 2.
3 Accreditation for Chemical Laboratories, Guidance on the interpretation of the EN 45001 series of standards and ISO/IEC Guide 25, EURACHEM Secretariat, Laboratory of the Government Chemist, Teddington, 1993, EAL–G4.
4 International Guide to Quality in Analytical Chemistry. An Aid to Accreditation, D. Holcombe, CITAC Secretariat, LGC, Middlesex, 1995.
5 Good Laboratory Practice—Primer, L. Huber, Hewlett Packard Company, Germany, 1993.
6 Validation of Computerised Analytical Systems, L. Huber, Interpharm Press, Germany, 1995.

CHAPTER 12

Experiences in the Implementation of Quality Assurance and Accreditation into the Food Analysis Laboratory: Administrative Aspects—Reports and the Chain of Documentation, Internal and External Quality Audits and Management Reviews and Continuous Improvement

In this chapter, specific areas of interest are addressed which have been shown to cause some problems in their introduction into a food analysis laboratory. The specific areas are concerned with some administrative aspects in the laboratory. The chapter is divided into three parts dealing with:

1. Reports and the chain of documentation.
2. Audits.
3. Importance of continuous improvement.

1 Reports and the Chain of Documentation

Introduction

This section discusses various ways to document important information in order for analytical work to be carried out and finalised to the satisfaction of the client.

The quality of every step of the chain of documentation must be controlled in order to ensure that the final product, the analytical result, is of an adequate quality. Such a chain of documentation in the laboratory will normally contain the following information links:

- the analytical task in the form of an experimental or sampling plan or a request from the client;
- the sample and information on the sample procedure used, receipt of sample, *etc.*;
- sample pre-treatment and analytical techniques (methods, instructions, validation and verification of data, *etc.*);
- equipment, instruments, *etc.* (calibration status, verification and quality control);
- raw data and calculated results;
- the analytical or study report; and
- the quality manual.

These have all been commented on in this book; the documentation as outlined above will in principle serve two alternative purposes:

- it will describe and direct the way in which analytical work is to be carried out; or
- it is generated as the analytical work is carried out (description of sample pre-treatment, raw data and final report).

Documents such as methods and instructions which describe the performance of analytical work must be included in the laboratory's document control; they should be typed, formally approved and have a documented distribution within the laboratory. Information on samples, specific analytical procedures, raw data, *etc.* is often documented only by hand as the data is generated and gathered. Irrespective of the way in which the information is noted, the documentation should meet the following requirements:

- it must be credible, *i.e.* it must be legible, comprehensible, not manipulated, checked and approved, *etc.*;
- it must be traceable to samples, methods, instruments, analyst, time/date, *etc.*;
- it must be complete enough so that the work could be repeated if required, and so that it is possible to confirm later that the work was carried out in a credible manner.

The amount and nature of documentation depends on the degree of details required in order to enable the laboratory to repeat the work, as well as the purpose of the analytical result. The requirements on credibility and complete documentation are most stringent in analytical work associated with official food control. In such cases, results and their documentation must be acceptable to a court of law. However, routine analytical work carried out, for example, for internal production control purposes, requires much less documentation. The competence of the staff also influences the extent of the required documentation. Experienced staff carrying out routine work may need less guidance and may not be required to describe work carried out in

Table 12.1 *Documentation in routine/non-routine analysis*

Extensive documentation	Typically laboratories investigating cases of food poisoning, elaborating new methods or studying new analytical problems	Routine laboratory involved, in *e.g.* official food control
Limited documentation	Typically research oriented laboratory offering routine analytical services within its field of competence	Routine laboratory performing analytical work, *e.g.* within quality control
No. of instructions	Limited number of instructions, *etc.* with document control	Large number of instructions, *etc.* with document control

detail. However a laboratory examining, for example, cases of food poisoning and having access to very few documented methods will find it necessary to carefully document all work carried out during the examination. In Table 12.1, the relation between routine/non-routine analyses and the extent of documented/not documented information is given.

Laboratories should decide in advance on the extent and degree of detail of documentation required in order for the analytical result to demonstrate adequate quality (from the client's point of view).

It is necessary for laboratories to evaluate the sufficiency and extent of their documented information when carrying out their internal audits. Flaws in the traceability, credibility and completeness in the analytical system are very often detected only after the work has been completed. Therefore all pertinent information must be filed so that it is retrievable. Files should be stored so that documents are protected from moisture, heat or unauthorised usage.

Laboratories should develop and implement instructions describing the routine document control, documentation of analytical data, reporting and archiving. It is important that the routine procedures are adjusted to the needs of the laboratory, and are not made unduly complicated.

Document Control of Analytical Methods, Instructions, etc.

Document control is a system which should ensure that correct and up-to-date documents are in the right place, and that outdated documents are not used.

Analytical methods, instructions, standard operating procedures, manuals and all other information in frequent use should be included in the control documentation. The instructions on control measures must describe the way in which these documents have been prepared and identified. It is important that the instructions also specify where and when information on document control must appear in the documents. The instructions on document control should cover all steps of the 'life' of a document, namely:

- its creation and discussions with future users;
- its approval;
- its distribution and copying;
- its implementation;
- any introduction of single corrections, and complete revisions; and
- any annulment and repeals.

The instructions should also include information on the relationship between various documents. Such information is usually available in the quality manual of the laboratory.

Correct Papers in the Correct Place

Creation of documents

In order for analytical methods, manuals, *etc.* to be correct, they must be drawn up in a competent way on the basis of known facts. The text of the document should contain references to literature; alternatively the source(s) may be described and archived elsewhere in such a way that they can be retrieved if necessary. Great care should be taken in preparing documents to ensure that future users of the document understand and are able to use the instructions. It is therefore recommended that drafts of documents are circulated among the staff concerned before documents are approved and taken into use. If it is practical, laboratories should appoint suitable members of staff to be responsible for elaborating documents. These 'document owners' then become the persons who understand the contents of the document best and therefore can explain to colleagues why the document has been elaborated in a certain manner.

Approval

Before document controlled information is taken into use it must be checked and approved by a responsible person. The checking is of great importance since errors and deficiencies not detected before the document is taken into use easily become established truths giving rise to losses, *e.g.* in the form of wrong analytical results. The laboratory must identify members of staff authorised to approve documents.

Distribution

Approved documents must be distributed in such a way that it is known who has received copies of the document. Larger laboratories often establish distribution lists in which documents are signed for. If only a few copies are required it may be practical to indicate the distribution in the document itself. It may be convenient to archive originals or master copies of all analytical methods and similar documents in one single place in the laboratory, this then providing sufficient information regarding their distribution.

Any distribution of formally approved documents outside the laboratory's document control is not to be recommended. If the laboratory wishes to distribute its documents outside the document control, such documents must be clearly labelled as unofficial, otherwise there is a significant risk that future readers are unaware that the document has undergone changes, or that older, outdated versions of documents are introduced and used long after they have been taken out of current use.

Implementation
There is no reason to elaborate documents, however good, in order to control the quality of work if there is no need for them. As mentioned previously, this may be avoided if the staff involved are given an opportunity to participate in their elaboration and that the documents are distributed in an efficient manner.

Sufficient time must be reserved for presentation and discussion of documents to be implemented. The 'document owner' should participate in this work. After the document has been taken into use, the staff concerned should be invited to submit comments. The document owner should collect any comments, and make use of them when changes are introduced.

Amendments
The laboratory's instructions regarding document control must identify the person to be contacted should documents need to be revised, and also describe how revisions or annulments are to be carried out. In all cases the person who approved the original document should be contacted, as well as the document owner, since that person probably has the best information on it.

If amendments are minor, *e.g.* in the form of an addition or a change of only a few words, and the changes are acceptable, a staff member authorised to approve documents may introduce changes by hand, dating and signing these directly into all distributed copies of the document. Alternatively, the owner can change and sign the master copy, which then is copied and distributed to everybody involved. In such cases it is important that the owner carefully recalls all earlier versions of the document. It should be made clear to the staff why changes must not be introduced by other than authorised staff. In the case of more extensive amendments, there may be a need to substitute one or more pages of the document, or revise all of it. This work should then be conducted in the same way as if elaborating a new document.

Annulment
When distributing an entirely revised document or revised single pages of documents, the laboratory must ensure that older versions are taken out of use and no longer are available in the laboratory. However, the laboratory should always keep on file one copy of any documents which have been annulled so that it later can be demonstrated which version of the document was in use at a particular time.

Format

Analytical methods, instructions, manuals, *etc.* may in principle be elaborated in any suitable way. However, it may be an advantage if the laboratory uses a uniform format since this may facilitate both the elaboration and comprehension of the document.

It must be indicated on the document whether it is approved or not and when it was approved. Additionally, the document may contain information on when it must be taken into use. Pages must be numbered, and the total number of pages given on every page. Documents also need to have a unique identification, such as a name or a code, which must be given on every page. Many laboratories have found it convenient to prepare documents with a header or a footer containing all the document control information. In addition to the information given above, the header or footer may also contain information on where the document is to be kept.

It is recommended that suitable parts of the document control information are given also in draft documents. Such practice makes it easier to distinguish various versions to be circulated prior to approval. As a minimum, drafts should be dated, have page numbers, an identification and contain the name of the document owner.

Document controlled information may be written on paper or be stored electronically. Some laboratories find it useful to have documents on papers of various colour, or paper with coloured logos so that official final versions may be distinguished from drafts. An often used way of indicating that a document has been approved is to give it a handwritten signature.

Storing of documents on computer media poses special requirements on the availability of printers, and also on the protection of document integrity. Systems must be introduced to ensure that unauthorised staff cannot introduce changes. In addition, the laboratory must have procedures to ensure that staff does not use print-outs of older versions.

Documentation of Variable Information

All important information concerning experimental work should be documented objectively and correctly. No information influencing the overall picture must be omitted. Raw data of analytical work is an example of 'variable information', and examples include:

- all forms of registration of measurement results, such as instrumental print-outs, chromatograms, registration in work books, standard curves, control charts, *etc.*;
- descriptions on procedure (including references to methods, instructions, *etc.*);
- copies of analytical reports, *etc.*; and
- other data (calculations, notes on service of instruments, *etc.*) needed for the description of the analytical chain in an understandable manner.

Laboratories may also document analytical methods and control information according to the instructions given below. An example of such a situation is when a method is used only rarely, *e.g.* in connection with examination of food poisoning incidences.

Immediate Documentation

Analytical work should be documented as the work is carried out and not later in a processed or refined form. The original registration should always be filed, since every transcription means there is a risk of errors being introduced. It is therefore not acceptable first to note the information and then to write it out in order for the data to look 'nice'.

Documentation of raw data, *etc.* may be done in handwriting, by automatic printouts or by indirect or direct registration in a computer. Irrespective of how raw data are documented they must be legible, and must always state who carried out the analytical work and what equipment was used. Raw data must be traceable to individual determinations or analytical runs. Registration on paper should as far as possible be made by the person who carried out the analytical work. Registration may be done:

- in bound laboratory journals;
- on pre-printed forms; or
- on already existing documents, such as a document accompanying a sample (the usual practice in sampling and when samples are received in the laboratory).

Bound laboratory notebooks are often used when registering analytical steps carried out at the same time on a large number of samples, for notes on service, controls of balances, *etc.* The notebooks must be labelled to aid traceability and the pages should be numbered. Pre-printed forms are used where appropriate, for example in the reading of ELISA micro titre plates or microbiological cultures. By providing specific lines or boxes to be filled in, such forms at the same time have the role of a check-list or memorandum of the various steps to be documented. In very strict cases it may become necessary to use pre-printed forms or laboratory journals equipped with the laboratory's unique identification or code, and which staff must sign for before use.

Automatic printouts can also be carried out in various ways, for example using thermal paper, paper rolls or individual sheets. It must be noted that thermal paper has very poor keeping quality. Solvent residues in the air, heat and sunshine may, in the worst case, erase the registered information within days or weeks.

Storage of data on computer may be done directly or indirectly. A discussion on questions relating to computers, LIMS and PC format, *etc.* is included in Chapter 11 on Computers. Registered information must normally be printed as hard copy onto paper.

Correct Data

As described previously, the original documentation of raw data should be kept on file and form the basis of further calculations. Transcription of raw data is one of the most common sources of errors in an analytical laboratory. Irrespective of the manner in which data is documented, it must be proof-read. All calculations and transcriptions should be checked and the checks must be dated and signed.

All information must be registered using permanent ink. Normally this means that ball-point pens filled with archive-resistant ink should be used. Staff should be warned against the use of the simplest types of pens; their ink may fade away and disappear within a short period of time.

Any corrections or amendments needed to be introduced into a text should be done in such a manner that the original (crossed-out) text remains legible. Unless the amendment is trivial (*e.g.* a typographical error) it should be dated and signed. If significant amendments are made, there should be an explanation or reasons for the change.

The Report—a Final Product?

The report may be seen as the final product, presenting the analytical result to the customer. It is important that 'Particular care and attention shall be paid to the arrangement of the test report, especially with regard to presentation of the test data and ease of assimilation by the reader' as is given in section 5.4.3 of the standard EN 45001. The arrangement of an analytical report may need to be varied because of different customers needs, which is not always appreciated by laboratories and accreditation agencies. The report is one part of the laboratory's final product, the other part being the analytical result itself.

Format

A report may contain one or more analytical results, or may consist of several part-reports from long experimental runs, larger analytical tasks or studies. Laboratories may wish to distinguish between preliminary reports and final reports. Particular requirements on the reporting should be stated in the experimental plan or in the analytical request. This section covers the information which should be included in all types of reports.

The content of an analytical report resembles to a great extent the information given in the sample submission form or instructions accompanying the sample. The quality models EN 45001 and ISO/IEC Guide 25 list a number of general requirements which reports have to meet irrespective of whether the laboratory is accredited or not. However, it should be noted that the above quality models are primarily drawn up for routine laboratories offering defined analytical testing, and therefore are not always directly applicable to more complicated reports. The following information should be given in a report:

- name and address of the laboratory;
- name and address of the client (both the receiver of the report and the invoice, if these persons or organisations are not the same);
- unique identification of the report (such as a serial number) and of each page, and total number of pages of the report;
- description and identification of the sample or samples;
- date of receipt of sample and date(s) of carrying out the analysis;
- description of sampling procedure, where relevant;
- any non-compliances, additions to or exclusions from the analytical task and any other information relevant;
- identification of all analytical methods or procedures used;
- analytical results (together with important information on them in the form of tables, diagrams, *etc.*);
- a statement on measurement uncertainty (where relevant; see previous discussion on measurement uncertainty);
- signature and title of person(s) accepting technical responsibility for the report and its date of issue;
- a statement to the effect that the results relate only to the samples tested; and
- a statement that the report shall not be reproduced except in full without the written approval of the testing laboratory.

Amendments or additions to an analytical report after issue may be made only by issuing a new report. It should be emphasised that the new report replaces the earlier report, *e.g.* by labelling it 'amendment/addendum to report' together with the identification of the first report. If reporting errors are detected, the laboratory should investigate the effect of the error on the analytical result and its use, and also inform all clients in cases when the error may have affected the way in which the analytical result was utilised.

Distribution of Reports

Analytical results may be reported to external clients, to other departments of the same organisation or to other laboratories in the case of subcontracted work. The requirements on the manner of reporting and format of results may vary between these different types of clients.

Reports may, for example, be printed on paper which is sent by post or telefax or be electronically stored and sent by electronic mail, *etc.* It is recommended that the laboratory always keeps on file a hard copy of the report containing the main part of the information given above even if the client does not always wish to receive all the information in his copy of the report. A situation in which the client only requires the analytical result is in the production control situation or in monitoring situations when long sample series are taken over a considerable period of time.

Electronic reporting *via* fax, e-mail, *etc.* should only be undertaken with some caution since the laboratory has little control over the form in which the report reaches the client. In particular, older types of fax machines may distort digits and other information in the report so that they become impossible to read.

The report is the final product of the laboratory; if it is not legible or fails to reach the client within the agreed time, all analytical work and quality assurance measures carried out may be wasted. A prerequisite for 'correct quality' is that the laboratory has a good system for reporting its results.

Archiving and Discarding of Information

Documents on finalised analytical assignments or projects must be archived. Requirements on archiving vary, and therefore the archiving procedures must be adapted to the needs of the laboratory and its clients. OECD's GLP model places the most stringent requirements on archiving. According to this model, access to archives must be limited to specifically authorised persons, and all transfers of material to and from the archives must be documented.

Premises for Archives

Documents may be archived in either local or central archive premises. Often substantial quantities of raw data such as chromatograms are archived locally in the laboratory, while copies of requests, received instructions and manifests and external reports are archived centrally under stricter circumstances.

Archiving should be carried out in a way that ensures safety against access by unauthorised persons, moisture, heat, solvents, electromagnetic disturbance and other factors which may affect and destroy information on paper or stored electronically. However, food laboratories need normally not have archive premises or cabinets meeting very strict fire-protection requirements, or resembling safes, such as is required in OECD's GLP rules. Nevertheless, the EN 45001 standard requires the laboratory to arrange such protection if the client so requests.

Handling of Information

Archiving may be done 'batch wise' by gathering all kinds of information related to a specific analytical task. This is fairly common in laboratories accepting infrequently carried out analytical determinations. In some cases, archiving needs to done on the basis of an analysis or an analyte; this is the case when several different samples are analysed at the same time in an analytical run and is documented in a laboratory journal. Archiving must always be carried out in such a way that the information is retrievable. In some cases, the archives must be equipped with a register to facilitate their use.

Corrections to archived raw data must not be made without a written explanation. In addition, the correction must be approved and signed by the appropriate manager.

In order for a food laboratory not to be overwhelmed by archived information, the archive needs to be assessed from time to time. The laboratory should prepare procedures for the discarding of documents, *etc.* stating the various archiving times that they should be kept. It is often practical to label the archived material with the archiving time before it is placed in the archive. The archiving time depends on the laboratory's own needs, requirements of clients and other users of the information of the laboratory (*e.g.* users of the laboratory's analytical methods) and various requirements by the authorities. The procedures of the laboratory must be adapted to all these needs and requirements.

In general, the archiving time is short in the case of simpler forms of production control data, whereas raw data from, for example, official food control may need to be archived for several years depending on national law. Results from analytical assignments relating to product safety regulations or to OECD's GLP rules may need to be archived for up to 10 or 25 years or longer. A laboratory active in several different analytical fields can normally not apply one single archiving time to all its information.

Analytical methods and instructions, validation and verification data, as well as information on the instruments used and the people involved in the work, *etc.* must be archived for at least the period of time the reported results are archived.

2 Internal and External Quality Audits and Management Reviews

Introduction

The activities of a laboratory will continuously change as a result of new needs and requirements of its customers, new regulatory demands, developments in analytical techniques, new staff or changed responsibilities for existing staff, *etc.* Its operations must therefore be continuously evaluated and adapted to new conditions.

Evaluations of the operations of a laboratory may, for example, be conducted in the form of internal and external quality audits, which together with handling of complaints, customer satisfaction studies, *etc.*, enable the laboratory to improve its operations on a continuous basis.

The following definitions should be used when considering audits:

Quality audit: the systematic and independent examination to determine whether quality activities and related results comply with planned arrangements and whether these arrangements are implemented effectively and are suitable to achieve objectives (ISO 8402:1994).

Management review: the formal evaluation by top management of the status and adequacy of the quality system in relation to quality policy and objectives (ISO 8402:1994).

Senior management plays a very significant role in audits and reviews, since senior management has the overall responsibility for the operation of the laboratory. It is the responsibility of management to review systematically the quality system and, on the basis of the results, decide on future strategies including identifying the required resources. However, it should always be appreciated that each member of the staff is responsible for the quality of his work. Staff must therefore be motivated and encouraged to introduce improvements as seen or felt necessary without awaiting the results of internal audits.

Guidelines on the conduct of audits are given in standards such as EN 45001[1] and EN 45002,[2] ISO/IEC Guides 25 and 58,[3,4] ELA-G3[5] and in ISO 10011: Parts 1 to 3.[6]

The Different Types of Audits and their Objectives

A laboratory audit may have different objectives. Internal audits are conducted on the laboratory's own initiative as a part of the control of operations. They aim to identify flaws (non-compliances) in operations which need to be corrected. Standards such as EN 45001 and EN ISO 9001 require that laboratories periodically and systematically conduct internal audits. It is very important that the laboratory collects information on its own operations in order to be able continuously to develop and improve its procedures.

External audits may be carried out by customers of the laboratory (second party audits) or by independent accreditation or certification bodies (third party audits). Third party audits are a part of accreditation or certification body assessments. Even if external audits are carried out as a result of the initiatives from outside the laboratory, it has to make use of the information resulting from external audits in its own improvement work. It is important that the laboratory and the external auditor mutually agree on the time and the manner in which the audit is to be conducted.

Audits carried out by customers are not unusual in laboratories offering services to private companies. Customer audits may be carried out as a part of the assessment of whether the analytical work has been performed in accordance with the order or contract. Audits may also be carried out prior to signing a contract, for example in cases when the customer needs further details on services offered in order to specify accurately the work wished to be carried out.

Laboratories which apply for, or which already have received, accreditation, or which operate as a part of a certified organisation, will regularly be subjected to third party audits. The objective of these assessments is to determine whether the quality system of the laboratory and its operations comply with the requirements of standards such as EN 45001 and EN ISO 9001 or guides such as ISO/IEC Guide 25. The result of a third party audit is usually an accreditation or certification body decision granting (or continuing existing) accreditation or certification, respectively.

None of these three forms of audits can replace each other. Internal audits are necessary, if not for other reasons, then in order to follow up the results of external audits. Second party (customer) audits are important aspects in the establishment of customer's needs and the customer's perception of the quality of the laboratory's analytical services. Third party audits often result in information which is valuable and useful both to the laboratory and to its customers.

Experience shows that all quality audits increase the motivation of staff. Their routine work receives a 'new dimension' and their work is recognised as being important. The quality system and the quality work becomes concrete and tangible, and feed-back is received which informs staff in which areas they have performed well and in which areas improvements could be made. Discussions between the auditor and the staff on audit findings often form an excellent training session. A quality audit carried out well usually affects the entire operations in a very positive way.

Summary of the Processes of Internal and External Audits

A summary of the processes involved in internal and external audits is given below:

Audit plan	Designed by the auditor in co-operation with laboratory representatives on the basis of the audit schedule
Document familiarisation	Standards, quality assurance plans, test plans, handbooks, methods, results, *etc.*
Opening meeting	Presence of all staff, presentation of scope, audit plan, *etc.*
Collection of facts	Documentation, discussion and signing, objective evidence, findings
Summary	Auditors only, discussion and possible classification of non-compliances and observations
Closing meeting	Auditors, manager, quality co-ordinator; reason oriented summary, acceptance, timeframes for actions
Audit report	Description of findings, references to standards, standard operational procedures, *etc.*; description of which part of the laboratory and its operations were assessed
Action reporting	From laboratory to the auditor or quality co-ordinator, within an agreed timeframe
Follow-up	Follow-up by auditor or quality co-ordinator; identification of any additional actions
Summary of all audit findings	General summary of all audits carried out during the year
Management review	Review of the quality system by management and quality co-ordinator, follow-up of all audits, handling of complaints

Planning of Internal Audits

The extent and the conduct of an internal audit will vary with the size and nature of the laboratory. Laboratories with a staff of only one person or a few people may have to make use of external auditors.

For an audit to be viewed positively within the laboratory and thus lead to satisfactory conclusions, it must:

- be well prepared;
- be carried out in a professional and objective manner;
- focus on essential parts of operations within the laboratory; and
- be documented, and recommendations later followed-up.

Schedules for Audits and Reviews

The laboratory must establish an audit schedule which describes over a longer period of time when and how the various operations of the laboratory will be audited. All elements of the quality system should be addressed and audited at least once a year.

Prior to each audit the auditor should design a plan for the audit. The plan is unique for each occasion and contains information on the elements included in the audit. The plan is a useful tool in the preparatory work and can be seen as a 'pathway' to be followed during the audit.

Background documents for an audit plan and for an audit include, for example:

- reports from earlier internal and external audits;
- the quality manual and other quality documents such as control charts, results from participation in proficiency tests, *etc.*;
- relevant regulatory requirements;
- operational and testing plans, including quality assurance plans;
- methods of analysis and standard operating procedures;
- documentation on complaints and reclamations;
- relevant quality standards or guidelines (EN 45001, EN ISO 9001, ISO/IEC Guide 25, *etc.*); and
- documentation on new staff or new equipment.

The auditor should, on the basis of the existing documentation, select the elements to be covered by the audit. It is important to focus only on elements which are essential to operations and the quality system of the laboratory. In addition, priority should be given to areas where the laboratory has encountered problems.

The plan may be written in a suitable format which will also serve as a 'checklist' to be followed during the audit. Such a checklist should in no way prevent a flexible conduct of the audit, something which may happen if previous internal lists, or checklists 'borrowed' from other organisations, are used as such, without revising them to take account of the laboratory's own needs. Examples of useful checklists for audits have been published.[7]

Internal audits should not be carried out without prior notice. The auditor should set the date of the audit and agree on the time schedule in consultation with the person responsible for the operations to be audited and other staff. It may be an advantage to carry out audits in several, recurrent and shorter sessions during the year. This will be beneficial for constructive discussion and may contribute to maintaining the interest of staff in quality matters. If the laboratory is subjected to many second or third party audits or assessments, it may be justified to reduce the number of internal audits. However, EN 45001 and ISO/IEC Guide 25 require that internal audits are carried out at least once a year. Too short an interval between audits means that there is a risk that staff find them to be repetitive and boring and so adopt a negative attitude towards them.

Auditors should aim at visiting all categories of staff, as a means of motivating everybody to produce quality work.

Horizontal and Vertical Audits

Audits may be carried out either horizontally or vertically. A horizontal approach means that one specific element of the quality system or one specific analytical technique is addressed throughout the laboratory. An audit of the competence, training and records of the staff is an example of a horizontal audit, as is the evaluation of all LC systems used in the laboratory.

The conduct of horizontal audits may be an excellent way of finding out if a flaw or defect identified, for example during an external audit, is repeated throughout the laboratory or if it was an isolated occurrence. The findings will in turn dictate the corrective actions to be taken.

A vertical audit may be carried out by evaluating an analytical task from order and sample receipt through to reporting of the results, disposal of the sample and filing of the report. Maximum information is obtained if the task is selected from tasks finalised some months prior to the audit since the staff will then not remember all details. It is appropriate to take as the starting point of vertical audits a selection of test reports on completed analytical tasks and to trace the information back to receipt of sample or sampling. If possible, the test reports should cover different types of reported analytical work. In a vertical audit, the sample's progress through the laboratory is followed back from the filed report. This kind of tracking audit should give answer to, for example, the following questions: 'who handled the sample?', 'was that person authorised to do so?', 'what analytical methods were used?', 'were the methods approved for the particular use?', 'which instruments were used?', 'what was the calibration status of the instruments?', 'were the raw data documented?', and 'were the calculations checked?', *etc.*

A clear advantage of the vertical approach as compared to the horizontal approach is that the vertical audit gives an overall picture of operations (*e.g.* staff, equipment, methods, procedures and routines) in their 'natural context'. Often horizontal and vertical audits are combined to give the best possible results.

The Auditor is in the Key Position

It is frequently assumed that auditors are asked by the management to check the way in which staff work, and catch them making mistakes. This is very wrong! The purpose of the auditor is to help the staff to help themselves; any identified flaws should be regarded as opportunities for improvements.

As mentioned previously, experience show that there are very positive effects from well-conducted quality audits. A prerequisite is that that the auditor has a positive view on the procedures of the laboratory

Staff should therefore view the auditor as:

- being impartial and having integrity;
- competent and objective; and
- interested in listening and discussing.

Impartiality and Integrity

An important rule-of-thumb is that staff cannot audit their own activities. It is always difficult to detect flaws in one's own work. Laboratories employing only a few people should therefore consider making use of external experts for their internal audits. Such external auditors should be well versed with the nature of the laboratory's activities, and the requirements of relevant standards or guides. In these cases, it may be sensible to agree in writing on confidentiality issues.

In larger laboratories, colleagues may audit each other's operations. However, the conduct of an internal audit is one of the most difficult auditing tasks. The auditor must have the courage to ask detailed and, if needed, inquisitive questions. If this is done in an aggressive manner, staff will easily become uncommunicative and negative in attitude. If, on the other hand, the auditor agrees with everything which is said, avoids profound scrutiny and subsequent discussion of findings, the audit is certain to fail. Not everybody can or should be assigned the role of internal auditor. Assigned auditors should receive training in carrying out internal audits.

Competence and Objectivity

The auditor must show an interest in operations and in an inquisitive way evaluate operations for the benefit of continuous improvements of laboratory activities. An internal auditor should be competent and preferably have knowledge and experience of work in the specific analytical area which is to be audited. The competence of the auditor will make it possible that the priorities in the planning and conduct of audits are the right ones. If, in addition, the members of the staff look upon the auditor as a person knowledgeable in their field of work, then there is an increased probability of objective and constructive discussions taking place.

Large companies may choose to divide the audit work into two parts, a horizontal audit covering the quality system and conducted by an expert in this field and a vertical audit conducted by analytical experts. Such a division of work is usual in third party audits, *e.g.* in the assessment of technical competence by accreditation bodies.

To Listen and to Discuss

In order to ensure the success of an internal audit, the dialogue between the auditor and members of the staff must be carried out in a positive and constructive atmosphere. It is not the role of the auditor to make decisions or to offer unequivocal opinions on how particular matters should be handled. The auditor should conduct the audit on the basis of the predetermined plan, but must at all times be sensitive and adjust to the way in which the staff experience the audit.

The auditor should take a respectful attitude to what he sees. The initial assumption must be that staff carrying out the work are experts. The staff being audited must have the right to ask why their work has to comply with specific quality requirements. The credibility of the auditor is strengthened if the auditor allows staff to question the comments of the auditor and is able to give satisfactory explanations to them. If some aspect appears to be a non-compliance, the auditor should carefully scrutinise that aspect and evaluate whether or not there in fact is a non-compliance.

The auditor should listen to all arguments from staff and be careful not to press them to hurry their opinions. If audits are not conducted in a positive and respectful manner, information may be concealed and the audit result thus misleading.

Conducting an Internal Audit

Introduction

Internal quality audits should be carried out in such a way that everybody involved understands the objective and the extent of the audit, as well as the content of the audit plan. Therefore the audit should start with a meeting during which these questions are discussed. If the entire staff are present at this opening meeting, the auditor also has the opportunity to stress the quality of work carried out in the laboratory in a positive manner.

After the opening meeting, the audit usually continues with the collection of information on the procedures of the laboratory, usually from a tour of the laboratory. When looking at the laboratory the auditor should be accompanied by one or several representatives of the laboratory, so that the auditor should have somebody with whom to discuss his findings with after the completed audit. One of the accompanying persons should be the quality co-ordinator (unless the quality co-ordinator is also the auditor).

In a medium-size laboratory, some hours need to be reserved for discussion of the various elements of the quality system (*e.g.* EN 45001, 5.3.3: equipment). It may be expected that for each analytical technique in operation, at least two hours are needed for discussions and scrutiny.

Questioning Technique

While collecting information about the procedures/operation of the laboratory activities under audit, it is important that this and the accompanying discussions take place in a calm and structured way. The auditor should speak as little as possible and avoid jumping to premature conclusions. Follow-up questions should be posed in case of insufficient clarity.

During the collection of information, all questions should be asked in a direct manner, for example:

- From where and in which way do you receive samples, analytical information, requests?
- What do you do with the samples?
- Which analytical methods, instructions, instruments, *etc.* do you use?
- How have you learnt? Do you know?
- Where do you send materials and reports?
- Show me!

The result of the audit will be better if the staff are allowed to contribute actively to the information being gathered.

Some examples of questions which may be asked are given below:

- Are the raw data from all analytical steps, the quality control, important equipment, sample preparation, sample receipt and analysis order retrievable ?
- Are the raw data traceable to analysts, dates, equipment, samples, *etc.*?
- Are the raw data credible (*e.g.* was permanent ink used, were changes authorised, *etc.*)?
- Do the raw data agree with reported data?
- Is it clear from the raw data which methods, instructions, *etc.* were used?
- Were the latest versions of methods, *etc.* containing no unauthorised additions or amendments used?
- Was the equipment used and/or the analytical system under statistical control at the time that the analysis work was carried out?
- Have actions been taken if the equipment or the analytical system were out of statistical control?

Documentation of Findings

The information collected should be documented in such a way that it is clear what was seen and where. Any misunderstanding in this documentation should, if possible, be immediately remedied.

The following points on the documentation of audits should be taken into account:

- immediately make a note of findings (use a permanent pen, date and sign);
- write legibly and in a way which is understood by others (this is of particular importance if those not present at the audit need to read the notes at a later date);
- write objectively and focus on facts, not on individuals;
- write in a way which makes it possible to trace the findings (what was seen and where);
- ask the staff involved to read through what was documented, discuss it, amend if required, and finally ask the responsible analyst to sign the document to indicate it is understood and agreed.

It is an advantage if the audit findings can be documented on check-lists prepared in advance or on forms similar to those used by accreditation and certification bodies. This will make it easier to obtain the right papers later as well as to facilitate evaluation against the audit plan.

Summary of Findings and the Closing Meeting

After touring the laboratory, the auditor(s) should, in private, summarise all audit findings before they are presented to the laboratory. It is very helpful in ensuring the quality of future work of the laboratory for findings to be graded/prioritised in relation to each other.

Findings may, for example, be graded as 'observations', 'non-compliances', 'major non-compliances' or 'minor non-compliances'. Findings which do not constitute a quality risk, but which are of interest from a quality assurance point of view, should be classified as observations. A non-compliance is a defect in relation to specified requirements. Major non-compliances indicate systematically occurring defects or a flaw which may result in sizeable analytical errors. For a non-compliance or a major non-compliance to be accepted as such, it should be objectively described, be focused on facts and not be directed towards a person.

Auditors must remember when preparing and then discussing the audit findings to:

- thoughtfully grade and classify non-compliances after the laboratory inspection (whilst an impression can be gained during the inspection, it is impossible to decide if a finding occurs systematically, or is serious or not);
- classify the finding as non-compliances only if there exists a traceable requirement which can be presented to the laboratory;
- justify the classification and allow the representatives of the laboratory to question it and then to take any comments into consideration.

Auditors must be flexible during the concluding discussions. It is usually appropriate that only the staff involved, the quality co-ordinator and the responsible analyst are present when discussing details. However, a summary should be written covering general findings. Such a summary may be very useful for all staff. At the closing meeting, an evaluation, the extent to which the objectives of the audit plan were reached, must be stated.

After the audit, the quality co-ordinator and the management should invite the staff to comment on the planning and conduct of it. Such comments should be taken into account in the attempt to improve future internal audit routines.

External Audits from the Laboratory's Point of View

External audits should in principle be designed and conducted as outlined above. The laboratory and its staff have a right to be informed about the objective and the extent of the audit, as well as on the content of the audit plan, irrespective of whether it is a second or third party audit. However, it should be remembered that the initiative and the responsibility concerning external audits lie respectively with the customer and the accreditation/certification body.

Audits Conducted by Customers

Not all customers of food laboratories have either the experience of audits nor of carrying any out. This is unfortunate as such occasions provide excellent opportunities to discuss the requirements of the customer, and the expectations and needs concerning the analytical service being offered.

It is the responsibility of a laboratory operating a well-functioning quality system to clarify the wishes of the customer. Under 6.1 of EN 45001 it reads 'The laboratory shall afford the client or his representative co-operation to enable him to clarify the client's request'. For example, the choice of analytical methods needs often to be discussed to a greater extent than presently normally occurs. Does the analytical request mean a continuous control of a food manufacturing process? Is it expected that the analytical result will be the subject of a dispute in a court of law? Can rapid methods with poorer specificity be used, or should the laboratory use a collaboratively tested reference method?

The laboratory should ensure that audits carried out by customers are performed out in a positive atmosphere. The quality co-ordinator and the management should do all they can for the operations/procedures in the laboratory to make a good impression. During the opening meeting it is important to describe the quality system of the laboratory and how it works. If the laboratory is accredited, or is a part of a certified organisation, a concise presentation of the scope of the accreditation or the relevant part of the certified operations should be given. This is important because customers may, for example, wrongly believe that accreditation in a limited area means that the laboratory is competent to perform all types of analyses.

It is important that the representative of the laboratory (*e.g.* the quality co-ordinator) takes detailed notes during the collection of information. The customer may be unfamiliar with conducting and documenting the collected data, and a report is not always written and made available to the laboratory.

Audits Conducted by Accreditation and Certification Bodies

Third party audits by accreditation and certification bodies are a part of an assessment carried out according to guidelines of international standards such as EN 45002 and 45003, as well as ISO 10011, Parts 1 to 3. These documents offer the laboratory opportunities to prepare itself for the audit. After a laboratory has decided to apply for accreditation or certification, the staff, and particularly the quality co-ordinator, should familiarise themselves with the relevant quality standards and guides (EN 45001 or ISO/IEC Guide 25, one of the EN ISO 9000 standards, respectively). If the quality co-ordinator is familiar with the accreditation or certification standard, this will contribute to the usefulness of the discussions during audits/assessments.

During third party audits, the laboratory should take its own notes. The staff should be well prepared and instructed to question if unclear in any situation. It is possible that the atmosphere during the first few visits may be very tense, which then easily leads to misunderstandings by both parties.

After an audit or assessment, the auditors will normally prepare detailed reports and make them available to the laboratory.

Follow-up of Audit Findings

A prerequisite for audits leading to tangible and lasting improvements is that all identified defects are evaluated and acted on as soon as possible. Irrespective of the type of audit, this evaluation starts early, at the closing meeting of the audit. For the follow-up to be efficient, the documentation from the collection of facts and the discussions of the meeting needs to be structured and traceable.

The Closing Meeting

The quality co-ordinator and representatives of the laboratory should be present at the closing meeting. However, it can be discussed whether it is practical for the entire staff to participate in the closing meeting of a second or third party audit, since the presence of many people may make it difficult to discuss details in a direct and clear manner.

The representatives of the laboratory should not hesitate to ask for any clarification during the closing meeting. Are there, for example, additional facts behind a deviation, which have not been considered earlier and which show that there in fact is no deviation? Is the classification/prioritisation of non-compliances reasonable? Has it been possible to identify the underlying reasons, and have these been justified by reference to relevant requirements in standards (or has the auditor just identified superficial defects in the system)?

Accreditation and certification bodies often wish the representatives of the laboratory to suggest corrective actions. This is suitable in the case of serious non-compliances. However, the laboratory should state the corrective actions only if they can be given in a meaningful way, comments such as 'must be taken care of' have little function.

A frequently occurring problem is that not enough time is reserved for the closing meeting, which may cause stress and misunderstandings. In the case of internal audits, it may sometimes be an advantage to schedule the meeting to take place a couple of days after the collection of facts, or to plan to have several meetings. This may give the staff involved time to think through the audit findings. In the case of external audits and assessments, the closing meeting is usually in direct connection to the audit.

Any conclusions from the meeting should be documented. It is recommended that representatives of the laboratory are given the opportunity to select the timeframe within which non-compliances should be corrected.

After the Audit

As soon as possible after the audit, the staff involved, the quality co-ordinator and the management should initiate actions to correct identified non-compliances. Acute problems must be solved immediately. Work on more long-term problems should be thoughtfully planned and carried out to ensure that actions lead to permanent results (see also Section 3 of this chapter regarding continuous improvement). In addition, the staff of the laboratory should evaluate whether identified non-compliances also occur within other groups in the laboratory or in connection with methods other than those in connection with which the defects were identified.

Work in order to plan and correct non-compliances should be documented. The documentation should include the name of the person responsible for actions and a time deadline. Any summary report should be distributed to everybody involved. If non-compliances are documented separately, these should be distributed to staff involved. The actions to be taken can be written directly on the notification forms.

It may be appropriate to arrange a meeting in the laboratory some time (a few days) after the audit when the agreed corrective actions are discussed. The improvement work will be successful only if all relevant staff are informed and involved in the work.

Reporting of Corrective Actions

The accreditation and certification bodies usually require laboratories to report the corrective actions taken. They require information on the actions, not only the information 'the deviation has been corrected'. Normally major non-compliances require more extensive information than minor non-compliances.

As in the case with internal audits, it may be appropriate for the auditor to request some form of reporting, even if this reporting does not need to be as detailed as that of an external audit.

In order to ensure that all relevant actions have been taken, the quality co-ordinator, together with the auditor, has the responsibility of following-up internal audits as soon as possible after the deadline for the corrective actions has passed.

Management Review

Senior laboratory management has the responsibility to ensure that management reviews are carried out at least once a year, and that their results are documented. Laboratory and section managers, the technical manager, the quality co-ordinator and others responsible for the quality system should participate in the review.

The aim of the management review is to consider which changes in the operations and quality system are needed in order to ensure that the quality as defined in the quality policy is achieved on a continuous basis. The considerations must be based on the results of internal and external audits, as well as on all information from participation in proficiency testing schemes, customer satisfaction studies, handling of complaints, *etc.* Any changes in the volume of work, staff and analytical techniques must be taken into account. The combined information may make it necessary for the laboratory to introduce changes, for example in the organisation, premises, staff, equipment, procedures and operations, as well as in the internal work load.

Any laboratory management which issues ambitious quality policies, but which then involves itself in quality work very infrequently, does not make itself credible to the staff of the laboratory. The role of the management as an active participant in the quality aspects of the laboratory cannot be over-emphasised.

The management can and must, at the annual review of the quality system, take the strategically important and maybe painful decisions necessary for effective operation in the laboratory. It is also important that management at the same time ensures that these decisions are implemented in the laboratory in a positive manner. The continual support of management for quality work must be appreciated by all staff.

3 Importance of Continuous Improvements

Achieving Quality by Continuous Improvements

The quality system of a laboratory will be successful only if all benefits that may be derived from 'quality work' are used. A systematic, co-ordinated and prioritised approach to improvements is a prerequisite for this. The PDCA

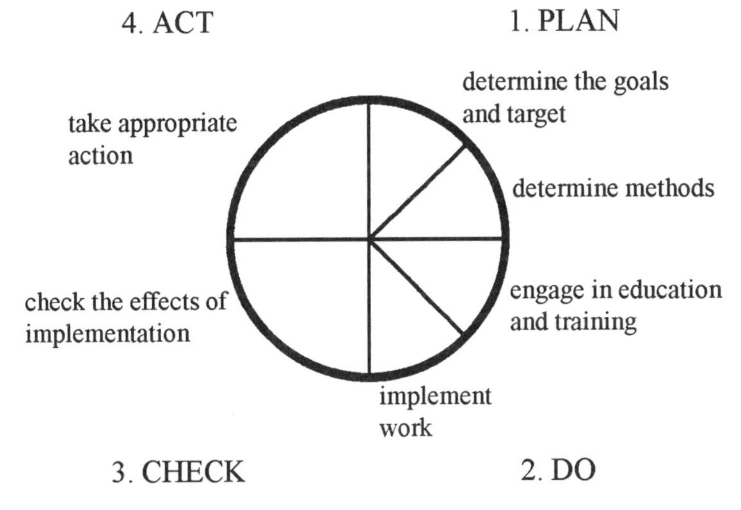

Figure 12.1 *The PDCA cycle*

(plan, do, check, act) cycle in its different variations can be used when introducing improvements in the laboratory. The cycle divides the improvement process into four steps (see Figure 12.1).

The philosophy of this cycle may be applied to all improvement work, independent of its nature.

An improvement process may also be described as follows:

1. *investigate* the real causes of identified non-compliances (do not act only on symptoms);
2. *analyse* the whole process where the non-compliance occurs to identify the potential sources;
3. *act* to prevent the non-compliance re-occurring (may lead to changes, *e.g.* in analytical procedures, organisation within the laboratory);
4. *monitor* the improvements to ensure that the non-compliance has been corrected; and
5. *implement* improvements and document them in the quality system (it is important that actions not only solve the problem but also harmonise with the overall quality system of the laboratory).

Sources of improvements can be both internal and external, for example:

- results from internal quality control and proficiency testing schemes;
- information on internal and external failure costs;
- customer complaints;
- results from internal and external audits, and management reviews;
- evaluation of analytical efficiency and timely reporting of results;
- suggestions from staff; and
- interviews and inquiries to customers.

It is important that all staff in a laboratory understand that documentation of failures, customer complaints, *etc.* are not personal criticisms. The major obstacle to improvements are negative, apprehensive attitudes. It has to be appreciated that the assembled information is an aid to help staff to improve, and that critical comments should be seen as constructive suggestions. It is recommended to use the term 'improvement' instead of 'corrective action', as given in quality models such as EN 45001, ISO/IEC Guide 25 and the EN ISO 9000 series.

'Adequate quality' of a food analysis or examination can be related to predetermined requirements, which continuously alter due to changes in customer needs. Staff and laboratory competence vary. Requirements for new analytical procedures within the laboratory re-emphasises the need for continuous improvements. A laboratory devoted to quality never completes its quality work as quality at one period of time may not be adequate at another. The goal should always be to achieve 100% quality, as acceptance of even a small number of unsatisfactory analyses may result in large costs.

It has been shown that the most important customer requirements when buying diagnostic laboratory services are:[8]

1. customer experience of overall quality of service;
2. reliability of the result;
3. overall ease of use of service;
4. prompt notification of results; and
5. provision of specialist advice.

Quality deficiencies in any of these five major requirements result in large costs for the laboratories and their customers. Quality improvements are essentially looking forward and aiming to achieve significantly higher levels of performance and with a reduction in cost. They must therefore focus on the customer, and the overall requirements for quality.

The quality of products can be measured continuously (*e.g.* using internal quality control procedures), periodically (*e.g.* evaluation of timely reporting), through spot sampling of information (*e.g.* internal audits) or by interviews and inquiries. In general, the same methods can be used to measure quality as those used for evaluation of customer requirements and needs.

Internal Quality Control and Proficiency Testing

Results of internal quality control measures and participation in proficiency testing schemes are important sources of information to aid improvement in the 'analytical-technical' quality of the laboratory.

To be of any use, internal quality control measures and participation in proficiency testing schemes must be an integrated part of the analytical work as previously discussed. Successful quality assurance work depends on immediate actions being taken when non-compliances have been identified. If this is not carried out, other work will be affected in the same way, and there is a risk that non-compliances will be totally neglected.

The following problems are regularly experienced when introducing quality assurance:

- quality control and proficiency testing results are documented, but not used;
- results are used as final inspection procedures and not as information to improve future work;
- results are documented in a computer that is not 'integrated into' the analytical system;
- evaluation of quality control and proficiency testing results are carried out long after the analysis has been performed; and
- non-laboratory staff carry out such evaluations.

Many laboratories document quality control results directly into control charts in personal computers. Since computerised information is more difficult to overview, evaluation of the quality control data becomes more complicated. Furthermore, the graphic evaluation of a control chart often requires that another mode other than purely electronic be used, which then complicates directly using the result. In addition, computers are not always placed with direct access to the analytical instrumentation. As a consequence, transference of results into charts is sometimes carried out manually hours or days after a control analysis has been completed and a long time after the analytical report has been issued.

It is recommended that quality control data are manually documented in control charts immediately as the control analytical result becomes available. Plans must be in place for direct action when analytical deficiencies are recorded. If these procedures are implemented, the laboratory has increased the possibility of taking appropriate actions whenever needed and reporting an increased number of correct analytical results. Staff will gain more confidence in their work as they can see that the analytical system is under statistical control.

Evaluation of results from proficiency tests long after the results have been reported is again evidence of poor procedures. It is also not appropriate for only one or a few persons to be involved in this evaluation.

Some organisers of proficiency tests are very slow in reporting the results, which makes the evaluation difficult for participant laboratories. However, proficiency tests are useful tools for improvement of analytical work provided the problems presented above are addressed; they are also excellent tools in the education of staff.

Costs of Internal and External Failures

The costs of quality and quality assurance can be divided into preventive and failure costs. Some laboratories, if introducing quality assurance measures, consider that 'preventive costs' are too high and that they therefore cannot afford quality assurance. The costs of non-compliances, however, represent by far the greatest part of the quality costs. Experience has shown that internal and external failure costs range from 5% to 30% of laboratory turnover. However, laboratories often do not recognise this and do not document and evaluate these costs.

Failure costs can be identified in many ways. Staff may discover an aspect of, for example, an analytical procedure that is often unsatisfactory. It is important that all findings of non-compliances are documented. As a consequence, the laboratory must have procedures for handling evaluation and documentation which are known and used by all staff and management. Once recognised, evaluated and documented, non-compliances become excellent opportunities to introduce improvements.

Complaints as Possibilities for Development

External and internal complaints often supply information on the functional quality of laboratory services. Complaints can be seen as one type of failure cost associated with analytical results as well as with other laboratory services delivered.

In general, the quality of services delivered, such as analytical reports, should be acknowledged by the customer. If the customer has any questions on the report, or has complaints on the service delivered, then the service is of an inadequate quality. It is therefore of fundamental importance to take seriously all questions, comments and complaints from the customer.

Quality models EN 45001 and ISO/IEC Guide 25 require that the laboratory has a documented policy and procedures for the resolution of complaints regarding the laboratories activities received from clients or other parties. Records shall be maintained of all complaints and of actions taken by the laboratory with regard to the complaints. It is recommended that the procedure is designed to work in a stepwise manner (as in the PDCA cycle):

1. *document* the comments and complaints and the supplier of the information;
2. *analyse* the comments together with the quality co-ordinator and a member of the management (complaints should be taken seriously);
3. decide what *direct actions*, if any, to be taken and appoint a person responsible for this;
4. evaluate if there are any long-time effects due to the comment, and appoint a person responsible for *preventive actions* (decisions and actions should be documented); and
5. *monitor* all actions to ensure that they are effective to overcome the problem.

Normally laboratories which operate a quality system or which are accredited initially claim that no complaints are received. However, complaints will be received, but the organisation is not sensitive enough to recognise them. To benefit from the experience from complaints, the attitude of staff may have to alter.

It sometimes happens that the person receiving a comment or complaint rejects it as being wrong, without considering or documenting the information. It also happens that individuals take the comments personally and so react with animosity, *e.g.* 'our laboratory never makes mistakes'. Such reactions will not aid future co-operation. Customers receiving competent and positive treatment are likely to stay as customers to the laboratory. Dissatisfied customers not only stop using the services of the laboratory, but may also communicate rumour of the complaint to other current and potential future customers.

Evaluation of Analytical Efficiency and Timely Reporting of Results

Evaluation of the efficiency and the timely reporting of results will suggest where improvements in the quality of different parts of the laboratory may be made. By selecting a specific subject for evaluation, it is possible for the management and staff to evaluate problems that may have been identified as a result of, for example, complaints or inquiries responded to by the laboratory.

Economy and efficiency are generally not viewed as quality matters in models such as EN 45001, ISO/IEC Guide 25 and the EN ISO 9000 series. However, those factors have an direct impact on the overall quality of the laboratory as poor economy and low efficiency have adverse effects on its analytical work.

There are examples of laboratories preparing analytical reports on all initiated analytical work and not only on work that is completed. In these laboratories it is possible to calculate the 'analytical efficiency' as a percentage, given by:

$$\frac{\text{No. of finalised analytical reports}}{\text{No. of initiated analyses}} \times 100$$

Research laboratories may accept a lower 'analytical efficiency', as defined above, than a routine laboratory, but it is recommended that the laboratory determines what is a reasonable analytical efficiency. Once staff and management have established a goal for the efficiency, work to improve it can be set in context. Examples exist where laboratories have increased the analytical efficiency, as measured above, by 10% annually over a five year time-period!

Measurement of the time from the receipt of an order to the delivery of a service provides important information on how procedures may be improved in a laboratory.

As with analytical efficiency, staff and management must evaluate and determine what is a reasonable throughput time for the laboratory. It is important to remember that this evaluation should also involve the customer. The evaluation starts with a description of the process of the service. Each step is documented and all persons involved are defined. The throughput time for each step is then determined, after which inefficient steps in the analytical process can be modified and thus efficiency improved.

Evaluation of the throughput time is of particular importance when a laboratory is involved in complex analytical procedures running over long periods of time. An even more beneficial technique is to evaluate throughput time and identify 'weak links' when laboratories are selling services other than analysis. Customers usually demand and appreciate prompt reporting of analytical results or delivery of non-analytical services such as professional judgements or opinions. Improved analytical efficiency and shortened throughput time also have positive effects on this aspect of the laboratory's work.

Internal and External Quality Audits

Internal and external quality audit results can be used as sources for improvement in all aspects of the quality system. By focusing internal audits on problems identified in internal quality control, customer complaints, *etc.*, laboratories may gain more information on potential sources of non-compliance.

The purpose of internal and external audits is to confirm that the quality system is being implemented as intended. Audits should be carried out in a positive and constructive way. A prerequisite for audits leading to actual and lasting improvements is that all identified defects are evaluated and remedied as soon as possible.

Frequent quality audits supply management with information on the quality of the laboratory work. Audit results should either be reported directly to management for immediate attention or be reported during regular scheduled management reviews. It is, however, important that audit results are not misused to punish or directly criticise staff. A majority of the non-compliances within an organisation such as a laboratory result from organisational deficiencies, which only management can solve. For this to happen, management must understand its responsibility to act and allocate resources for improvements.

Quality audits provide information that can be used both for creating a more efficient organisation and identifying stages of the analytical process which need revision. Audits must be viewed positively and give a motivating effect to staff.

Interviews and Inquiries

Interviews and inquiries generally supply information on the functional quality of analytical results and other services. It is recommended that the planning of interviews and inquiries is based on the most important customer requirements presented previously (*i.e.* overall quality, reliability, ease of use, prompt notification, specialist advice, *etc.*). Recognised specific problems or identified errors may also be used to aid the planning of interviews.

It is important for the laboratory to be able to determine if customers have confidence that a product (analytical result) or service meets its requirements for quality. Different techniques can be used, *e.g.* systematic customer interviews or the use of inquiry forms sent together with analytical reports.

When planning an interview the following aspects should be considered:

1. Before the interview

- what information is needed?
- which questions shall be asked?
- how shall documentation of the answers be undertaken?
- will follow-up questions be needed (and make a list of those if there are expected to be any)?
- how shall evaluation of answers, and presentation of results, be carried out?
- is there any need for confidential handling of information?

2. Interviewing

- must be performed in an undisturbed environment;
- enough time for follow-up questions must be reserved;
- answers must be documented and checked for correct interpretation;
- when required, treat all information in a confidential manner.

Inquiries have to be planned in a similar way. Information from inquiries can be evaluated and orderly presented with respect to customer requirements.

Management Responsibility and Employee Involvement at all Levels

Successful improvement work depends on good leadership. Management is responsible for setting of time schedules and allocating resources. Management must stimulate and encourage staff to aim for quality work to be produced and for quality assurance measures to be introduced. In particular, the following should be emphasised:

- customer satisfaction and customer-oriented management;
- fast reactions;
- active leadership for quality;
- staff involvement at all levels and teamwork;
- process orientation;
- the fact that quality begins with education and ends with education;
- quality work has to be given time and priority;
- improvements are based on facts; and
- continuous improvements are important.

References

1 General Criteria for the Operation of Testing Laboratories, EN 45001, CEN/CENELEC, Brussels, 1989.
2 General Criteria for the Assessment of Testing Laboratories, EN 45002, CEN/CENELEC, Brussels, 1989.
3 General Requirements for the Competence of Calibration and Testing Laboratories, ISO, ISO/IEC Guide 25, Geneva, 1990.
4 Calibration and Testing Laboratory Accreditation Systems—General Requirements for Operation and Recognition, ISO/IEC Guide 58, ISO, Geneva, 1993.
5 Internal Audits and Management Reviews for Laboratories, European Cooperation for Accreditation of Laboratories, The Netherlands, 1996, Guide EAL-G3, 2nd edition, 1996.
6 Guidelines for Auditing Quality Systems, ISO 10011, ISO, Geneva; Part 1, 1990; Parts 2 and 3, 1991.
7 Food and Drink Laboratory Accreditation—A Practical Approach, S. Wilson and G. Weir, Chapman and Hall, London, 1995.
8 The Journey to Quality and Continuous Improvement: The First Steps, in Quality Assurance and TQM for Analytical Laboratories, C.A. Gibson and L.R. Stephens, ed. M. Parkany, RSC, UK, 1995.

CHAPTER 13

Experiences in the Implementation of Quality Assurance and Accreditation into the Food Analysis Laboratory: Sensory Analysis

In this chapter, sensory analysis is considered, this being an area of increasing interest but where problems have been identified in its introduction into a food analysis laboratory. In some sectors, *e.g.* the olive oil sector, sensory analysis is now required by legislation.[1]

1 Introduction

Analysts involved in the sensory analysis of foods have, for a long time, had difficulty in getting their work accepted as being objective, scientific and 'equal' to chemical and microbiological analysis. However, it is now accepted that sensory analysis is indeed a scientific discipline used to measure and interpret reactions to characteristics of foods and other materials as they are perceived by the senses of sight, smell, taste, touch and hearing. Sensory analysis includes both qualitative and quantitative measurements, and may be applied, for example, in shelf-life studies, product matching, specification and quality control, taint and off-odour/flavour identification and product quality.

In the sensory laboratory, a clear distinction should be made between sensory analysts and sensory assessors. The sensory analyst organises, conducts and reports the results of sensory tests. The sensory assessor takes part in a sensory test as a member of a panel, a group of assessors chosen to participate in a sensory test. The division of responsibilities between various staff categories is described in the standard ISO/CD 13300 'Sensory analysis. General guidance for the staff of a sensory evaluation laboratory. Part 1: Staff structure and responsibilities'.[2]

2 Different Types of Sensory Tests

Laboratories can seek formal recognition (accreditation) for objective sensory methods, *i.e.* methods in which the effects of personal opinions are minimised, and in which a panel consisting of trained sensory assessors is used. Examples of objective sensory methods are:

Discriminative tests
Triangle test: a method of difference testing involving the simultaneous presentation of three samples, two of which are identical; the assessor is asked to identify the odd one out.
Paired comparison test: a method of difference testing in which samples are presented in pairs for comparison on the basis of one or more defined characteristics.
Duo-trio test: a method of difference testing in which the control is presented first, followed by two samples, one of which is identical to the control sample; the assessor is asked to identify the sample which is either the same as, or different from, the control.
Ranking test: a method of classification in which the assessor is asked to place a series of samples in order of intensity or degree of a given characteristic.

Descriptive tests
Intensity measurements: the assessors are asked to use numbers or words to express the intensity of a perceived attribute, or asked to react to the attribute.
Profile/quantitative descriptive analysis: the assessors are asked to detect and describe the perceived sensory attributes of a sample and to rate the intensity of each attribute.

As in chemical and microbiological testing, all test methods used in the sensory laboratory must be fully documented and validated, be kept up to date and be available to the staff. It is important that the laboratory, before taking any method, including national or international standard methods, into routine use, validates them in-house. It must test and document the method to confirm that it may be used competently by its laboratory staff, *i.e.* is suitable for its intended use, especially as regards its trueness and precision.

Affective tests, such as consumer preference tests and hedonic tests (tests which are relating to likes or dislikes), are used to assess the personal, subjective response (preference and/or acceptance of a product). At the present time, it is not considered possible to seek formal recognition for the competence (accreditation) for tests of this nature.

3 The Sensory Analyst

The general rules for personnel in a laboratory apply equally to the sensory analyst, *i.e.* minimum requirements on qualification, expertise and experience have to be defined, and staff authorised to organise, conduct and evaluate, as well as report the results of tests. Staff should also receive adequate training and records of the competence of each member of the staff should be kept.

In addition to more general requirements, sensory analysts should have a basic knowledge of the different types of sensory methods and should also be knowledgeable with respect to the recruiting, selection, screening, training and monitoring of sensory assessors. Expertise in experimental design and knowledge in statistics is also required, especially if no statistical support or advice is available in-house to the sensory analyst. Thus, in sensory analysis, it is particularly important that the analyst responsible for the organisation and conduct of the tests is fully acquainted with the general sets of rules for this particular type of test, for example:

- the minimum number of assessors in a panel;
- the importance of not providing any prior information on the samples;
- the impact of the environmental conditions and equipment used;
- the way in which sensory samples are prepared, coded and distributed; and
- the importance of distributing the samples in a randomised order.

Sensory analysts must take into consideration and give priority to the health and safety of sensory assessors. The sensory analyst is obliged to give full consideration to the welfare of the assessors and to instruct assessors to immediately report any ill effects. In addition, the sensory analyst must inform the assessor of any food to be included in the test which deviates from established growing or processing conditions, or which has been modified by genetic engineering, and thus to allow assessors, if they so wished, not to participant in such a test.

The sensory analyst should be aware of sensory fatigue, *i.e.* a form of sensory adaptation in which a decrease in sensitivity occurs. Sensory adaptation is considered to be a temporary modification of the sensitivity of a sense organ due to continued and/or repeated stimulation. If such effects, or any session fatigue or assessor discomfort, are noted in the monitoring of performance, the laboratory should pay careful attention to experimental designs, to the balanced presentation of samples and to allow sufficient time between tests. Sometimes the laboratory must reduce the number of samples per session, or the number of sessions per day.

The work and duties of 'the sensory analyst' can be—and often are—divided between the responsible sensory analyst (technical manager of the sensory laboratory), a panel leader and/or a panel technician. In such cases the distribution of responsibilities should be defined and documented.

Useful guidance on the qualifications of sensory analysts is given in ISO/CD 13300 'Sensory analysis. General guidance for the staff of a sensory evaluation laboratory. Part 2: Recruitment and training of panel leaders for descriptive analysis'.[2]

4 The Sensory Assessor

A sensory analysis panel should be regarded as a 'true measuring instrument', with the result of a test directly depending on its members and their competence. It is critical that the selection, training and monitoring of performance of sensory assessors are carried out with great care. Standard ISO 8586, Parts 1 and 2, give detailed guidance on these aspects.[3,4]

Naturally, it is a prerequisite that any sensory assessor candidate is interested and motivated to perform sensory assessments. Uninterested, inactive assessors, or assessors who are too busy to concentrate, should not be used in sensory analysis.

A number of recommended procedures for the selection, training and the regular monitoring after training of the performance of individual sensory assessors are given in the following sections. However, sensory laboratories need to adapt the procedures to their own specific needs.

Common to all of the recommendations is that all procedures are carefully and fully documented and compared to the pre-set requirements of the laboratory. Also the screening and the monitoring of results of each individual assessor need to be documented in order for the laboratory to be able to demonstrate that all sensory tests have been performed using sensory assessors competent to perform the test in question.

Recruitment, Preliminary Screening and Basic Training

The first requirement is that sensory assessors should have normal primary senses. It needs to be confirmed that the assessor candidate recognises and perceives odours and primary tastes, for example by training assessor candidates in order that they learn a number of basic odours and to identify the basic tastes: sweet, sour, bitter and salty. Where relevant, colour vision and the detection of specific taints/odours should also be confirmed. If the laboratory carries out descriptive tests, it is important to confirm that the sensory candidate has the ability to describe product characteristics. Finally, the candidate's ability to work in a team should be taken into consideration.

After a person has been found suitable to fulfil the basic requirements of an assessor, he or she should be declared qualified to act as a sensory assessor. It is good practice to record any allergies, preferences/strong dislikes, colour or flavour blindness, *etc.*, and to take these into account when selecting individual assessors to be included as members in a panel.

Training in General Principles and Methods

Candidate assessors should be trained in the use of senses, and be thoroughly familiar with the various methods in use. It is important that terminology and any scales in use are clearly explained. They should be made aware of the effect of extraneous/disturbing factors such as eating spicy food or smoking prior to a test, and the use of scents.

The laboratory should design and document its selection and training programme to ensure that assessors are adequately trained before they are included in a sensory panel. In order to evaluate the competence of an assessor under training, the trainee may participate 'as an outsider' in real tests, and results of the trainee may be plotted on the control chart used to monitor the performance of the regular panellists. A trainee result lying within the observed variation of the regular panel should be considered acceptable. Laboratories must decide and document how many 'acceptable performances' a candidate assessor should produce before the candidate is considered competent to be a regular sensory panellist. It is usual that laboratories require a minimum of two acceptable performances before declaring a candidate competent to be a regular panellist.

Training for Particular Purposes

For many foods the sensory assessors need to receive specific training, *e.g.* the analysis of fish. Laboratories performing sensory tests on such foods must confirm that the assessors participating in such assessments are able to meet the specialised requirements. This can be done by careful selection of assessors on the basis of, for example, results obtained by altering the concentration of the relevant constituent and by the analysis of replicate samples.

Monitoring of Individual Assessors

After a sensory assessor has been found to have the experience and competence to be a member of a sensory panel, the individual performance of the assessor should be monitored on a regular basis. Results, along with the date and information on the product being assessed, should be recorded, for example on a control chart, where individual results are plotted against averages. Deviating results produced by individual panellists should be dealt with and any corrective actions considered immediately. Laboratories should decide on criteria for the acceptance/rejection of an assessor, and, in the case of rejection, on further training programmes of the individual assessor. Re-training, and criteria for re-training, may also become necessary in cases where a sensory assessor has not carried out a test for a significant period of time.

For easy access and clarity, it is sensible to keep records separately and individually for each assessor. These records should include confirmation of the senses, history of training regarding various methods in use and performance when acting as a sensory assessor.

Quality Control of Methods and Panel Performance

Monitoring of panel performance, *i.e.* the quality control of the analytical system including performance of methods, procedures and measuring instrument (*i.e.* the panel), varies according to the type of samples, methods and frequency of determinations. Monitoring is important and laboratories should select the most appropriate method for their particular needs. The level should be sufficient to demonstrate the validity of results, and it naturally varies widely. As a guide, the level should, according to EAL (European Co-operation for Accreditation of Laboratories), be not less than 5% of the sample throughput, while 10% is considered appropriate for routine work. For non-routine work a greater percentage may be required. When the laboratory has documented evidence which supports the reliability of the method and the panel, the level of quality control may be reduced.

Criteria for accepting results of a panel should in all cases be set, as well as clear rules on which actions to take if a panel does not carry out its assessments reliably.

Panel performance may be established by carrying out on different occasions replicate analysis of samples as a defined percentage of the total number of samples analysed. However, this is not always possible or meaningful in the food laboratory since foods are spoiled and other materials, such as food contact materials, change with time.

Many laboratories see it as more useful to introduce random repeat samples into the sample analysis system which then enables the repeatability to be monitored.

It is not easy to obtain commercially available control materials having specific characteristics, nor is it easy to prepare such materials within the laboratory. If such a material is available, it could periodically be included in the tests on different occasions and the results recorded, *e.g.* on a control chart. This chart should contain warning and action limits, in the same way as charts for use in chemical or microbiological analysis. If the results obtained for the control material are outside the action limit, then the laboratory should consider rejecting the results obtained for the test samples on that particular session.

Successful participation in proficiency testing schemes would be an excellent way of demonstrating that a sensory panel achieves reproducible results. However, interlaboratory tests are infrequently carried out in sensory analysis, at least at the international level. Laboratories should, whenever possible, seek to co-operate with other laboratories performing the same types of tests. Analysing another laboratory's samples and a comparison of the results is at least a partial compensation for the lack of proficiency testing schemes.

Health Factors

Health and related factors that may affect the performance of the sensory assessors should be recorded. Such factors may include strong dislikes, allergic reactions, colds, upset stomachs, toothache, pregnancy, certain medications and psychological stress. Consideration should be given to not using assessors in tests if health factors are found to influence the testing.

5 Accommodation and Environment

Environmental conditions are particularly important in sensory work as they have an effect on the test results. If possible, a specific area should be set aside. The laboratory must provide environmental conditions and controls appropriate to the various tests in use.

Ideally the area in which sensory analysis is carried out should be a quiet, odourless area free from distractions, with controlled temperature, lighting, ventilation and humidity. Bright colours should be avoided in the testing area. The area should be equipped with partitions to allow assessors to concentrate without visual contacts with other assessors. Special attention should be given to the house-keeping of the testing area.

There should be separate sample preparation premises to which the sensory assessors do not have access. If at all possible, the testing area and the sample preparation room(s) should be near to each other in order to minimise transportation of samples during which, for example, the serving temperature may change. The design for test rooms for sensory analysis is described in the appropriate ISO standard.[5]

If it is not possible to fulfil all of the above criteria, *e.g.* on factory sites, the laboratory must be able to demonstrate that environmental conditions do not jeopardise the validity of the test results.

Critical environmental conditions should be monitored and documented together with action limits for the conditions and the instruction on which actions to take if limits are exceeded. Conditions must include temperature, and in many cases humidity, since a too-dry environment will dry the mucous membranes of the assessors, thus influencing test results.

6 Equipment

General rules apply to most of the equipment normally in use in the sensory laboratory. Special attention should be given to the cleaning of sample preparation equipment in order to avoid cross-contamination and tainting. Refrigerators in which samples are stored may need to be cleaned more often than other refrigerators in the laboratory.

Heating units and their performance should be determined and monitored; it may be necessary to investigate the temperature distribution within ovens.

The equipment used for serving test samples, irrespective of whether they are made of glass, pottery or paper, must be identical in a sensory session. Sample serving equipment must be thoroughly cleaned and be kept solely for the purpose of sensory analysis. It must be checked that any utensils used in the tests do not impart a taint. The use of marker pens which give off a strong odour is to be avoided when coding the sample containers.

7 Methods

The general rules regarding methods in use in a food laboratory equally apply in the sensory laboratory. It is the responsibility of the sensory analyst to select the most appropriate method for each analytical task.

Methods for sensory testing should include sections on:

- sensory assessor training requirements;
- sample preparation and presentation;
- sensory panel composition; and
- any special environmental conditions.

8 Records

In addition to the normal records to be kept in the food laboratory, additional records need to be kept on sensory tests. Records of sensory tests that have been carried out should include the following extra information:

- copies of the instruction sheets, questionnaires or forms given to the sensory assessors;
- the order of presentation of the samples to each individual assessor;
- the identity of all sensory assessors who participated in the test;
- reference to the selection and training records of all sensory assessors who participated in the test; and
- if needed, results of or reference to monitoring of environmental conditions.

9 Evaluation of Assessors and the Reliability of Results in Sensory Analyses

The importance of sensory analysis can also be confirmed by the following information which is to be used in the EC Milk Market Support sector.[6]

Determination of the 'Repeatability Index'

At least ten samples will be analysed as blind duplicates by an assessor within a period of 12 months. This will usually happen in several sessions. The results for individual product charateristics are evaluated using the following formula:

$$W_I = 1 + \frac{\sum(x_{i1} - x_{i2})^2}{n}$$

where W_I is the repeatability index, x_{i1} is the score for the first evaluation of sample x_i, x_{i2} is the score for the second evaluation of sample x_i, and n is the number of samples

The sample to be evaluated should reflect a broad quality range. W_I should not exceed 1.5 (on a 5 point scale).

Determination of the 'Deviation Index'

This index should be used to check whether an assessor uses the same scale for quality evaluation as an experienced group of assessors. The scores obtained by the assessor are compared with the average of the scores obtained by the assessor group. The following formula is used for the evaluation of results:

$$D_I = 1 + \frac{\sum[(x_{i1} - \bar{x}_{i1})^2 + (x_{i2} - \bar{x}_{i2})^2]}{2n}$$

where x_{i1} and x_{i2} are as defined previously, $\bar{x}_{i1}$ and $\bar{x}_{i2}$ are the average scores of the assessor group for the first and second evaluation respectively of sample x_i, and n is the number of samples (at least 10 per 12 months).

The samples to be evaluated should reflect a broad quality range. D_I should not exceed 1.5 (5 point scales).

Comparison of the Results Obtained in Different Regions of a Member State and in Different Member States

Where applicable, at least one test per year is organised which allows the comparison of the results obtained by assessors from different regions. If significant differences are observed, Member States are required to take the necessary steps to identify the reasons and to arrive at comparable results.

Member States are encouraged to organise tests which allow the comparison of the results obtained by their own assessors and by assessors from neighbouring Member States. Significant differences should lead to an in-depth investigation with the aim of arriving at comparable results.

Member States shall report on the results of these comparisons.

References

1 Commission Regulation 2568/91 of 11 July 1991 on the Characteristics of Olive Oil and Olive-Residue Oil and on the Relevant Methods of Analysis, *O. J., L248*, 5.9.91.

2 Sensory analysis—General guidance for the staff of a sensory evaluation laboratory. Part 1: Staff structure and responsibilities. Part 2: Recruitment and training of panel leaders for descriptive analysis, ISO/CD 13300, ISO, Geneva, DN/PF 05.11.96.

3 Sensory analysis—General guidance for the selection, training and monitoring of assessors—Part 1: Selected assessors, ISO 8586–1, ISO, Geneva,1993.

4 Sensory analysis—General guidance for the selection, training and monitoring of assessors—Part 2: Experts, ISO 8586–2, ISO, Geneva, 1994.

5 Sensory analysis—General guidance for the design of test rooms, ISO 8589, ISO, Geneva, 1988.

6 Guidelines for the Interpretation of Analytical Results and the Application of Sensory Evaluation in Relation to Milk and Milk Products under the Common Market Organisation, DG VI, Commission of the European Communities, Brussels, Document VI/2721/95 Rev. 1 of 20.07.95.

CHAPTER 14

Vocabulary, Terminology and Definitions

The definitions for terms that are used throughout this book and which are of direct concern to food analysts, together with the appropriate reference, are given below. In some instances a number of definitions for a single term are given as different organisations working in the sector are not agreed on the final wording.

The terms are divided into two parts, the first dealing with terminology which has recently been adopted by the Codex Alimentarius Commission[1] and the second the terminology of interest frequently used by various International Organisations of importance.

1 Analytical Terminology for Codex Use

Introduction

The purpose of this document is to provide harmonised analytical terminology specifically related to the work of the Codex Committee on Methods of Analysis and Sampling. It was recognised that various disciplines, *e.g.* analytical chemistry and metrology, use the same terms in different ways and this causes confusion. Even within the chemical field the meaning of the same term may vary among commodity areas. Therefore, the Codex Committee on Methods of Analysis and Sampling/Codex Alimentarius Commission needs to harmonise core terminology with clear definitions to ensure that analysts mean the same thing when using a particular term. The terminology was submitted to the Codex Alimentarius Commission and was adopted in June 1997. The adopted text is given here.

VOCABULARY[2]

RESULT: The final value reported for a measured or computed quantity, after performing a measuring procedure including all subprocedures and evaluations. {IUPAC-1994}

[1] See 'Report of the 22nd Session of the Codex Alimentarius Commission', FAO, Rome, 1997, ALINORM 97/37.

[2] The source of the definition, if identical or used with minor editorial revisions, is given in curly braces.

NOTES: {VIM}

1. When a result is given, it should be made clear whether it refers to:
   - the indication [signal]
   - the uncorrected result
   - the corrected result

   and whether several values were averaged.
2. A complete statement of the result of a measurement includes information about the uncertainty of measurement.

ACCURACY (AS A CONCEPT): The closeness of agreement between the reported result and the accepted reference value.

NOTE:
The term accuracy, when applied to a set of test results, involves a combination of random components and a common systematic error or bias component. {ISO 3534–1} When the systematic error component must be arrived at by a process that includes random error, the random error component is increased by propagation of error considerations and is reduced by replication.

ACCURACY (AS A STATISTIC): The closeness of agreement between a reported result and the accepted reference value. {ISO 3534–1}

NOTE:
Accuracy as a statistic applies to the single reported final test result; accuracy as a concept applies to single, replicate, or averaged value.

TRUENESS: The closeness of agreement between the average value obtained from a series of test results and an accepted reference value.

NOTES:

1. The measure of trueness is usually expressed in terms of bias. {ISO 3534–1}
2. Trueness has been referred to as 'accuracy of the mean'.

BIAS: The difference between the expectation of the test results and an accepted reference value. {ISO 3534–1}

NOTES:

1. Bias is the total systematic error as contrasted to random error. There may be one or more systematic error components contributing to bias. A larger systematic difference from the accepted reference value is reflected by a larger bias value. {ISO 3534–1}
2. When the systematic error component(s) must be arrived at by a process that includes random error, the random error component is increased by propagation of error considerations and reduced by replication.

PRECISION: The closeness of agreement between independent test results obtained under stipulated conditions. {ISO 3534–1}

NOTES: {ISO 3534–1}

1. Precision depends only on the distribution of random errors and does not relate to the true value or to the specified value.
2. The measure of precision is usually expressed in terms of imprecision and computed as a standard deviation of the test results. Less precision is reflected by a larger standard deviation.
3. 'Independent test results' means results obtained in a manner not influenced by any previous result on the same or similar test object. Quantitative measures of precision depend critically on the stipulated conditions. Repeatability and reproducibility conditions are particular sets of extreme conditions.

REPEATABILITY [REPRODUCIBILITY]: Precision under repeatability [reproducibility] conditions. {ISO 3534–1}

REPEATABILITY CONDITIONS: Conditions where test results are obtained with the same method on identical test items in the same laboratory by the same operator using the same equipment within short intervals of time. {ISO 3534–1}

REPRODUCIBILITY CONDITIONS: Conditions where test results are obtained with the same method on identical test items in different laboratories with different operators using different equipment. {ISO 3534–1}

NOTE:

When different methods give test results that do not differ significantly, or when different methods are permitted by the design of the experiment, as in a proficiency study or a material-certification study for the establishment of a consensus value of a reference material, the term 'reproducibility' may be applied to the resulting parameters. The conditions must be explicitly stated.

REPEATABILITY [REPRODUCIBILITY] STANDARD DEVIATION: The standard deviation of test results obtained under repeatability [reproducibility] conditions. {ISO 3534–1}

NOTES: {ISO 3534–1}

1. It is a measure of the dispersion of the distribution of test results under repeatability [reproducibility] conditions.
2. Similarly, 'repeatability [reproducibility] variance' and 'repeatability [reproducibility] coefficient of variation' could be defined and used as measures of the dispersion of test results under repeatability [reproducibility] conditions.

REPEATABILITY [REPRODUCIBILITY] LIMIT: The value less than or equal to which the absolute difference between two test results obtained under repeatability [reproducibility] conditions may be expected to be with a probability of 95%. {ISO 3534–1}

NOTES:

1. The symbol used is r [R]. {ISO 3534–1}
2. When examining two single test results obtained under repeatability [reproducibility] conditions, the comparison should be made with the repeatability [reproducibility] limit r [R] = 2.8 s_r[s_R]. {ISO 5725–6, 4.1.4}
3. When groups of measurements are used as the basis for the calculation of the repeatability [reproducibility] limits (now called the critical difference), more complicated formulae are required that are given in ISO 5725–6:1994, 4.2.1 and 4.2.2.

APPLICABILITY: The analytes, matrices and concentrations for which a method of analysis may be used satisfactorily to determine compliance with a Codex standard.

NOTE:
In addition to a statement of the range of capability of satisfactory performance for each factor, the statement of applicability (scope) may also include warnings as to known interference by other analytes, or inapplicability to certain matrices and situations.

SPECIFICITY: The property of a method to respond exclusively to the characteristic or analyte defined in the Codex standard.

NOTES:

1. Specificity may be achieved by many means: it may be inherent in the molecule (*e.g.* infrared or mass spectrometric identification techniques), or attained by separations (*e.g.* chromatography), mathematically (*e.g.* simultaneous equations) or biochemically (*e.g.* enzyme reactions). Very frequently, methods rely on the absence of interferences to achieve specificity (*e.g.* determination of chloride in the absence of bromide and iodide).
2. In some cases, specificity is not desired (*e.g.* total fat, fatty acids, crude protein, dietary fibre, reducing sugars).

SENSITIVITY: Change in the response divided by the corresponding change in the concentration of a standard (calibration) curve, *i.e.* the slope, s_i, of the analytical calibration curve.

NOTE:
This term has been used for several other analytical applications, often referring to capability of detection, to the concentration giving 1% absorption

in atomic absorption spectroscopy, and to the ratio of found positives to known, true positives in immunological and microbiological tests. Such applications to analytical chemistry should be discouraged.

NOTES: {IUPAC-1987}

1. A method is said to be sensitive if a small change in concentration, c, or quantity, q, causes a large change in the measure, x; that is, when the derivative $\mathrm{d}x/\mathrm{d}c$ or $\mathrm{d}x/\mathrm{d}q$ is large.
2. Although the signal s_i may vary with the magnitude of c_i or q_i, the slope, s_i, is usually constant over a reasonable range of concentrations. s_i may also be a function of the c or q of other analytes present in the sample.

RUGGEDNESS: The ability of a chemical measurement process to resist changes in results when subjected to minor changes in environmental and procedural variables, laboratories, personnel, *etc.* {IUPAC-1995}

The following set of definitions are from {IUPAC-1944–2}, with some minor revisions for clarity:

INTERLABORATORY STUDY: A study in which several laboratories measure a quantity in one or more 'identical' portions of homogeneous, stable materials under documented conditions, the results of which are compiled into a single document.

NOTE:
The larger the number of participating laboratories, the greater the confidence that can be placed in the resulting estimates of the statistical parameters. The IUPAC-1987 protocol (*Pure Appl. Chem.*, 1994, **66**, 1903) requires a minimum of eight laboratories for method–performance studies.

METHOD–PERFORMANCE STUDY: An interlaboratory study in which all laboratories follow the same written protocol and use the same test method to measure a quantity in sets of identical test samples. The reported results are used to estimate the performance characteristics of the method. Usually these characteristics are within-laboratory and among-laboratories precision, and when necessary and possible, other pertinent characteristics such as systematic error, recovery, internal quality control parameters, sensitivity, limit of determination, and applicability.

NOTES:

1. The materials used in such a study of analytical quantities are usually representative of materials to be analysed in actual practice with respect to matrices, amount of test component (concentration), and interfering components and effects. Usually the analyst is not aware of the actual composition of the test samples but is aware of the matrix.
2. The number of laboratories, number of test samples, number of determi-

nations, and other details of the study are specified in the study protocol. Part of the study protocol is the procedure which provides the written directions for performing the analysis.
3. The main distinguishing feature of this type of study is the necessity to follow the same written protocol and test method exactly.
4. Several methods may be compared using the same test materials. If all laboratories use the same set of directions for each method and if the statistical analysis is conducted separately for each method, the study is a set of method–performance studies. Such a study may also be designated as a method-comparison study.

LABORATORY–PERFORMANCE (PROFICIENCY) STUDY: An interlaboratory study that consists of one or more measurements by a group of laboratories on one or more homogeneous, stable, test samples by the method selected or used by each laboratory. The reported results are compared with those from other laboratories or with the known or assigned reference value, usually with the objective of improving laboratory performance.

NOTES:
1. Laboratory–performance studies can be used to support accreditation of laboratories or to audit performance. If a study is conducted by an organisation with some type of management control over the participating laboratories—organisational, accreditation, regulatory, or contractual—the method may be specified or the selection may be limited to a list of approval or equivalent methods. In such situations, a single test sample is insufficient to judge performance. It is expected that the results from one of every 20 tests will be outside the value for the calculated mean, $\pm$ twice the standard deviation, due solely to random fluctuations.
2. Sometimes a laboratory–performance study may be used to select a method of analysis that will be used in a method–performance study. If all laboratories, or a sufficiently large subgroup of laboratories, use the same method, the study may also be interpreted as a method–performance study, provided that the samples cover the range of concentration of the analyte.
3. Separate laboratories of a single organisation with independent facilities, instruments and calibration materials, are treated as different laboratories.

MATERIAL-CERTIFICATION STUDY: An interlaboratory study that assigns a reference value ('true value') to a quantity (concentration or property) in the test material, usually with a stated uncertainty.

NOTE:
A material-certification study often utilises selected reference laboratories to analyse a candidate reference material by a method(s) judged most likely to provide the least-biased estimates of concentration (or of a characteristic property) and the smallest associated uncertainty.

Source Materials

1. International Vocabulary of Basic and General Terms in Metrology, 2nd Edition, International Organization for Standardization, Geneva, Switzerland, 1993, 58 pp.
2. ISO 3534–1:1993, Statistics—Vocabulary and Symbols—Part 1: Probability and general statistical terms, International Organization for Standardization, Geneva, 47 pp.
3. ISO 3534–2:1993, Statistics—Vocabulary and Symbols—Part 2: Statistical quality control. International Organization for Standardization, Geneva, 33 pp.
4. ISO 5725–6:1994, Accuracy (trueness and precision) of measurement methods and results—Part 6: Use in practice of accuracy values, International Organization for Standardization, Geneva, 41 pp.
5. IUPAC-1987, Compendium of Analytical Nomenclature. Definitive Rules 1987, H. Freiser and G.H. Nancollas, 2nd Edition, Blackwell Scientific Publications, Oxford, 1987.
6. IUPAC-1994, Nomenclature for the Presentation of Results of Chemical Analysis, (IUPAC Recommendations 1994), L.A. Currie and G. Svehla, *Pure Appl. Chem.*, 1994, **66**, 595.
7. IUPAC-1994–2, Nomenclature of Interlaboratory Analytical Studies (IUPAC Recommendations 1994) W. Horwitz, *Pure Appl. Chem.*, 1994, **66**, 1903.
8. IUPAC-1995, Nomenclature in Evaluation of Analytical Methods Including Detection and Quantification Capabilities. (IUPAC Recommendations 1995) L.A. Currie, *Pure Appl. Chem.*, 1994, **67**, 1699.

2 Terminology of Interest Frequently Used by Various International Organisations of Importance

ACCURACY: Closeness of the agreement between the result of a measurement and a true value of the measurand.

NOTES: 1. Accuracy is a qualitative concept.
2. The term *precision* should not be used for *accuracy*.

(*From 'International vocabulary for basic and general terms in metrology', 2nd Edition, ISO, Geneva, 1993.*)

ANALYTICAL SYSTEM: Range of circumstances that contribute to the quality of analytical data, including equipment, reagents, procedures, test materials, personnel, environment and quality assurance measures.

(*From 'The harmonised guidelines for internal quality control in analytical chemistry laboratories', Pure Appl. Chem., 1995, **67**, 649.*)

ANALYTICAL RUN: See 'Run'.

ASSIGNED VALUE: The value to be used as the true value by the proficiency test co-ordinator in the statistical treatment of results. It is the best available estimate of the true value of the analyte in the matrix.

(*From 'The International Harmonised Protocol for the Proficiency Testing of (Chemical) Analytical Laboratories', Pure Appl. Chem., 1993,* ***65****, 2123.*)

BIAS: Difference between the expectation of the test results and an accepted reference value.

NOTE:
Bias is a systematic error as contrasted to random error. There may be one or more systematic error components contributing to the bias. A larger systematic difference from the accepted reference value is reflected by a larger bias value.

(*From 'Statistics, vocabulary and symbols—Part 1: Probability and general statistical terms', ISO 3534–1:1993, ISO, Geneva (E/F)*)

BIAS OF THE MEASUREMENT METHOD: The difference between the expectation of test results obtained from all laboratories using that method and an accepted reference value.

NOTE:
One example of this in operation would be where a method purporting to measure the sulfur content of a compound consistently fails to extract all the sulfur, giving a negative bias to the measurement method. The bias of the measurement method is measured by the displacement of the average of results from a large number of different laboratories all using the same method. The bias of a measurement method may be different at different levels.

(*From 'The International Harmonised Protocol for the Proficiency Testing of (Chemical) Analytical Laboratories', Pure Appl. Chem., 1993,* ***65****, 2123.*)

CERTIFIED REFERENCE MATERIAL: Reference material, accompanied by a certificate, one or more of whose property values are certified by a procedure which establishes its traceability to an accurate realisation of the unit in which the property values are expressed, and for which each certified value is accompanied by an uncertainty at a stated level of confidence.

(*From 'International vocabulary for basic and general terms in metrology', 2nd Edition, ISO, Geneva, 1993.*)

CONTROL MATERIAL: Material used for the purposes of internal quality control and subjected to the same or part of the same measurement procedure as that used for test materials.

(*From 'The harmonised guidelines for internal quality control in analytical chemistry laboratories', Pure Appl., Chem. 1995,* ***67****, 649.*)

CO-ORDINATOR: The organisation with responsibility for co-ordinating all of the activities involved in the operation of a proficiency testing scheme.

(*From 'The International Harmonised Protocol for the Proficiency Testing of (Chemical) Analytical Laboratories', Pure Appl. Chem., 1993,* ***65****, 2123.*)

EMPIRICAL METHOD OF ANALYSIS: A method which determines a value that can be arrived at only in terms of the method *per se* and serves by definition as the only method for establishing the measurand. (Sometimes called 'defining method of analysis'.)

(*From the draft 'Harmonised Guidelines for the Use of Recovery Information in Analytical Measurement', IUPAC, Research Triangle Park, NC, USA, 1997.*)

ERROR: Result of a measurement minus a true value of the measurand.

(*From 'International vocabulary for basic and general terms in metrology', 2nd Edition, ISO, Geneva, 1993.*)

FITNESS FOR PURPOSE: Degree to which data produced by a measurement process enables a user to make technically and administratively correct decisions for a stated purpose.

(*From 'The harmonised guidelines for internal quality control in analytical chemistry laboratories', Pure Appl. Chem., 1995,* ***67****, 649.*)

INTERLABORATORY TEST COMPARISONS: Organisation, performance and evaluation of tests on the same items or materials on identical portions of an effectively homogeneous material, by two or more different laboratories in accordance with pre-determined conditions.

(*From 'The International Harmonised Protocol for the Proficiency Testing of (Chemical) Analytical Laboratories', Pure Appl. Chem., 1993,* ***65****, 2123.*)

INTERNAL QUALITY CONTROL: Set of procedures undertaken by laboratory staff for the continuous monitoring of operations and the results of measurements in order to decide whether results are reliable enough to be released. IQC primarily monitors the batchwise accuracy of results on quality control materials, and precision on independent replicate analysis of test materials.

(*From 'The harmonised guidelines for internal quality control in analytical chemistry laboratories', Pure Appl. Chem., 1995,* ***67****, 649.*)

LABORATORY BIAS: The difference between the expectation of the test results from a particular laboratory and an accepted reference value.

(*From 'The International Harmonised Protocol for the Proficiency Testing of (Chemical) Analytical Laboratories', Pure Appl. Chem., 1993,* ***65****, 2123.*)

LABORATORY COMPONENT OF BIAS: The difference between the laboratory bias and the bias of the measurement method.

NOTES:

1. The laboratory component of bias is specific to a given laboratory and the conditions of measurement within the laboratory, and it may also be different at different levels of the test.
2. The laboratory component of bias is relative to the overall average result, not the true or reference value.

(*From 'The International Harmonised Protocol for the Proficiency Testing of (Chemical) Analytical Laboratories', Pure Appl. Chem., 1993,* ***65****, 2123.*)

MEASUREMENT UNCERTAINTY: Parameter associated with the result of a measurement, that characterises the dispersion of the values that could reasonably be attributed to the measurand.

NOTES:

1. The parameter may be, for example, a standard deviation (or a given multiple of it), or the half-width of an interval having a stated level of confidence.
2. Uncertainty of measurement comprises, in general, many components. Some of these components may be evaluated from the statistical distribution of results of a series of measurements and can be characterised by experimental standard deviations. The other components, which can also be characterised by standard deviations, are evaluated from assumed probability distributions based on experience or other information.
3. It is understood that the result of a measurement is the best estimate of the value of a measurand, and that all components of uncertainty, including those arising from systematic effects, such as components associated with corrections and reference standards, contribute to the dispersion.

(*From 'Guide to the expression of uncertainty in measurement', ISO, Geneva, 1993.*)

NATIVE ANALYTE: Analyte incorporated into the test material by natural processes and manufacturing procedures (sometimes called 'incurred analyte'). Native analyte includes incurred analyte and incurred residue as recognised in some sectors of the analytical community. It is so defined to distinguish it from analyte added during the analytical procedure.

(From the draft 'Harmonised Guidelines for the Use of Recovery Information in Analytical Measurement', IUPAC, Research Triangle Park, NC, USA, 1997.)

PRECISION: Closeness of agreement between independent test results obtained under prescribed conditions.

NOTES:

1. Precision depends only on the distribution of random errors and does not relate to the accepted reference value.
2. The measure of precision is usually expressed in terms of imprecision and computed as a standard deviation of the test results. High imprecision is reflected by a larger standard deviation.
3. 'Independent test results' means results obtained in a manner not influenced by any previous result on the same or similar material.

(From 'Terms and definitions used in connections with reference materials', ISO Guide 30:1992, ISO, Geneva.)

PRECISION PARAMETERS: REPEATABILITY

r = Repeatability, the value below which the absolute difference between two single test results obtained under repeatability conditions (*i.e.* same sample, same operator, same apparatus, same laboratory, and short interval of time) may be expected to lie within a specific probability (typically 95%); $r = 2.8 \times s_r$.

(From 'AOAC International'.)

s_r = Standard deviation, calculated from results generated under repeatability conditions.

(From 'AOAC International'.)

RSD_r = Relative standard deviation, calculated from results generated under repeatability conditions $[(s_r/\bar{x}) \times 100]$, where $\bar{x}$ is the average of results over all laboratories and samples.

(From 'AOAC International'.)

PRECISION PARAMETERS: REPRODUCIBILITY

R = Reproducibility, the value below which the absolute difference between single test results obtained under reproducibility conditions (*i.e.* on identical material obtained by operators in different laboratories, using the standardised test method), may be expected to lie within a certain probability (typically 95%); $R = 2.8 \times s_R$.

(*From 'AOAC International'.*)

s_R = Standard deviation, calculated from results under reproducibility conditions.

(*From 'AOAC International'.*)

RSD_R = Relative standard deviation calculated from results generated under reproducibility conditions $[(s_R/\bar{x}) \times 100]$.

(*From 'AOAC International'.*)

PROFICIENCY TESTING SCHEME: Methods of checking laboratory testing performance by means of interlaboratory tests. It includes comparison of a laboratory's results at intervals with those of other laboratories, with the main object being the establishment of trueness.

(*From 'The International Harmonised Protocol for the Proficiency Testing of (Chemical) Analytical Laboratories', Pure Appl. Chem., 1993,* ***65****, 2123.*)

QUALITY ASSURANCE: All those planned and systematic actions necessary to provide adequate confidence that a product or service will satisfy given requirements for quality.

(*From 'Quality assurance and quality management—vocabulary', ISO 8402:1994, ISO, Geneva.*)

QUALITY ASSURANCE PROGRAMME/SYSTEM: The sum total of a laboratory's activities aimed at achieving the required standard of analysis. While IQC and proficiency testing are very important components, a quality assurance programme must also include staff training, administrative procedures, management structure, auditing, *etc.* Accreditation bodies judge laboratories on the basis of their quality assurance programme.

(*From 'The International Harmonised Protocol for the Proficiency Testing of (Chemical) Analytical Laboratories', Pure Appl. Chem., 1993,* ***65****, 2123.*)

RATIONAL METHOD OF ANALYSIS: A method which determines an identifiable chemical(s) or analytes(s), for which there may be several equivalent methods of analysis available.

(From the draft 'Harmonised Guidelines for the Use of Recovery Information in Analytical Measurement', IUPAC, Research Triangle Park, NC, USA, 1997.)

RECOVERY: Proportion of the amount of analyte, present in or added to the analytical portion of the test material, which is extracted and presented for measurement.

(From the draft 'Harmonised Guidelines for the Use of Recovery Information in Analytical Measurement', IUPAC Research Triangle Park, NC, USA, 1977.)

REFERENCE MATERIAL: Material or substance, one of whose property values is sufficiently homogeneous and well established to be used for the calibration of an apparatus, the assessment of a measurement method, or for assigning values to materials.

(From 'International vocabulary for basic and general terms in metrology', 2nd Edition, 1993, ISO, Geneva.)

REPEATABILITY CONDITIONS: Conditions where independent test results are obtained with the same method on identical test items in the same laboratory by the same operator using the same equipment within short intervals of time.

(From 'Statistics, vocabulary and symbols—Part 1: Probability and general statistical terms', ISO 3534–1: 1993, ISO, Geneva.)

RUN (ANALYTICAL RUN): Set of measurements performed under repeatability conditions.

(From 'The harmonised guidelines for internal quality control in analytical chemistry laboratories', Pure Appl. Chem., 1995, 67, 649.)

SURROGATE: Pure compound or element added to the test material, the chemical and physical behaviour of which is taken to be representative of the native analyte.

(From the draft 'Harmonised Guidelines for the Use of Recovery Information in Analytical Measurement', IUPAC, Research Triangle Park, NC, USA, 1977.)

SURROGATE RECOVERY: Recovery of a pure compound or element specifically added to the test portion or test material as a spike. (Sometimes called 'marginal recovery', IUPAC, Research Triangle Park, NC, USA, 1977.)

(*From the draft 'Harmonised Guidelines for the Use of Recovery Information in Analytical Measurement'.*)

TARGET VALUE FOR STANDARD DEVIATION: A numerical value for the standard deviation of a measurement result, which has been designated as a goal for measurement quality.

(*From 'The International Harmonised Protocol for the Proficiency Testing of (Chemical) Analytical Laboratories', Pure Appl. Chem., 1993, **65**, 2123.*)

TESTING LABORATORY: A laboratory that measures, examines, tests, calibrates or otherwise determines the characteristics or performance of materials or products.

(*From 'The International Harmonised Protocol for the Proficiency Testing of (Chemical) Analytical Laboratories', Pure Appl. Chem., 1993, **65**, 2123.*)

TRACEABILITY: Property of the result of a measurement or the value of a standard whereby it can be related to stated references, usually national or international standards, through an unbroken chain of comparisons all having stated uncertainties.

(*From 'International vocabulary for basic and general terms in metrology', 2nd Edition, 1993, ISO, Geneva, 1993.*)

TRUE VALUE: The actual concentration of the analyte in the matrix.

(*From 'The International Harmonised Protocol for the Proficiency Testing of (Chemical) Analytical Laboratories', Pure Appl. Chem., 1993, **65**, 2123.*)

TRUENESS: Closeness of the agreement between the average value obtained from a large series of test results and an accepted reference value.

NOTE:
The measure of trueness is usually expressed in terms of bias.

(*From 'Statistics, vocabulary and symbols—Part 1: Probability and general statistical terms', ISO 353–1: 1993, ISO, Geneva.*)

Subject Index

Accreditation, 231–294
  apparatus and equipment, 231–238
  confidentiality, 217
  EN 45000 series, 5, 17–20
  food control laboratories, 8–9
  housekeeping, 218
  ISO Guide, 7, 17–21, 25–26
  laboratory,
    facilities, 216
    staff, 214–215
  quality co-ordinator, 213–214
  requirements,
    management, 210–211
    methods, 42–44
    staff, 212
    waste disposal, 219
    working environment, 216
Additional Measures Official Control of Foodstuffs Directive, 5–8
Apparatus and reagents, 231–238
  equipment,
    measuring, 236
    volumetric, 235
  general requirements, 233–235
  glass- and plasticware, 237
  reference standards and reference materials, 236
  water, chemicals and gases, 237

Calibration, 238–246
  balances, 242
  complete analytical procedure, 244
  different degrees, 239
  frequency of checks, 241
  pipettes, 243
  programme, 240–241
  thermometers, 243
  user certified reference materials, 245
Certification – EN ISO 9000, 21–24
Collaborative trials, 56–61
  micro valve protocol, 64
  number of materials, 57
  number of replicates, 58
  other procedures for method validation, 62–64
  outlier treatment, 60
  participants, 56
  replicas, 58
  sample,
    homogeneity, 57
    type, 57
  statistical analysis, 59
  statistics, 98–105
Computers, 246–253
  archiving, 251
  backing-up procedures, 252
  critical considerations, 246
  data, 250
    identification of, 251
    storage of, 252
  equipment, 250
  quality assurance, 250–253
  software programs, 248
  storage media, 252
  validation, 247
    before use, 248
  viruses, 250
Criteria approach for methods of analysis, 51–56

Definitions, 295–308
  accuracy, 296, 301

Definitions, *continued*
analytical system, 301
applicability, 298
assigned value, 302
bias, 302
certified reference material, 302
collaborative trial, 299
control material, 303
co-ordinator, 303
empirical method of analysis, 303
error, 303
fitness for purpose, 303
inter-laboratory test comparisons, 303
internal quality control, 303
laboratory,
bias, 304
component bias, 304
performance study, 300
measurement uncertainty, 304
native analyte, 305
precision, 296, 305
proficiency testing scheme, 306
quality assurance, 306
rational method of analysis, 307
recovery, 307
reference material, 307
repeatability, 297, 305
conditions, 307
limit, 297
standard deviation, 297
reproducibility, 297, 306
standard deviation, 297
result, 295
ruggedness, 299
run, 307
sensitivity, 298
specificity, 298
study,
inter-laboratory, 299
laboratory performance, 300
material certification, 300
method performance, 299
proficiency, 300
surrogate, 307
recovery, 308
target value for standard deviation, 308
testing laboratory, 308
traceability, 308
trueness, 296, 308
true value, 308

Documentation, 254–264
analytical methods, 254–257
care, 257–260
reports, 261–264

Eco Management and Audit Scheme (EMAS), 29
Environmental management standards, EN ISO 14000, 28–29

Food,
analysis, 2
analysis laboratories – European Union, 4–6
examination, 2

Good Laboratory Practice, 24–27

Horwitz ratio, 64–66

In-house method validation, 66–97
AOAC International, 93–97
EURACHEM, 88–92
Inspectorate for Health Protection, The Netherlands, 97
NMKL, 68–87
protocols, 66–67
Internal quality control, 13, 148–171
basic concepts, 149–150
blank determinations, 163
choice of analytical method, 153, 184
material from proficiency testing, 161
preparation of control materials, 159–160
proficiency testing, 154
recommendations, 167–169
scope, 151–152
Shewart control charts, 163–167
estimation of statistical parameters, 164–166
interpretation, 166
traceability, 163
true value,
by analysis, 161
by formulation, 161
uncertainty, 152–153
use of reference materials, 159–160
within-run precision, 157,
Internal Harmonised Guidelines,

for internal quality control, 7, 148–169
for recovery corrections, 108–124
International Harmonised Protocol,
for proficiency testing, 7, 174–195
for validation of methods of analysis, 56

Laboratory quality assurance, 34–35
Legislative requirements,
for food analysis laboratories, 4
Codex Alimentarius Commission, 6–8

Measurement uncertainty, 12, 41–42, 125–147
accreditation agencies' views, 129–130
alternative approaches, 129
Codex approach, 132–133
ISO/EURACHEM approach, 126–129
MAFF project, 130
NMKL approach, 135–147
Methods of analysis, 36–107
competency, 39
precision characteristics, 64–66
validation, 38
verification, 39
Methods of analysis formal quality requirements,
AOAC, 44
CEN, 48–49
Codex Alimentarius Commission, 45–47
European Union, 47–48
Microbiological methods, 50
Mutual recognition, 2

National accreditation agencies, 30–33
Needs of customers, 10–11
New approach, 2

Official Control of Foodstuffs Directive, 4

Proficiency testing, 13, 172–202
assigned value, 182–183
estimation of, 186
choice of analytical method, 153, 184
combination results, 188–190
commercial samples, 201
commercial schemes, 195–198
essential elements, 175–176
estimation of assigned value, 186
examples, 192–195
FAPAS, 198
food control laboratories, 9–10
formation of z-scores,
by perception, 187
by prescription, 187
by reference to generalised model, 188
by reference to validated methodology, 188
framework, 178
interpretation of z-scores, 188
ISO Guide 43, 27
limitations, 177
NFMPT, 199
organisation, 178
quality manual, 175
ranking, 192
running scores, 190–192
statistical procedures, 186–192
test materials, 178–181
Proficiency testing and accreditation, 174

Quality assurance measures, 11
Quality audits, 264–276
auditor, 269
conducting an internal audit, 270
continuous improvements, 276–283
PDCA cycle, 276–278
corrective action, 275
cost of internal/external failures, 279–281
different types, 265
employee involvement, 283
evaluation of analytical efficiency, 281
external audits from the laboratory's point of view, 273–274
follow-up, 274–276
horizontal and vertical, 268
internal quality control and proficiency testing, 278–279
management responsibility, 283
management review, 276
planning, 267
processes of auditing, 266
timely reporting of results, 281

Recovery corrections, 12, 108–124
arguments against corrections, 115

Recovery corrections, *continued*
  arguments for corrections, 114
  empirical procedures, 115
  estimation, 116–118
  internal standards, 113
  isotope dilution, 112
  matrix mismatch, 113
  reference materials, 111
  spiking, 112
  surrogates, 111
  uncertainty, 118–123
Respecting a limit, 14, 203–209
  analytical result,
    from indirect analysis, 204
    from routine methods, 205
    from single analysis, 204
  results of analysis in dispute, 208
  using validated methods, 203

Sampling, 15, 220–221
  anonymous, 227
  checking of sample description, 225
  equipment and packaging, 223
  labelling, 223
  laboratory samples, 223
  legal action, 227
  official control, 224
  packaging and transportation, 225
  pre-treatment,
    for chemical analysis, 228
    for microbiology examination, 229
    for sensory evaluation, 229
  receipt in the laboratory, 225
  representative samples, 222
  storage, 226
Sensory analysis, 285–293
  accommodation and environment, 291
  analyst, 286–287
  comparison of results, 293
  deviation index, 293
  different types, 286
  equipment, 291
  evaluation of assessors, 292
  health factors, 290
  methods, 292
  monitoring of individual assessors, 289
  preliminary screening of basic training of assessors, 288
  quality control of performance, 289–290
  records, 292
  reliability of results, 292
  repeatability index, 292
  sensory assessor, 288
  training, 288
Standard Operating Procedures (SOP), 40–41

Total Quality Management (TQM), 29–30

50 natural ways to cure a headache

50 natural ways to cure a headache

Raje Airey

LORENZ BOOKS

contents

INTRODUCTION 6

HEADACHE TREATMENTS 10

LIVE IN HARMONY...

1) Refreshing water 12
2) Regular exercise 13
3) Easy neck stretches 14
4) Relaxing yoga positions 15
5) Healthy eating 16
6) Quick-fix snacks 17
7) Identifying food intolerance 18
8) Green leaf cleanser 19
9) Have more sex 20
10) Get a good night's rest 21
11) Keep a pet 22
12) Laughter is the best medicine 23

...AND IN PEACE

13) Crystal healing 24
14) Headache healing spell 25

50 natural ways to...
cure a headache

15) Blue colour therapy 26
16) Green colour therapy 27
17) Relax your body 28
18) Relax your mind 29
19) Protective bubble visualization 30
20) Gentle candlelight 31
21) Clear the clutter 32
22) Healing homeopathy 33
23) Reiki headache treatment 34

NATURE'S MEDICINE CHEST
24) Healing herb tub 35
25) Gentle lavender & rosemary oils 36
26) Soothing bath water 37
27) A relaxing atmosphere 38
28) Soothing away eyestrain 39
29) Feverfew 40
30) Evening primrose 41
31) Herbal steam inhalant 42
32) Comforting compress 43
33) Lime blossom & lemon balm tisane 44
34) Wood betony infusion 45
35) Herbal sedatives 46
36) Lavender tincture 47
37) Hangover headache treatments 48
38) Bach flower remedies 49

SMOOTHING HEADACHES AWAY
39) Calming head massage 50
40) Neck & shoulder easer 51
41) Shoulder reliever 52
42) Anxiety calmer 53
43) Shiatsu massage 54
44) Foot reflexology 55
45) Scalp massage 56
46) Head revitalizer 57
47) Migraine easers 58
48) Calming sleep massage 59
49) Draw out the pain 60
50) Tension reliever 61
INDEX 62

introduction

Almost everyone knows what it is like to have a headache. It is thought that more than 90 per cent of the population will have experienced a headache at one time or another, and unfortunately for many people they are almost a routine part of life.

There are hundreds of different causes of headaches, both psychological and physiological, and many different types of headache, ranging in severity from a crippling migraine which may last several days to a hangover headache which can clear up in a few hours. In some cases, headaches may indicate a major disorder such as a brain tumour or a life-threatening illness such as meningitis, but this is extremely rare. Most headaches seen by the family doctor are known as "benign recurring headaches"; the vast majority of these are described as "tension headaches".

muscle restriction

Tension headaches are so-called because they are usually caused by some type of physical tension in the muscles of the shoulders, neck and head, and by constriction or congestion of the blood vessels in the head. The pain typically arises from the base of the skull (occiput) and extends up over the back of the head to the forehead and temples. The pain results from the continuous, partial contraction of muscles attached to the scalp and can affect the whole head.

◂ *Use essential oil to help relieve a headache. Put a few drops in your bath water, on a handkerchief, or use an aromatherapy oil burner.*

Some people wake up with a headache, which then lasts all day with varying degrees of severity, ranging from a general dull ache to sudden jabbing pains in a particular spot. Other people experience this type of headache as a feeling of pressure, like a tight band around the head, or as a persistent throbbing. Although tension headaches are not associated with visual disturbance, many sufferers dislike bright light and find it hard to concentrate.

migraines

Many regular headache sufferers describe their condition as "having a migraine". However, a migraine is a specific medical condition and is not the same as an everyday tension headache. Migraines are a fairly common neurological disorder, with three times as many women as men suffering. Many women's migraines occur premenstrually and are linked to hormonal imbalances.

The severe pain of a migraine headache is thought to be caused by the dilation (swelling up) of the blood vessels in the head, causing a disturbance in the flow of blood to the brain. This follows a brief period of constriction of the vessels which partly accounts for the visual disturbances (known as "aura") that many people experience prior to the headache. Migraines cause chemical changes in the body and typical symptoms include aversion to light (photophobia), nausea, vomiting and diarrhoea. The headache itself is often one-sided and is marked by severe pain in the forehead or temples, or rising up from the back of the neck. A migraine attack can last up to 72 hours and becomes extremely debilitating. It can take a day or two after an attack to get back to normal.

bouts of pain

Cluster headaches are often confused with migraine as the severe pain tends to be centred around the eye area and is typically one-sided. These headaches occur in bouts during a 1–2 month period. In an attack, the headache will come on suddenly and

▸ *If you have a busy lifestyle, just a few minutes spent meditating each day will reduce the number of headaches you have.*

last up to an hour. Several attacks may be experienced in a day, often waking the sufferer from sleep, or causing them to pace about as the pain is so intense. Cluster headaches are less common than migraines; they rarely occur in anyone under 30 and most sufferers are men. Drinking alcohol during a bout can bring on an attack.

a symptom of another illness

Other types of headache arise as secondary symptoms of problems elsewhere in the body. This may include mechanical injuries, such as whiplash or head injuries, or general wear and tear on the body, caused by failing eyesight, poor posture, or arthritis in the neck, for instance. Similarly, headaches associated with premenstrual syndrome (PMS), high blood pressure, blocked sinuses, inflammation of the middle ear and viral infections are also fairly common.

headache triggers

Most headaches do not come out of the blue, but are triggered by certain factors. Migraines, cluster and tension headaches are typically linked to stress, overwork and negative emotional states such as worry, anxiety, depression and held-in anger and resentment. After stress, food allergies and/or intolerances are one of the most common causes of migraine and tension headaches. Certain foods, such as red wine, cheese and chocolate, are well-known triggers. Others include low blood sugar, caffeine withdrawal, lack of sleep and toxicity in the body, caused by poor digestion or over-indulgence in sugar and alcohol, for instance. Long-distance driving, too much sun, changes in the weather and sensitivities to environmental triggers, such as perfume, car exhaust fumes, cigarette smoke and paint fumes, can also play a part.

nature's remedies

Conventional treatment for a headache is standard painkillers, either available on prescription for migraine and severe headaches, or over-the-counter for everyday tension headaches. The drugs often combine painkillers, such as aspirin or paracetamol, with other drugs which have a sedative, antispasmodic action. Because these drugs are so common and easily available, we tend to think they are harmless and that we can take them every day. While these drugs have their place, they also have many potentially harmful side effects, particularly if taken on a regular basis. It is believed that many people are addicted to painkillers, and that many headaches are caused as the result of taking too many pills. For this reason, increasing numbers of people are turning to natural medicine as they look for effective treatments which are non-habit forming and non-toxic.

This book contains information and ideas for how to treat headaches without using drugs. In natural medicine, pain is seen as the body's way of telling you that something is

wrong; it has a protective function, acting as an "early warning system". Consequently, many of the treatments are based on dealing with the underlying cause of the pain and recognizing the importance of diet, lifestyle and psychological factors. Natural remedies can be used to help the body work through its own healing process with minimum discomfort. Use it as a guide to find a treatment that is just right for you.

▲ *Most people appreciate the soft and subtle scent of flowers. Their gentle perfume can help relieve a headache.*

headache treatments

The treatments in this book are based on holistic principles. This means they have an underlying assumption that good health depends on a balance between physical, mental, emotional and spiritual wellbeing. The treatments are drawn from a wide variety of natural-healing traditions; some of these practices have been used for centuries as an aid to health and wellbeing. They are all based on therapeutic techniques which help to stimulate the body's own natural healing ability, and include hands-on therapies such as massage, shiatsu and reflexology as well as treatments based on herbal remedies, aromatherapy and nutrition. There are also treatments to relieve stress and tension, such as meditation and yoga, and subtle energy healing methods using reiki, crystals and colour.

The treatments in this book are safe to use in the manner in which they are described. They are not intended as a substitute for any medicines prescribed by your doctor. If you are taking prescribed medication, you should always check with your doctor first before trying other treatments.

1

refreshing water

The body of a healthy adult is made up of about 75 per cent water. Water is vital for life, yet many common health problems, such as headaches, are linked with dehydration.

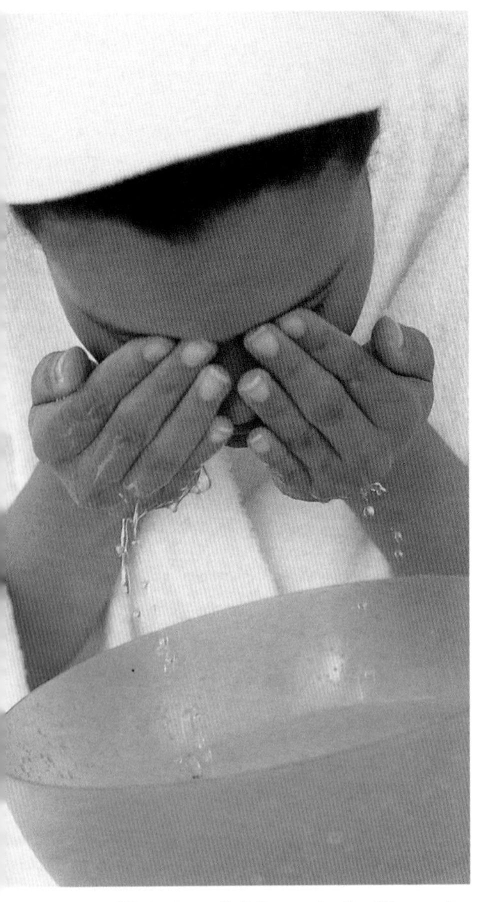

▲ *Water is revitalizing and refreshing and an excellent panacea for a headache.*

Start the day with a glass of water. This flushes out your kidneys and detoxifies your system. Water is best drunk half an hour before eating and between meals, to allow for the flushing action and to avoid interfering with the body's digestive processes.

At the first sign of a headache, drink a couple of glasses of water. Often this is enough for it to lift without the need for further treatment. Add a slice of lemon for a refreshing tang and a burst of vitamin C to kick-start the liver. Drinking water when stressed or anxious also helps to keep your body fluids flowing smoothly and can help to calm you down.

To dispel headaches brought on by eyestrain, try splashing the eyes and forehead with warm and then cold water. This stimulates the circulation and refreshes tired eyes.

KEEP HYDRATED

Alcohol, tea, coffee and fizzy drinks are diuretics; for every drink it is recommended to drink at least one glass of water to counter its effect.

2

regular exercise

Exercise is one of the best stress-busting methods around. During exercise, the body burns off excess adrenalin, so if you suffer from tension headaches, make exercise an everyday part of life.

When we are stressed, the body responds by producing extra adrenalin as its "fight or flight" mechanism comes into force. If this adrenalin is not used, it overloads the system and gets stored in the body, creating tense, tight muscles, which leads to headaches. No matter what your age, weight or physical build, there will be some form of exercise that is right for you.

keep focused

One of the hardest things about exercise is staying motivated. Enthusiasm may be high at the beginning of a new regime, but typically wanes as the weeks go by. Most importantly, first find a method of exercise that you like: if you don't like it, you won't do it. This could be running, yoga, swimming, walking, gardening, t'ai chi, playing a team sport or joining the gym. Then build exercise into your normal routine so that it becomes as much of an everyday activity as eating or washing.

If you are very busy, find opportunities for exercise in regular activities: take the stairs instead of the elevator; cycle or walk rather than use the car; or use a manual lawnmower to mow the grass. If exercise doesn't come naturally, remember that it can be fun as well as beneficial.

◂ Swimming is a good all-round exercise to do. The water supports your body so the exercise remains gentle and won't add to your headache.

3

easy neck stretches

Many tension headaches begin with a feeling of pressure in the head or at the base of the neck. A few simple stretches can help to relieve this muscular tension before the headache kicks in.

head and neck stretch

1 Turn the head to one side, then slowly rotate it in a semicircular movement, letting the chin drop down across the chest. Repeat in the opposite direction.

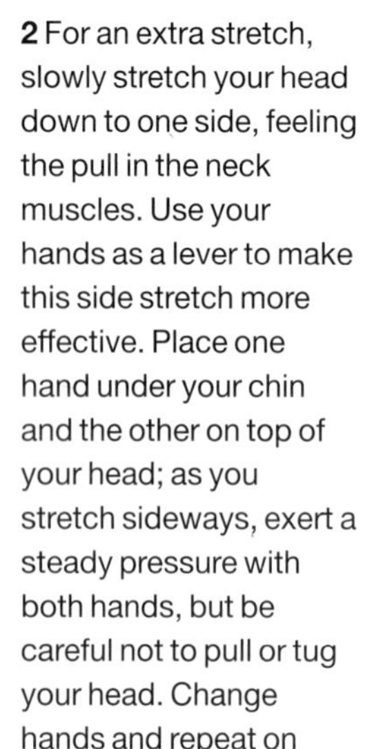

2 For an extra stretch, slowly stretch your head down to one side, feeling the pull in the neck muscles. Use your hands as a lever to make this side stretch more effective. Place one hand under your chin and the other on top of your head; as you stretch sideways, exert a steady pressure with both hands, but be careful not to pull or tug your head. Change hands and repeat on the other side.

face work-out

Another area where you hold a lot of tension is the face. This simple exercise is excellent for stretching the delicate facial muscles to release tension. To do it, find a quiet spot where you are unobserved. Open your mouth as wide as possible and push out your tongue. At the same time, open your eyes into as wide a stare as you can manage. Hold for a moment or two then relax. Repeat a couple of times.

4

relaxing yoga positions

Yoga relaxes the muscles, slows the breath and brings stillness to the mind, making it a useful therapy for treating stress-related disorders. These two poses can be undertaken by anyone.

standing bend (right)
Place a headrest on a stool. Stand with your feet parallel and hip width apart. Breathe out and bend forwards from the hips slowly. Rest your arms over your head on the stool and relax for 1–2 minutes.

lying flat (below)
Lie on a flat surface, with your legs and arms releasing out to the sides. Use a cushion if you feel more comfortable. Close your eyes and relax for 10–15 minutes.

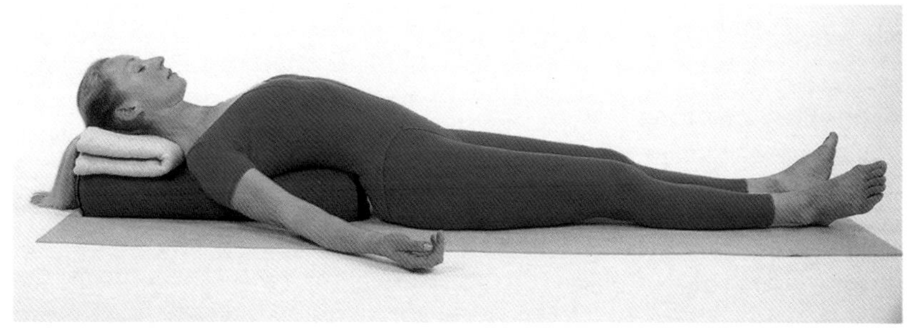

5 healthy eating

Food fuels the body. Frequent headaches may be linked to a poor diet, so it is worth spending time preparing meals from good quality, fresh ingredients rather than using convenience foods.

A well-balanced diet is rich in fibre, vitamins and minerals; this means eating plenty of fresh fruit and vegetables, wholegrains such as rice, millet and barley, and wholegrain bread and cereals. For protein, eat a little lean meat, poultry, fish, cheese, nuts and soya products.

B-complex vitamins

Having enough of the B-complex vitamins is necessary for the healthy functioning of the nervous system. They get used up more rapidly when we are under stress. Studies have shown that niacin (B_3) can help prevent and ease the severity of migraines. The best natural sources are in liver, kidney, lean meat, wholegrains, brewer's yeast, wheat germ, fish, eggs, roasted peanuts, the white meat of poultry, avocados, dates, figs and prunes.

HEALTHY CHOICES

If you are prone to headaches, cut down on your intake of processed foods, sugar, salt, refined carbohydrates, tea, coffee, fizzy drinks and alcohol.

▲ *Choose organic products as these are free from potentially harmful toxic residues.*

Many common foods contain chemicals that can trigger neural and blood vessel changes in the brain, causing migraines or severe headaches in susceptible people. Common migraine triggers are chocolate, citrus fruits, cheese, coffee, bacon and alcohol, particularly red wine.

6 quick-fix snacks

Many headaches are caused by a low blood-sugar level. As your energy level drops, don't be tempted to go for an instant fix with caffeine or sugar, instead choose healthy snacks.

There are many steps you can take to help break an unhealthy cycle. Avoid the temptation to snack unhealthily or to miss meals and "eat on the run". A few nuts or seeds or a piece of fresh fruit is a good substitute for a chocolate bar or bag of crisps (potato chips).

healthy food rules

Eating little and often is a good habit to cultivate. Eat regular light meals, based on fresh, whole ingredients, every 3–4 hours. This will help to stabilize your blood sugar level and prevent excessive energy swings. You should feel better and notice an improvement in your headaches.

Monitor your intake of caffeinated drinks and limit yourself to no more than one or two cups of tea or coffee a day. There are many replacements, such as herbal teas and cereal coffees made from barley, rye, chicory or acorns for instance, which are available in good health stores. Replace sugar with a little fructose (fruit sugar) or honey, either of which is preferable to an artificial sweetener or refined sugar.

▾ *Complex carbohydrates are the best source of energy. Nuts are filling and release energy slowly.*

7 identifying food intolerance

Most migraine sufferers are aware that certain foods and drinks can trigger an attack. Frequent headaches may also be a symptom of widespread and recognized food allergies or intolerances.

Food intolerance is when the body becomes hypersensitive to certain foods; the immune system perceives the substance as harmful, and sets off a chain reaction in the body which produces various symptoms, including sneezing, itchy rashes, sinus problems, lethargy, an uncomfortable bloated feeling and headaches. The onset of food intolerance can occur at any age and to a substance that was previously tolerated. The only way of finding out if you have a food intolerance is to eliminate the suspect food/s from your diet, one at a time, and see whether your symptoms disappear. Common offenders include products made from cow's milk, wheat, corn, yeast, eggs, nuts and shellfish.

▲ *Try cutting out cow's milk or eggs to help your headache symptoms.*

▲ *Red wine, cheese and chocolate can bring on migraines for some people.*

plan ahead

If you think you may have a food intolerance and want to try an elimination diet, make sure you plan ahead so that you don't run out of suitable foods. Base your diet on fresh foods and do not skip meals. Avoid eating out, or if you do, choose plainly cooked dishes. Always check the labels on any manufactured foods, in case they contain the foods that you want to eliminate.

8 green leaf cleanser

Toxicity in the body is one of the principal causes of headaches. Often this is the result of digestive problems such as constipation and/or the absorption of incompletely digested foods.

A sedentary lifestyle combined with a diet low in fibre and water and high in processed foods and caffeinated drinks makes constipation a common health problem. Additionally, regular use of antibiotics and other drugs, alcohol and/or a high intake of sugar can lead to inflammation of the walls of the small intestine, causing intestinal permeability or a "leaky gut". This means that toxic waste is reabsorbed through the intestinal wall back into the bloodstream, causing headaches, fatigue, skin problems and bad breath.

To improve digestion, increase your daily intake of fibre-rich foods, such as raw fruit and vegetables, brown rice and wholegrains. You should also increase your fluid intake. Filtered water is a must, but for a detox include fresh green vegetable juices.

green juice

All fresh juices have a cleansing effect on the digestive system and are gently laxative. Celery juice is effective against headaches, and combines well with dark green vegetables (such as kale, watercress and spinach) which are rich in B vitamins and minerals. To make enough for one serving, you will need 6 large spinach leaves, 2 sticks of celery, plus 2 or 3 tomatoes for flavour. Wash the ingredients and put them through a juice extractor. Serve immediately as fresh juices lose their potency if they are left. If you prefer, you may dilute the juice with water. Drink up to three glasses a day between meals.

▾ *Fresh vegetable juices help the body to detoxify and the cells to regenerate and repair themselves.*

9 have more sex

During sex, the body **relaxes**, easing muscular tension and dissolving **energy blocks**, which are often the source of headaches. It is one of the best **natural therapies** there is.

10 get a good night's rest

Sleep is one of nature's great healers. During sleep the cells of the body renew and repair themselves. Sleep deprivation can lead to all kinds of physical and emotional problems including headaches.

Sleeplessness is a common response to stress as your mind and body cannot let go enough to give you the rest you need. If you have headaches caused by stress, learning to switch off and relax is essential for promoting restful sleep. Make sure you have a healthy diet, take regular exercise and have a calming routine to wind down before bedtime.

SLEEP DO'S AND DON'TS

- Get plenty of fresh air and exercise on a regular basis.
- Don't sleep in late, but get up early and get yourself moving.
- Avoid drinking caffeinated drinks such as cola, tea and coffee at bedtime. A herbal tea or a warm, milky drink is a better alternative.
- Sleep in a well-ventilated room, preferably with the window open.
- Avoid heavy meals late at night.
- Essential oils such as lavender, chamomile and marjoram all have sedative properties. Add a couple of drops of one of them to a warm bath before bedtime, or else put a couple of drops on to a paper tissue and place under the pillow.
- Hops can be dried and used to fill a "sleep cushion" for the bed. Alternatively, they can be brewed and made into a tea, to be drunk before you go to bed.

◂ *If you can't sleep at night but find yourself falling asleep in the middle of the day, it's time to rethink your daily routine.*

11 keep a pet

For animal lovers, keeping a pet can be **therapeutic**. If you feel a headache coming on, take the **dog** for a walk or sit and stroke your **cat** – you may find that the headache disappears.

12 laughter is the best medicine

Laughter is nature's tonic. It eases muscle tension, deepens breathing, improves circulation and releases headache-relieving endorphins to the brain to give you a natural "high".

If you feel you spend too much time working and not enough time having fun, or too much time on your own and not enough time with friends or family, try to redress the balance. Research shows there is a strong link between happiness and good health, so balance the stress of daily life by spending time regularly with friends and family.

▾ *Laughter makes the world look brighter.*

13 crystal healing

Crystals and stones magnify and transform energy, making them effective for healing purposes. There are many different crystals, but amethyst is very useful for soothing tension headaches.

amethyst healing pattern

This exercise focuses on freeing up the energy pathways between the neck and the head. You will need three or four washed amethyst points. Lie on your back on the floor in a warm place. Put one amethyst point on each side of the base of the neck, just above the collarbones, pointing up towards the top of the head. Place a third stone, pointing up, in the centre of the forehead on the brow chakra (third eye). A fourth amethyst can be placed, point outwards, at the top of the head, above the crown chakra. It is important to place the amethysts so that the points face upwards. This directs the flow of energy up through the neck and head and encourages the headache to lift.

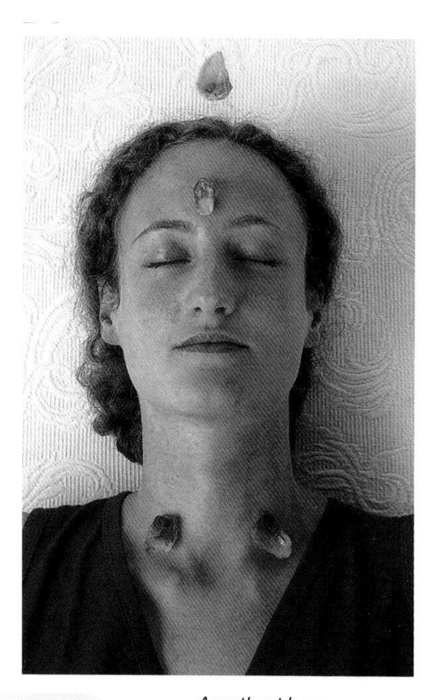

▲ *Amethyst has a calming, protective quality and is helpful for mental disturbances. Its quietening effect makes it an excellent aid to meditation.*

CRYSTAL CLEANSING

Because crystals act as energy transmitters, it is important to keep them clean. Before using them for healing they should always be washed in salt water. Ideally they should be left overnight, covered in salted water; the salt has a purifying action, helping to draw out any negativity which is being "held" in the stones. Always pour the water away.

14 headache healing spell

Try this ancient incantation to cure all types of headache. Before you make the spell, find a suitable tree to bury it under – ash, birch, juniper, orange and cedar trees all have healing powers.

you will need
gold candle and match
gold pen
15cm (6in) square of natural paper
knife
lime
gold cord
15cm (6in) square of orange cloth
spade

1 Light the candle and invite your guardian angel or spirit helper to support the healing. Make up your own words, or say the following:

I light this flame to honour
your presence and ask you to
hear this prayer.

2 Write your name clearly with the pen on the paper, at the same time visualizing a protective bubble of health and wellbeing surrounding you. Keep yourself focused.
3 Cut the lime lengthways into two. Fold the paper three times and place it between the two lime halves. Bind the lime halves together with gold cord, while saying the following invocation (prayer):

Powers of lime,
Health is mine,
Cleanse the body,
Cleanse the mind,
Spirit pure,
Fill my being with health,
With health,
With health.

4 Place the bound lime in the orange cloth and bind the cloth with gold cord. Blow out the candle and say farewell to your higher self, guardian angel or spirit helper.
5 Bury the parcel in the earth under your chosen tree. Ask the tree to help you return to good health and thank the tree.

15

blue colour therapy

Colours can affect our mood and we can tap into this power to use colour for healing purposes. Blue is one of the best colours for calming and soothing frayed nerves.

To ease a tension headache, look for colours which have a calming and cooling effect on the mind and emotions. When you need rest and healing, blue is a good colour to choose. Blue is the colour of the seas and skies and is associated with peace and tranquillity. Soft and soothing blue is a perfect antidote to the stresses, strains and tensions of modern living.

You can work with colour in a variety of ways. The clothes you wear, the decor of your room and your personal possessions are some of the most obvious ways to bring colour into your life.

changing the environment

Bathe yourself in coloured light using coloured films in combination with a free-standing spotlight. To do this, place a sheet of coloured film over the light, making sure that it is not touching the hot bulb. Turn off all the other lights and turn on the spotlight. If it is daytime, draw the curtains or blinds. Sit in the path of the spotlight's ray and bathe in the coloured light for an instant, on-the-spot therapy.

COLOUR CHOICES

You may have noticed that your preference for particular colours varies over time. Start paying attention to which colours attract and which ones repel you. It may mean that you need more of the ones you are attracted to, and a bit less of the others.

◂ *Blue is restful on the eye and is excellent for slowing things down.*

16 green colour therapy

In the spectrum of colours, green lies centrally between red and blue. Green can have a balancing, harmonizing effect. The green of the natural world is nourishing to the spirits.

Take a walk in a leafy green forest and feel the colour green refresh and revitalize you. A walk in the country, taking in fresh air, has the same effect. For headaches arising from nervous exhaustion and debility, green is a good colour to work with. Green is the colour of harmony; it is an excellent tonic for the nerves and helps to restore stability when you are out of balance. Green is all around us in so many tonal values. Light, vibrant green in particular has an uplifting effect on the spirit and is useful when you are feeling depressed.

▲ *When you are feeling drained, spend some time in a quiet park, garden or green fields and notice the effect on your energy levels.*

colour vibration

One way of treating yourself is to have a selection of coloured silks which you can use to wrap yourself in. By wrapping coloured silk around your body, you envelop yourself in pure colour vibration. Choose the colour to which you feel most attracted for your treatment. Turquoise has a soothing and calming effect on the central nervous system.

STRESS BUSTER

If you don't have a large piece of cloth, a small, green silk square placed behind your head in a chair can relieve tension and pressure.

17 relax your body

Relaxation is as important for your health and wellbeing as exercise and a nutritious diet. If you do not switch off from the tensions of everyday life, you are more likely to suffer frequent headaches.

Breathing is something you do unconsciously, but when you are relaxed and calm your breathing pattern is different from when you are tense, anxious or negative. At times of great tension and stress, breathing is usually irregular and shallow, and does not completely fulfil your need for oxygen. If you learn to control your breathing it will help you to stay relaxed even in the most tense or stressful situations.

breathe deeply

Working with the breath is one of the best ways to relax both mind and body. This technique is often used at the end of yoga or exercise sessions. Here is a simple breath control strategy that you can practise at any time. Learning to focus your attention on just your breathing and nothing else will enhance your body awareness and control and help to make you feel calm and centred.

1 Place your hands with your palms under your chest, on your ribs, and your fingers loosely interlocked. Inhale slowly and continuously through your nose, to a count of four. Do not strain, keep yourself relaxed.

2 As you inhale, concentrate on allowing your ribs to expand laterally: your fingers should gently part. Don't let your ribs jut forward. Exhale slowly, expelling all the breath from the lungs, then repeat.

18 relax your mind

A short meditation break in the day will help your mind to unwind and help you return to your activities with a clear head. Meditating at night will help you to relax and prepare for sleep.

visualization exercise
Sit down in a comfortable spot where you won't be disturbed. Close your eyes and allow your mind to drift to a pleasant, peaceful place. A special place where you can relax... completely. A safe... secure... place ... where nothing can ever bother you. It may be a place that you know ... or one that you imagine. Perhaps a garden... or a place in the countryside ... or maybe a room. But it is a place where you feel safe and able to let go... completely... a place that is unique and special to you.

When you are in your place... notice the light: is it bright, natural or dim? Notice also the temperature level... hot, warm or cool? Be aware of the colours that surround you... the shapes... and textures... the familiar objects that make that place special.

Continue to relax in your special place... enjoying the sounds... the smells... the atmosphere... with nobody wanting anything from you ... just you in your special place where you can truly relax.

To end the meditation, slowly bring your attention back to the room. When you are ready, open your eyes.

▲ *Candlelit rooms help create the right mood for meditation.*

CHANGE YOUR BRAINWAVES
Meditation is one of the best forms of relaxation there is; in meditation, the brain-wave pattern changes to relaxed alpha waves, similar to the pattern shown in sleep.

19 protective bubble visualization

Tap into the creative power of your imagination to increase your health and wellbeing and help remove the stress that causes headaches. Creative visualization is simple to learn and effective.

A lot of the stress of daily living comes from trying to satisfy the needs, demands and expectations of others. When the pressure becomes too much, it is easy to react by snapping angrily or by swallowing feelings of resentment and hostility. This is a classic scenario for a thumping headache. To protect yourself from outside pressure, try this visualization which involves creating a protective bubble or shield around yourself.

▲ *"Thinking yourself well" has many beneficial effects.*

calm and clarity

Sit comfortably, close your eyes and imagine yourself in the kind of stressful situation that typically leads to a headache. Picture yourself and any other people who may be involved. Now notice a slight shimmer of light between yourself and the other people ... a protective bubble around you. Learn to believe that this bubble only allows positive and helpful energies to reach you and reflects any negativity back to its source... leaving you free to get on with your life feeling calm and inwardly strong.

While you are in your bubble of light, imagine talking to someone who has been causing pressure to build. See yourself communicating with that person in a calm and clear way until they understand the position. Next, find unhelpful emotions such as past resentments and hurts and imagine pushing them out through the bubble where they can no longer limit or harm you. As you finish the visualization, remind yourself that the bubble stays with you, protecting you. Use this technique next time you feel a headache coming on.

20 gentle candlelight

Practically all headaches feel worse if you are surrounded by harsh bright light. The warm soft glow of candlelight creates a comforting and soothing ambience. Use it to help you unwind.

choosing candles

To help a headache, combine some aroma and colour therapy and choose candles with healing scents and hues. Simple white candles are effective, or look for pale, soft colours, such as pinks and mauves which have a healing effect on the emotions. Avoid large candles in shades of vibrant red and orange, acid green and yellow or dark purple and black.

soothing aromas

Scent is largely a matter of personal preference, but when you have a headache sickly sweet smells such as vanilla or heavy scents such as musk are probably best avoided. Some people like light floral scents, while others may prefer hints of fresh citrus. Sandalwood is also a good choice; this fragrance is traditionally used as a therapeutic aid to meditation, as it helps the mind to relax. Frankincense is another scent which is used in meditation; it has a calming effect on the nerves and slows down the breathing. If you don't like perfumed candles but would like to use scent, you could burn incense sticks or vaporize essential oils instead.

▲ *A warm and uncluttered room is a good place to relax by candlelight. Make sure that the candles are in a suitable container to protect your furniture from hot candle wax.*

21 clear the clutter

Too much paraphernalia in our lives makes us overburdened, depleting our energy and leaving us open to illness of all kinds. Keeping your space clutter-free keeps the energy pathways clear.

Books, papers and toys left lying around, untidy cupboards or work-spaces, anything that's kept "just in case", and any unfinished tasks or jobs which need to be done are all examples of clutter. When you start to accumulate junk, it's a fact that you will always add to it. Having piles of debris lying around, and items wrongly filed or waiting to be put away, will eventually wear you down and hinder your movement around a room. Once a whole house becomes cluttered, the effect is debilitating and depressing, leading to illness. Notice the effect on your energy levels after you have had a good clear-out.

TASK LIST

Make a list of all the jobs that need doing and put them in order of priority; make a point of tackling something on your list each day.

- **Go through your cupboards and have a clear out at least twice a year. Throw out anything that you are not using and no longer need.**
- **Don't forget to clear places like the loft, garden shed and cellar. They are typical dumping grounds for clutter.**
- **Always keep your desk and work area clear.**
- **Deal with correspondence quickly and don't let things lie in your in-tray for too long.**
- **If you find it difficult to throw things away, then ask a friend to help you. They won't have an emotional attachment to your things, and will be able to offer you a more objective opinion.**

◂ *Make a list of any items you can sell, any jobs that need finishing or items that need mending, as you work through your space.*

22 healing homeopathy

Homeopathic remedies are prepared by diluting the original substance until what is left is a vibrational essence. These headache remedies stimulate the body to heal itself from within.

There are hundreds of remedies which are suitable for treating headaches, but below are some of the most widely available. Choose the remedy that most closely matches your symptoms. Take it three times a day in the 6C potency or once a day in the 30C potency until your symptoms have improved.

Belladonna for a throbbing, hammering headache that is worse at the temples, and the headache may be accompanied by fever.
Euphrasia for a headache accompanied by painful, watering eyes, where the sufferer is unable to bear bright light.
Hypericum for a pain that is lessened by bending the head backwards.
Nat. Mur for a migraine-type headache, which is preceded by misty vision or flickering lights in the eyes.
Nux Vomica to lessen the pain of a hangover headache.
Pulsatilla for a headache brought on by overwork, or associated with pre-menstrual tension or the menopause.
Silica for a headache that starts at the base of the neck and spreads up over the scalp, settling over the eyes.
Sulphur for a throbbing headache, which is improved when lying on the right and when gentle pressure is applied to the head.

▸ Homeopathic remedies are made from plant, mineral and animal substances, some of which are highly poisonous in their original form. Because the remedies are diluted many times they are safe to use and have no harmful side effects.

23 Reiki headache treatment

Reiki is a form of Japanese spiritual healing whereby chi, or "life energy", is channelled, in the case of a headache treatment, through the practitioner's hands on to the head of the sufferer.

1 Stand behind your partner and place both (warm) hands firmly on the sides of the head at the back, with your fingers coming up on to the top of the head. This cradling action feels very supportive and helps to dispel tension rising from the neck, balancing energy in the brain. Hold the position for a few minutes.

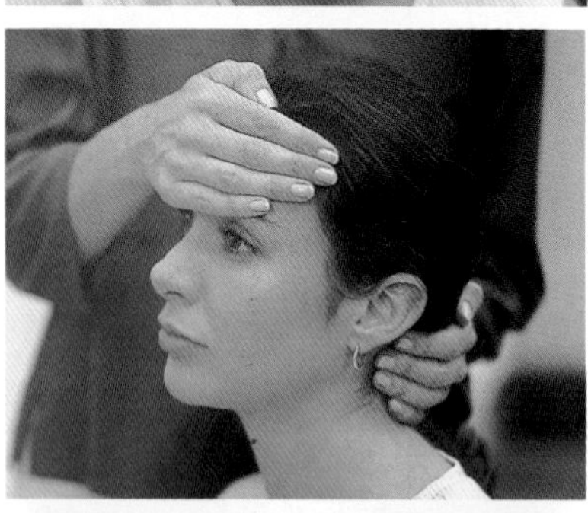

2 Move to the side, and place one hand firmly on the forehead and the other at the base of the skull. Reiki works by putting the hands in certain positions on the body and then allowing the healing energy to flow through them. The aim of the treatment is to dissolve energy blocks and rebalance the body.

3 Finish by placing one hand lightly over the eyes and the other on top of the head. This is very relaxing. After giving healing, you should always wash your hands. The person receiving reiki should drink plenty of water after the treatment to help flush out toxins. A reiki treatment can bring rapid relief to a headache.

24 healing herb tub

Many of the remedies suggested in this book use plants. For quality and freshness, nothing beats growing your own. An attractive way of doing this is to plant up a container of healing herbs.

Use a large container to give the plants room to grow and site it in a sunny spot near the house for easy access.

you will need
Half-barrel
Bricks
Drainage material
Soilless compost (growing medium)
Sharp sand
Watering can

plants
Feverfew (*Tanacetum parthenium*)
Lavender (*Lavandula* 'Hidcote', *L. stoechas*)
Marjoram (*Origanum vulgare* 'Variegatum')
Rosemary (*Rosmarinus officinalis*), prostrate and upright forms
Lemon balm (*Melissa officinalis*)

1 Rest the tub on a few bricks to raise it off the ground and promote better drainage. Cover the base with a layer of drainage material such as broken pots, broken polystyrene plant trays or horticultural grit. Almost fill the tub with a 50/50 mixture of compost and sharp sand.

2 Arrange the herbs, still in their pots, on the surface of the compost. When you are happy with the arrangement, make holes and plant them, firming the compost down and topping up if necessary. Leave a gap below the rim of the tub to allow for watering. Water the plants in well.

▾ *Keep the plants well trimmed and replace the top layer of compost annually. Feed regularly in the summer months.*

25 gentle lavender & rosemary oils

There are many essential oils which can help treat a headache, but two of the most popular and effective are lavender and rosemary, which are both from the same plant family.

Lavender is particularly good for headaches that are related to stress and tension, while rosemary is useful for ones brought on by mental fatigue and nervous exhaustion; either is useful for headaches associated with depression. A quick and convenient headache treatment is to mix a drop of either oil in a teaspoon of carrier oil, such as almond. Rub the mix into your temples. Or put a drop of neat oil on a handkerchief and inhale the aroma.

▲ *Lavender has many healing properties. It is an excellent headache remedy as it is a natural analgesic.*

▼ *Rosemary helps to regulate blood pressure and can reduce pain.*

HOW OILS WORK

Inhalation is the fastest way of enjoying the benefits of aromatic oils, as nerve pathways lead directly from the lining of the nose to the brain, having an immediate effect on the central nervous system.

CAUTION: Do not use rosemary oil if you are pregnant or suffer from epilepsy.

26 soothing bath water

Make up this bubble bath mix and keep it on standby for when you need to relax at the end of a long, stressful day – it will ease a headache and promote restful sleep.

lavender bubble bath
The recipe uses dried lavender flowers as well as lavender oil for extra strength. The mixture will keep for several months in a cool, dark place.

you will need
medium-sized bottle of clear, mild, unscented shampoo
45ml/3 tbsp dried lavender flowers
5 drops lavender essential oil
wide-necked, screw-topped glass jar
fine sieve
glass or plastic jug (pitcher)
squeezable plastic bottle

1 Pour the shampoo, the lavender flowers and the lavender oil into the glass jar. Replace the lid and shake vigorously to mix all the ingredients thoroughly together.
2 Leave this mixture to stand in a warm place, such as a sunny windowsill, for up to two weeks, shaking and turning the jar occasionally. The lavender flowers will infuse the shampoo.
3 Strain off the lavender liquid into the jug and then pour it into the squeezable plastic bottle. Discard the dried lavender flowers.

▸ *Let the day's tensions melt away with the delicate fragrance of lavender. Lavender is a recognized cure-all for many common ailments, including tense, nervous headaches.*

27 a relaxing atmosphere

Scenting the environment can soothe a headache and help you regain equilibrium quickly. Try using a vaporizer with a few drops of your favourite essential oil.

The bedroom and bathroom are both ideal places to relax, while the use of soft music, candles and essential oil burners is a popular way of creating a peaceful and soothing atmosphere. Essential oil burners have a small dish to hold water and a source of gentle heat underneath, often a night-light candle. The heat needs to be fairly low, but warm enough to heat up the water. A few drops of essential oil are added to the warm water, which then slowly evaporates, giving out its scent. Alternatively, oil drops can be added to a bowl of hot water if you do not have a burner.

Any of the following oils will help to create a relaxing atmosphere. Add two or three drops to a burner.

Clary Sage for stress and tension headaches.
Geranium for headaches caused by premenstrual syndrome.
Lavender for migraine and tension headaches.
Neroli for headaches caused by mental exertion.
Rose to soothe and calm the nerves.
Sweet Marjoram for headaches caused by anxiety and insomnia.

▲ *Essential oils help speed up the healing process. Use them as an aid to recovery, as well as to scent your room.*

BE PREPARED
Carry a small bottle of your favourite essential oil with you and try dabbing this on a handkerchief before reaching for the painkillers.

28 soothing away eyestrain

Many headaches are caused by eyestrain. Watching a lot of television, working in poor light, long-distance driving or sitting for long periods in front of a computer screen can trigger headaches.

If your eyes are feeling tired and ache and you feel a headache coming on, you can give them a treat by covering them with cucumber slices. Or, if you are at work, you could try splashing cold water on to your eyes for a similar effect.

Alternatively, make eye-packs out of chamomile tea bags. Chamomile has a very gentle, anti-inflammatory, calming action. It helps alleviate sore, tired eyes and headaches arising from tension and anxiety. Boil a kettle and pour the hot water on to two tea bags. Leave to infuse for 10 minutes, then take them out of the water. Let them cool down and squeeze out any excess water. Lie back and place the bags over your eyes for a soothing effect.

COMPUTER USERS

If you use a computer, there are some simple steps you can take to reduce the likelihood of getting headaches from eyestrain:

- **Take a short break away from your desk every 20–30 minutes.**
- **Make sure the computer has a good quality display and is fitted with an anti-glare filter. Adjust the brightness level on the screen to suit you.**
- **Set the computer screen at eye-level and position it so that it does not reflect glare from any other source of light, such as a window behind you, for instance.**

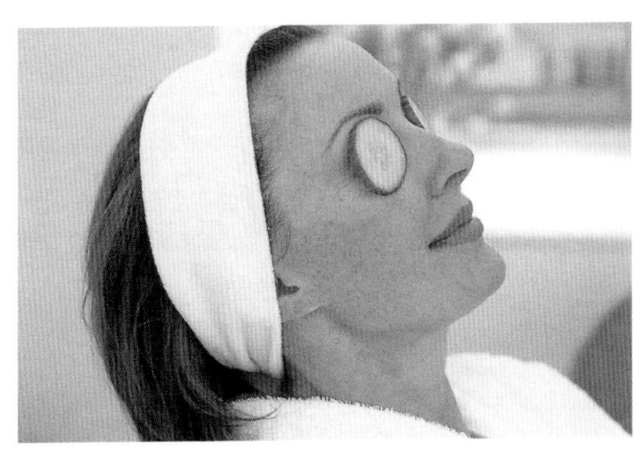

◂ *Cucumber has a cooling, refreshing effect and helps to increase the circulation to the eye area.*

29 feverfew

This bitter, edible plant has long been recognized as one of the most effective natural treatments for headaches and migraine. Feverfew works best as a prophylactic, taken over a period of time.

migraine cure

Science attributes the action of feverfew to a natural chemical which seems to inhibit the release of the hormone serotonin, which is thought to be a migraine trigger. A three-month course of feverfew can reduce the number and severity of attacks.

Feverfew can be taken in tablet form. Check the label and make sure the pills contain at least 0.2 per cent of the active ingredient, parthenolide. A daily dose of 200mg is usually sufficient. Alternatively, many headache and migraine sufferers have been helped by eating the fresh leaves. Eat two to three leaves daily in a brown bread sandwich. The bread makes the bitter leaves more palatable. You may also use a little honey to sweeten them.

CAUTION

- Avoid feverfew during pregnancy and while breastfeeding.
- Too much fresh feverfew can cause side effects, including mouth ulcers, stomach pain and swollen lips.
- If you are taking other medication, discuss with your doctor first.

▲ *Feverfew is easy to grow and can be used fresh or dried to treat headaches and migraine. The dried plant is also used to make feverfew tablets, available in good pharmacies and health stores.*

30

evening primrose

For many women, headaches are a recognized symptom of PMS and of the menopause, and some women notice more headaches when taking the pill or hormone replacement therapy.

hormonal regulator

Evening primrose is a traditional North American Indian medicine and enjoys a reputation as one of nature's most valuable and versatile remedies. A great deal of research has been done on the medicinal effect of the oil, which is extracted from the plants' seeds. It is a good source of Omega-6 fatty acids, vital for the healthy functioning of the immune, nervous and hormonal systems. In particular, the oil contains gamma-linoleic acid (GLA), which is especially helpful to counter hormonal problems.

Increasingly, many women regularly take a supplement of evening primrose oil, which is available in capsule form in pharmacies and health stores. The oil is not only helpful for treating PMS, but also migraine and menopausal problems. Medicinal doses range from 500–1500mg a day or as directed by a healthcare professional.

◂ *The evening primrose plant has fragrant yellow cup-shaped flowers which open at dusk.*

▴ *Women who suffer from migraine are most likely to suffer attacks around the time of their periods.*

HANGOVER CURE

Evening primrose oil has also been shown to counter the effects of alcoholic poisoning. For a hangover headache, try taking a couple of capsules instead of a painkiller.

CAUTION

Evening primrose should be avoided by women with breast cancer and sufferers of epilepsy.

31 herbal steam inhalant

A stuffy nose and blocked sinuses can sometimes be the cause of a headache. Steam inhalations are a good way of clearing the head. The addition of fresh herbs will help to relieve the headache.

1 To clear the sinuses and ease a congestive headache, gather together a large handful of lavender, rosemary, peppermint, sage, thyme and eucalyptus. Put the selected herbs in a bowl.

2 Pour on 1 litre/ 1¾ pints/4 cups of boiling water. Lavender and rosemary are natural painkillers, while peppermint and eucalyptus are good for clearing blocked noses and throats.

3 Lean over the bowl, with a towel draped over your head and shoulders to form a tent. Breathe in the steamy vapour as deeply as possible. The inhalation will decongest blocked nasal passages, kill any germs and clear the headache.

TIPS

- If fresh herbs are not available, then an inhalation using essential oils of peppermint, eucalyptus, lavender and rosemary may be tried instead. Use 2 drops of each oil to 1 litre/1¾ pints/4 cups of boiling water.

- The hormonal action of thyme can help lift depression and the symptoms of fatigue.

CAUTION

If you have high blood pressure or asthma, seek medical advice before using steam inhalants. Eucalyptus and peppermint may interfere with homeopathic remedies.

32 comforting compress

Many headache sufferers find a compress helps their symptoms. A compress is a cloth which has been soaked in water with a few drops of essential oil added, wrung out, then applied to the head.

Of all the essential oils used in the treatment of headaches, lavender and peppermint seem to be the most effective. Although they can be used separately, these oils work well together. Both are painkillers and complement each other: lavender is a sedative while peppermint is a stimulant. This dual action is found in many commercial headache remedies, which include a stimulant (such as caffeine) to counteract the slightly sedative, and even depressing effect of the painkiller. The important difference is that essential oils do not just suppress the pain but get to work on the cause of the headache. A compress will be most effective if it is used immediately at the onset of a headache.

you will need

600ml/1 pint/2½ cups cold water
bowl
2–3 drops lavender essential oil
2–3 drops peppermint essential oil
piece of soft cotton fabric

1 Pour the water into the bowl. When the water is still add the essential oils and gently stir the water to disperse the oils.
2 Fold the soft cotton fabric into a loose pad. Place it on the surface of the water and let it soak up the essential oils. Wring it out lightly.
3 Place the compress across the forehead to relieve the headache. As soon as it gets warm, soak it again in the water and re-apply.

TIPS

- During a migraine attack, some sufferers cannot tolerate the smell of peppermint, in which case try alternating hot and cold lavender compresses.
- A cold compress on the back of the neck will ease a tension headache, and one on the forehead is best for a thumping headache.

33 lime blossom & lemon balm tisane

A simple but effective way of curing a headache is to drink a herbal tea. Lime blossom and lemon balm (*Melissa*) leaves make a delicately fragranced drink for soothing tension headaches.

herbal infusions

Making herbal tisanes from fresh ingredients increases their potency and medicinal value. Lime flowers have a relaxing and cleansing effect in the body. They can help with high blood pressure, and are effective in treating headaches, including migraine. Lime flowers make a good remedy for any condition associated with tension, including depression, and back, neck and shoulder pain. Lemon balm is useful for calming tension and promoting feelings of wellbeing; it can also relieve headaches and migraine. The two herbs are complementary.

To make this tisane, gather a handful each of lemon balm leaves and lime blossom – pick lime blossom when the flowers are just opening. Use five or six fresh flowers and leaves per cup, and drop them into near-boiling water. Cover and leave to steep for three or four minutes. Strain off the liquid and drink hot or cold, three times a day. The tea has a fresh, lemony taste.

◂ *Lime blossom can also be mixed with peppermint for a more uplifting effect.*

CAUTION
Although natural, herbal teas can be potent and should be taken no more than three times a day and for no more than two weeks at a time.

34 wood betony infusion

Once considered a panacea for all ills, wood betony may be used alone or combined with lavender or rosemary to make a soothing herbal infusion to ease your headache.

The leaves and pink flowering tops of wood betony are used in medicinal preparations. The plant has a stimulating effect on the circulation and is also a relaxant, making it helpful for both congestive and tension headaches. It is also a tonic to the nervous system, helping to ease anxiety, lift depression and soothe pain.

The infusion may be made with either fresh or dried ingredients. Fill a cup with near-boiling water and add 5ml/1 tsp dried wood betony and 2.5ml/½ tsp dried lavender or rosemary. Double these quantities if you are using fresh herbs. Cover and leave to steep for 10 minutes. Strain and drink, sweetened with a little honey if required.

▲ *Wood betony was highly prized as a medicinal herb in Roman times. The Anglo-Saxons believed it capable of driving away despair.*

◀ *Fresh rosemary is invigorating and refreshing. It is excellent for clearing congestive headaches, and combines well with wood betony to relieve nervous tension.*

35 herbal sedatives

Headaches caused by stress and nervous exhaustion are generally linked with an inability to switch the mind off and relax, leading to insomnia and restless nights.

calming the nerves

There are several herbs which have a strong sedative action and which are useful for headaches associated with nervous exhaustion and overactivity. Among the most commonly used are valerian, vervain and sweet marjoram: any one of these can be made into an infusion that will help you to relax and promote restful, healing sleep. Do not take all three together.

For every 250ml/8fl oz/1 cup of near-boiling water you will need 10ml/2 tsp of fresh valerian, vervain or sweet marjoram. If using dried herbs, halve the quantity. Drop the herbs into the water, cover and leave to infuse for 5–10 minutes. Strain off the liquid, sweeten with a little honey, and drink before going to bed.

CAUTION
Sweet marjoram and vervain should not be taken during pregnancy. Valerian should not be taken by anyone with liver disease.

▲ *Vervain is a wonderful tonic for the nervous system, calming the nerves and easing tension. It protects against stress, and is useful for treating headaches caused by anxiety, depression and insomnia.*

▲ *Valerian has a calming effect on the mind and helps to relax tense muscles – use it to ease a tight neck and shoulders. It is a strong sedative, and forms the basis of the pharmaceutical drug, valium.*

36 lavender tincture

Tinctures are an effective way to extract the active ingredients of plants. They are made with fresh or dried herbs which are steeped in a mixture of alcohol and water. This one will cure a headache.

you will need
15g/½oz dried lavender
250ml/8fl oz/1 cup vodka, made up to 300ml/½ pint/1¼ cups with water
dark glass jar with an airtight lid

1 Put the dried lavender into a glass jar and pour in the vodka and water mixture. It will almost immediately start to turn a beautiful lavender blue.
2 Put a lid on the jar and leave in a cool, dark place for 7–10 days (no longer), shaking the jar occasionally. The tincture will eventually turn dark purple.
3 Strain off the lavender through a sieve lined with a paper towel before pouring into a sterilized glass bottle. Seal with a tight-fitting lid and store in a cool, dark place for future use.

As tinctures are highly concentrated medicinal extracts, take no more than 5ml/1 tsp, three or four times a day, as a headache treatment. The tincture may be diluted in a little water or fruit juice. Alternatively, a few drops may be added to a compress.

MEDICINE CHEST

- Among herbal remedies, tinctures have a relatively long shelf-life; properly stored, they will keep for up to two years, as the alcohol acts as a preservative.
- Lavender tincture is a useful remedy to have on standby as it can be used to treat many common health problems, such as burns, muscular aches and pains, coughs and colds, as well as headaches of all kinds.

37 hangover headache treatments

Waking up with a hangover takes all the fun out of the night before. Before dashing off for the black coffee, try some natural remedies. There are many simple things you can do which are effective.

fruit sugar

Hangover headaches seem to result from metabolic disturbances in the brain as a result of drinking too much alcohol. Experts think this may cause a type of "brain hypoglycemia" or low blood sugar, which is why some people recommend eating a snack before going to bed that is high in fructose (fruit sugar). Research suggests that fructose helps metabolize the chemical products of alcohol that cause headaches and hangovers. Fructose is found in carrots, and in all fruits, especially dates. Fructose bars are also available in health stores.

rehydrate

It's important to drink plenty of water and to increase your intake of vitamin C. Alcohol dehydrates the body; drinking water counters this effect, and helps to flush out toxins. Vitamin C is needed for more than 300 metabolic processes in the body.

Kiwi fruit and all citrus fruits are a good natural source. Make a drink from freshly squeezed orange or lemon juice, sweeten it with honey and it will help the headache.

▲ *Oranges and kiwi fruit are high in vitamin C.*

You could also keep a supply of homeopathic nux vomica at home. Sometimes, a single dose is enough to clear up nausea and a thumping headache. Take it in a 6C potency every hour for up to eight doses.

38 Bach flower remedies

The Bach flower remedies work on any underlying emotional or psychological cause of a headache, treating the negative mental and emotional states which lead to pain and tension in the body.

Dr Edward Bach discovered the healing energies of selected flowering plants and trees, by "tuning in" to their subtle vibrations. He noted that the plants affected him on a mental and emotional level, and devised a system of healing based on these states. The remedies match certain personality types and are chosen accordingly.

Bach flower essences are gentle and safe to use and are available in most pharmacies and health stores. They may be taken separately or in a combination of up to six remedies. They may also be taken in conjunction with other treatments. Add a couple of drops of each essence to a glass of water and sip at frequent intervals.

▲ *There are 38 Bach essences.*

FLOWER ESSENCES

To treat a headache, select from the following remedies:

- **Beech** if you are critical and intolerant of others and yourself.
- **Cherry Plum** if you have repressed anger and feel as if you might explode, or feel fearful.
- **Crab Apple** if you try to be "superman/woman", or are driven by the need to be perfect.
- **Gorse** if you have feelings of hopelessness and are resigned to fate.
- **Holly** if you suffer from feelings of jealousy, hatred, resentment and frustration.
- **Impatiens** if you feel tension due to rapid mental activity, or are always in a hurry.
- **Pine** if you have feelings of guilt and inferiority, if you always blame yourself or are always apologizing.
- **Vervain** if you live on your nerves, are keyed up and unable to relax.
- **Vine** if you are rigid in your thinking, if you are inflexible or ambitious, or if you think you are always right.
- **White Chestnut** for thoughts that go round and round.

39

calming head massage

Gentle massage of the temples and forehead can help to stop a tension headache from getting a tight grip and encourage the body to relax. Use essential oils for additional benefit.

Everyone can benefit from the comforting touch of massage. The sense of touch is a powerful tool of communication and can be used to benefit the recipient on an emotional, physical and mental level. It helps relaxation, relieves aching muscles and reduces pain making it a useful treatment for headaches.

Add any of the essential oils in the quantity stated in the box, right, to 30ml/2 tbsp of a light vegetable oil, such as almond or grapeseed, or to an unscented cream or lotion.

CHOOSING AN OIL

- **Rosemary** for a congestive headache. Rosemary is uplifting and clears the mind. Use 4 drops.
- **Peppermint** if the head feels too hot (peppermint has a cooling action). Use 4 drops.
- **Lavender** if warmth feels as though it is helpful. Use 6 drops.
- **Chamomile** for either headaches or migraine. Use 4 drops.
- **Marjoram** when the mind simply won't switch off. Use 4 drops.

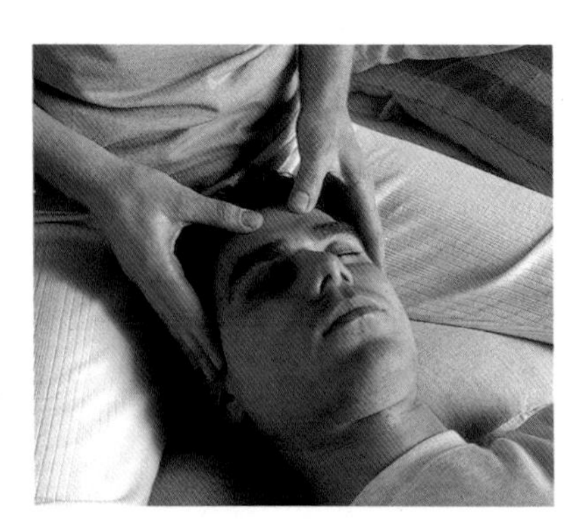

1 With your thumbs, use steady but gentle pressure to stroke the forehead and work the oils into the skin. The strokes will help ease tension.

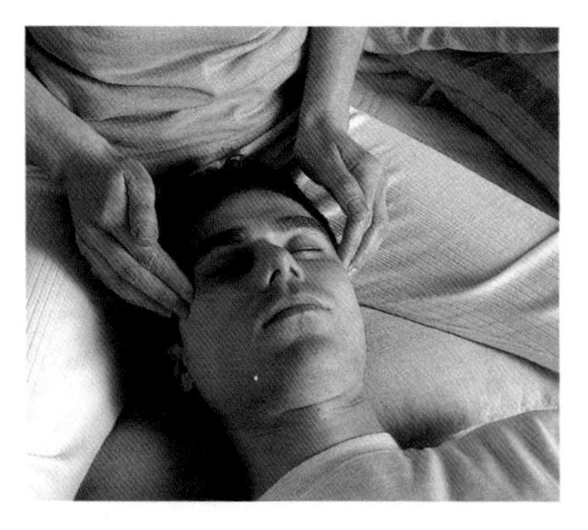

2 Gently massage the temples with the tips of your fingers to release tension and stress. Work the oils into the skin as before.

40 neck & shoulder easer

Sitting hunched over a desk for long periods, driving, or carrying heavy bags are just a few of the occupational hazards that create tension in the shoulders, neck and below the ridge of the skull.

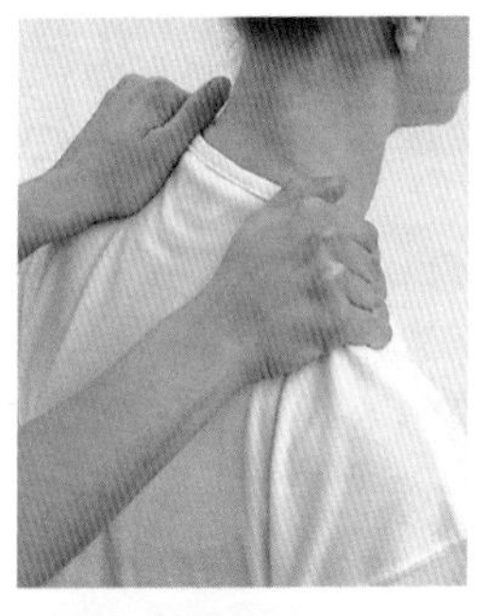

1 Anchor your fingers over the shoulders. Roll and squeeze the muscles in a kneading action, working out to the edge of the shoulders and down the tops of the arms.

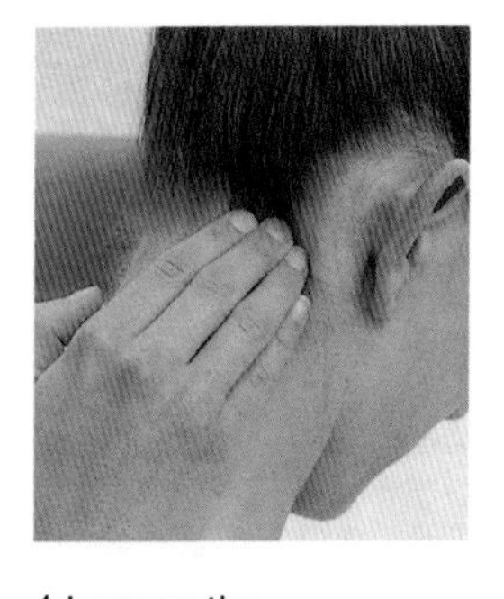

2 Place one hand across the forehead and the other across the nape of the neck. Squeeze the neck muscles between the fingers and heel of your hand.

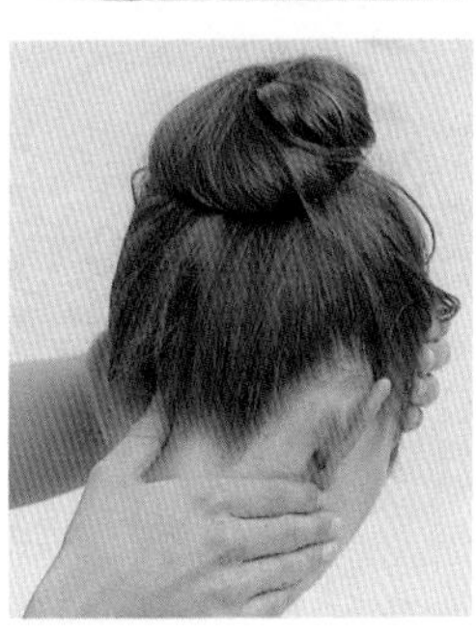

3 Supporting the forehead, use the thumb pad of the other hand to press upwards into the hollow at the top of the spine below the skull. Apply gentle upward pressure for a steady count of five, then release.

4 Loosen the constricted muscles under the base of the skull by massaging beneath the bony ridge, working from the top of the spine to the outer edge of the skull.

5 Change hands to massage beneath the other side of the skull. Ease scalp tension by rotating the fingertips of both hands in small circles all over the head.

41 shoulder reliever

Having a shoulder massage is one of the best ways of releasing muscular tension. It not only feels good, but can help to prevent or ease a headache. Practise this routine with a friend or your partner.

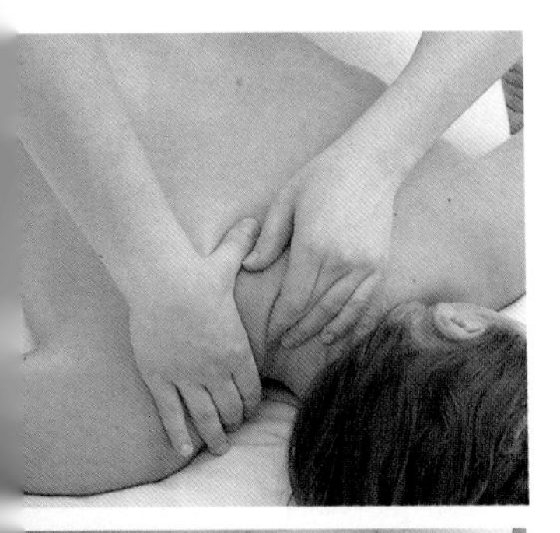

1 Make sure your partner faces away from you. Place both your hands on one shoulder, and with alternate hands squeeze your fingers and thumb together. Repeat on the other side.

2 Place your thumbs on each side of the spine on the upper back, with the rest of each hand over each shoulder. Squeeze your fingers and thumbs together, rolling the flesh between them.

3 Let your thumbs smoothly move out across the shoulder muscles, and release the pressure of the thumbs as you stretch the shoulder blades outwards, away from the spine, with your hands.

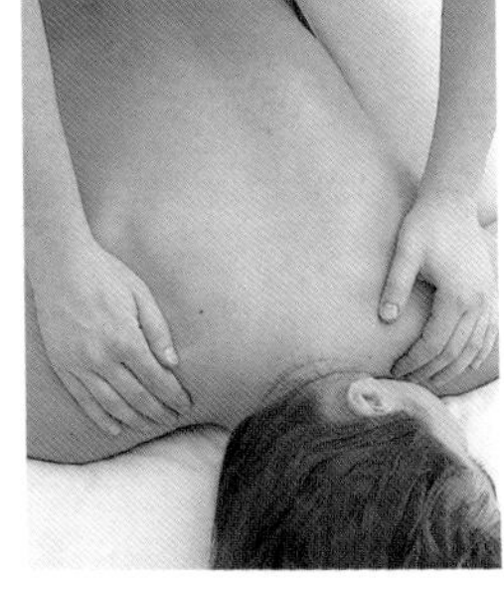

4 Return your hands to the centre and repeat this movement with a firm kneading action.

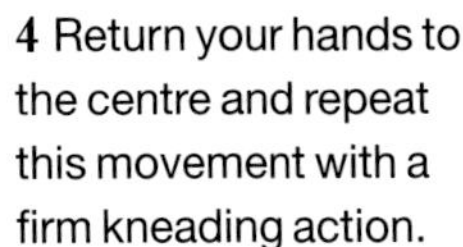

RELAX

A typical response to feeling burdened by life's responsibilities is to tighten up in the trapezius, the large muscle in the shoulders.We hold a lot of tension in this area; when you feel "uptight" it is often the shoulders that bear the load. Make a conscious effort to relax your shoulders.

42 anxiety calmer

When you are feeling anxious and upset, the muscles of the face tense up, making you look fraught. Having the face gently massaged is very relaxing, and a great way to soothe a headache.

Ask a friend to practise this routine with you. It is best done when you have an opportunity to relax afterwards. For a headache, it is a good idea to use essential oils, blended in a little massage oil. Choose a light vegetable oil, such as sweet almond or grapeseed, as a base. Use four or five drops of essential oil to 30ml/2 tbsp massage oil, but take care not to put too much on at a time, as most people don't like a greasy feeling on the face. If you prefer, you may use an unscented lotion or cream as a base.

1 Ideally have the person lying down with the head on a cushion. With your fingertips, smooth the essential oil blend into the face. Pay particular attention to the temples and forehead, using small circular movements and light brushing strokes.

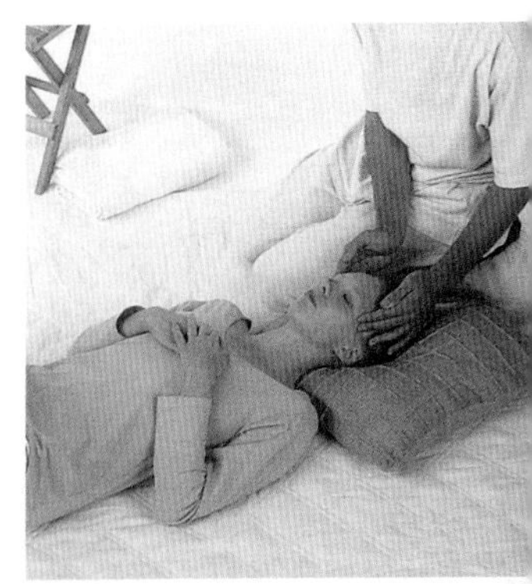

2 Using your thumbs one after the other, stroke tension away from the centre of the forehead. Finish the routine by holding your hands still on each side of the face; this feels very calming and reassuring.

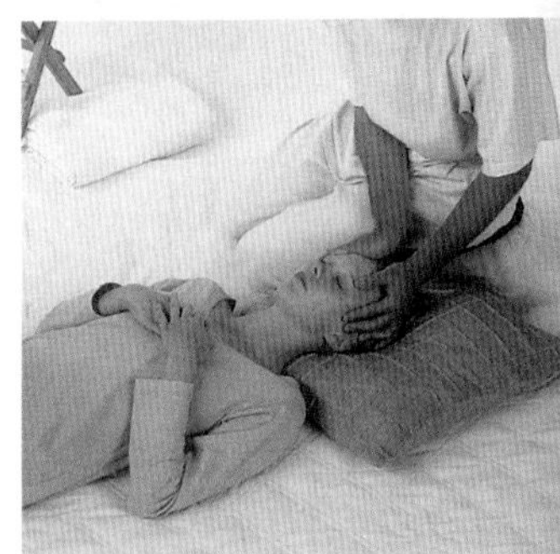

ESSENTIAL OILS

- **Chamomile** or **marjoram** for headaches that are the result of overwork, anger or worry.
- **Lavender** for migraine and tension headaches.
- **Tea tree** for clearing the head.

43 Shiatsu massage

This traditional Japanese healing system is based on applying pressure to the body, and using stretching and holding movements. This routine is excellent in the treatment of tension headaches.

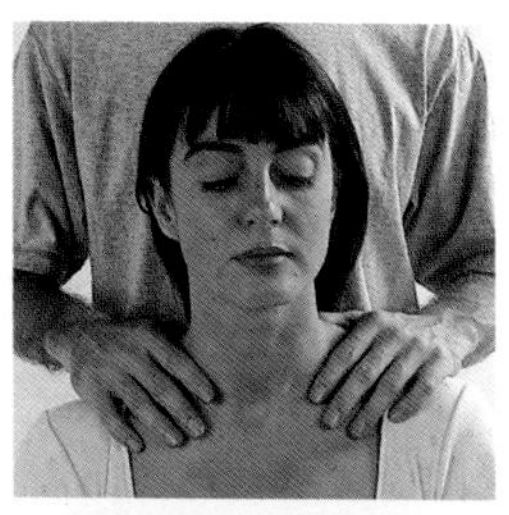

1 Standing behind your partner, place both hands loosely on each side of the neck. Gently massage the shoulders to help relax the breathing.

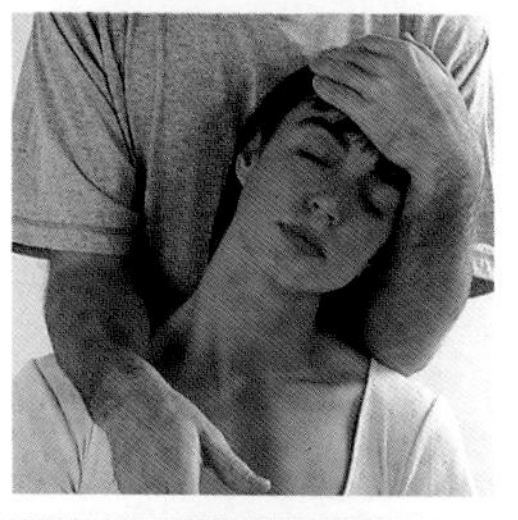

2 Tilt the head sideways and support with your hand. Place the forearm across the shoulder and apply downward pressure. Repeat on the other side.

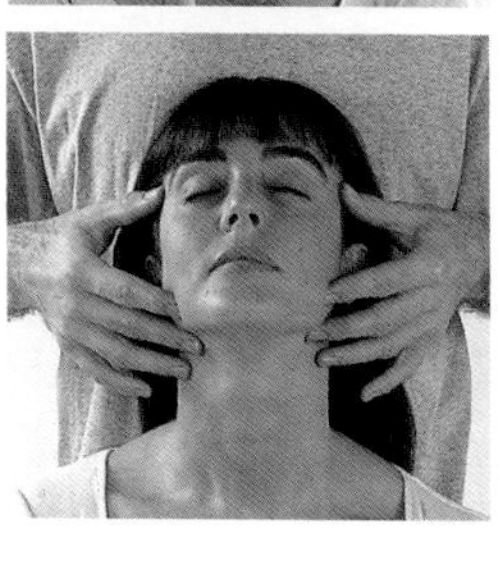

3 Apply gentle pressure with the thumb and forefinger from the base of the neck to the nape. Hold at the nape for 5 seconds, then release. Tilt the head back, supporting it. Place your thumbs on the temples with the fingers resting on each side of the face. Rotate the thumbs in small forward movements.

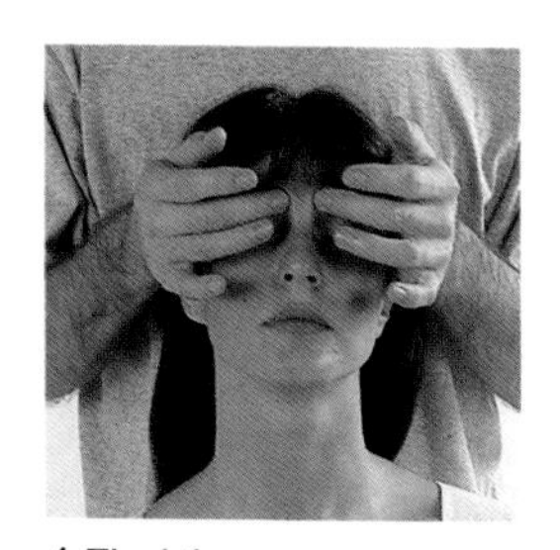

4 Find the pressure points just above the inner corner of each eye. Apply gentle pressure with the middle fingers to help disperse the pain. Hold for 5 seconds.

5 Position your thumbs about 5cm/2in apart on each side of the head, just above the hairline, with the palms pressed flat along the sides of the face. Press the thumbs evenly back along the top of the head.

POSTURE
Sit on an upright chair with good back support.

44 foot reflexology

The science of reflexology believes that our bodies are reflected in miniature in our feet. If we treat the specific area of the foot that represents the head, we can massage away a headache.

an ancient art

Having a reflexology treatment is relaxing and can treat specific health problems. It is effective for treating tension headaches as well as for migraines.

Your head is represented on the toes; the right side of your head lies on the right big toe and the left side on the left big toe. In addition, the eight other toes contain the reflexes to specific parts of your head, for fine tuning. By applying gentle pressure to these exact points, reflexology stimulates the body to heal itself.

Reflexology can be an excellent preventive therapy. If the headache is a symptom of another illness, a different part of the foot to that suggested here would need to be treated first.

1 Work the hypothalmus reflex first, as this controls the release of endorphins for the relief of pain.

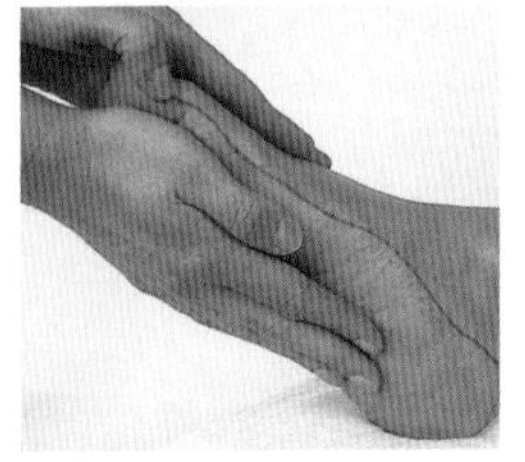

2 Work down the spine to take pressure away from the head. This will draw energy down the body and ground it.

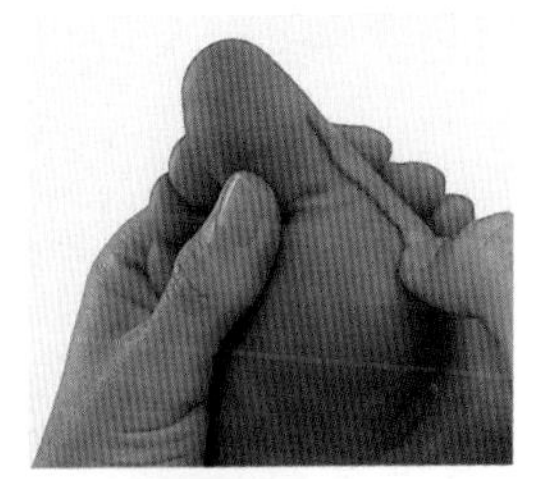

3 Work the cervical spine on the big toe. Work the neck of all the toes to relieve tension.

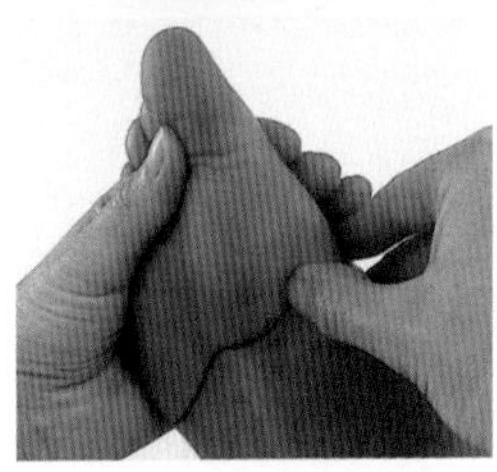

4 Work the diaphragm to encourage freer breathing. Repeat the reflexology treatment on the other foot.

45 scalp massage

A scalp massage is deeply relaxing. If you are suffering from stress-related headaches, use this treatment on a regular basis to reduce tension. It only takes a few minutes.

If you are stressed, the scalp muscles tighten. One side effect is that the roots of your hair become starved of nourishment and your hair will start to thin out and weaken. This massage stimulates the hair roots. The advantage of this massage is that it is one you can do for yourself.

▸ *Lank, lifeless hair may be an early sign of stress.*

1 Place the thumbs at the top of the ears and "glue" the fingers to the scalp, moving it firmly and slowly over the bone beneath.

2 Place the hands on another part of the scalp and repeat. Carry on until the whole scalp has been covered. Repeat steps 1 and 2 several times.

46 head revitalizer

This simple self-massage sequence will help to ease headaches, whatever their cause. You can also use it to increase your vitality and help you to focus your mind throughout the day.

1 Use small, circling movements with the fingers, working steadily from the forehead down around the temples and over the cheeks.

2 Use firm pressure and work slowly to ease tensions out of all the facial muscles. Use your fingers to gently press around the eye sockets, by the nose.

3 Smooth firmly around the arcs of the eye sockets. Work across the cheeks and along each side of the nose, then move out to the jaw line where a lot of tension is held.

4 Try not to pull downwards on the skin – let the circling movements help to smooth the stresses away and gently lift the face as you work.

USE AN OIL

If you are feeling overwrought, make up an aromatherapy blend of 4 drops lavender and 2 drops ylang ylang in 30ml/ 2 tbsp of a light massage oil, such as sweet almond or grapeseed.

47 migraine easers

Having a migraine is one of nature's ways of shutting the body down when things get too much. Rather than trying to "fight it off", respect what your body is saying and look after yourself.

There are several essential oils which can help a migraine, but use them sparingly and carefully. Many migraine sufferers have a heightened sense of smell at the onset of an attack, and may find any aroma intolerable.

massage mix

As soon as you feel a migraine coming on, try a massage blend of 2 drops rosemary, 1 drop marjoram, and 1 drop clary sage oil, diluted in 30ml/2 tbsp of a light vegetable oil, such as grapeseed or sweet almond. You may use an unscented lotion or cream base rather than oil if you prefer.

Alternatively, use a drop of each oil in a bowl of warm water, and apply a warm compress to the forehead. If this blend smells too strong, you could just try lavender on its own, 3–4 drops in the base oil or cream as before.

1 Using the massage mix, gently rub the temples with small circular movements, using the tips of your fingers.
2 If touching the head does not make the pain worse, ask a friend to give you a gentle head massage. This can feel very soothing and comforting and help you to relax.

◂ *Migraine headaches are not the same as everyday tension headaches, and if you suffer from these you should seek professional advice.*

48 calming sleep massage

The gentle wave-like strokes of this massage wash over and down the body with a deliciously hypnotic and sedative effect – ideal to ease away headaches before going to sleep.

1 Place one hand over the chest and the other over the back of the shoulder. As you breathe in, pull your hands steadily outwards and down to the edge of the shoulder. Pause briefly as you exhale, lightly cradling the top of the arm.

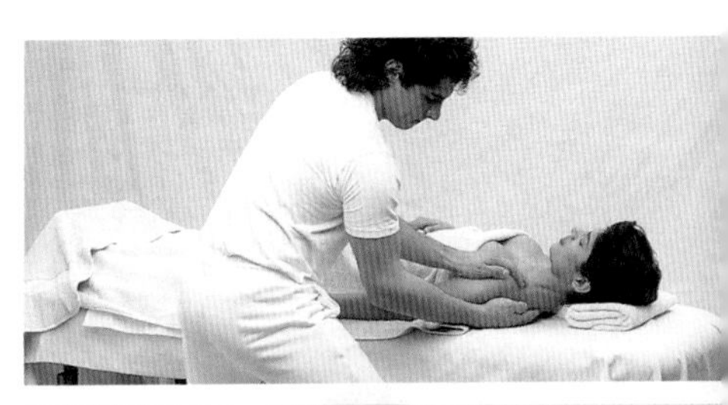

2 Continue the pulling motion down the length of the arm. As you breathe in, pull both hands down to just below the elbow joint. Relax as you breathe out, then continue the slide down the forearm and below the wrist.

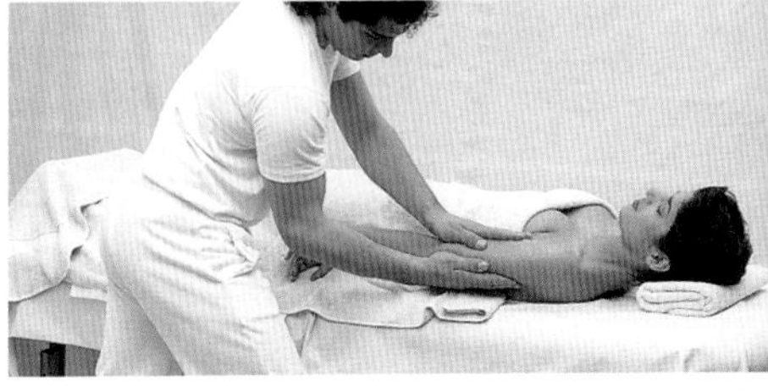

3 Draw your hands over both sides of your partner's hand and fingers, taking your stroke out beyond the body as the hand settles back on to the mattress. Repeat steps 1–3 on the other side of the body.

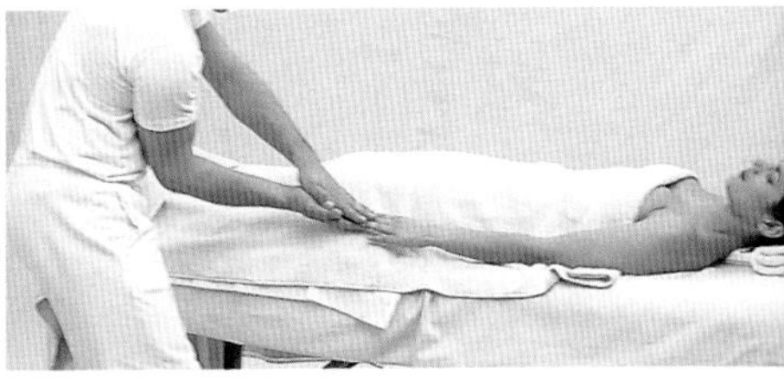

4 Pull your hands down over the hips and down the leg to just below the knee. Continue this wave-like motion down the lower leg to the ankle, then pull gently and steadily out over the toes. Repeat this sequence of strokes on the other side of the body. Repeat each movement up to five times.

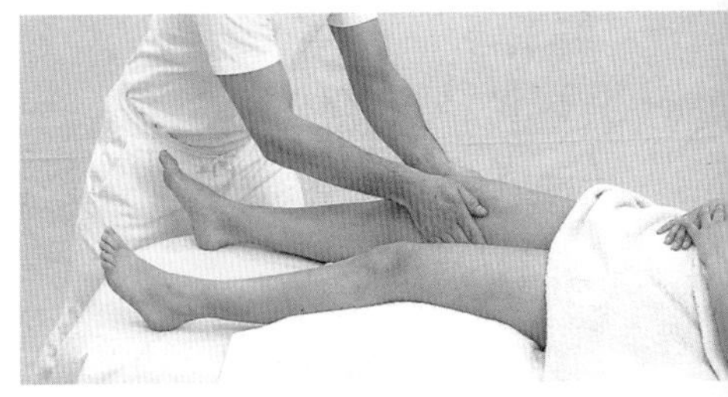

49 draw out the pain

Certain techniques can help to release the pain of a tension headache. Those shown here all involve applying pressure then releasing it. This helps to relax the muscles and draw the pain away.

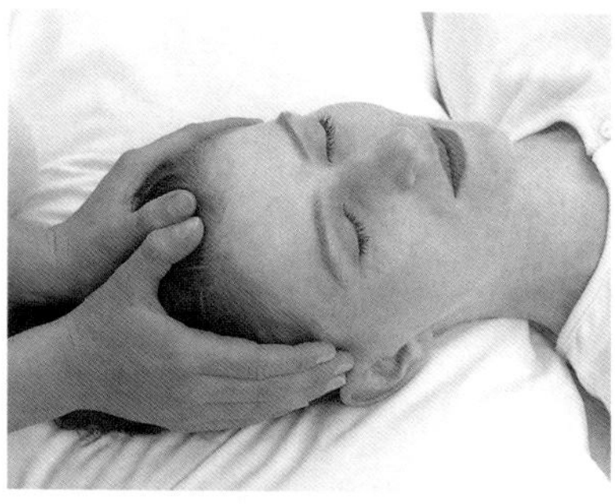

1 Settle your hands lightly around your partner's scalp for a few moments. Keeping your hands in the cupped position, lift them slowly away from the head as if they are drawing out the pain. Cup your hands around the head again, placing your thumbs between the eyebrows. Apply gentle pressure with your thumbs for a count of five, then release.

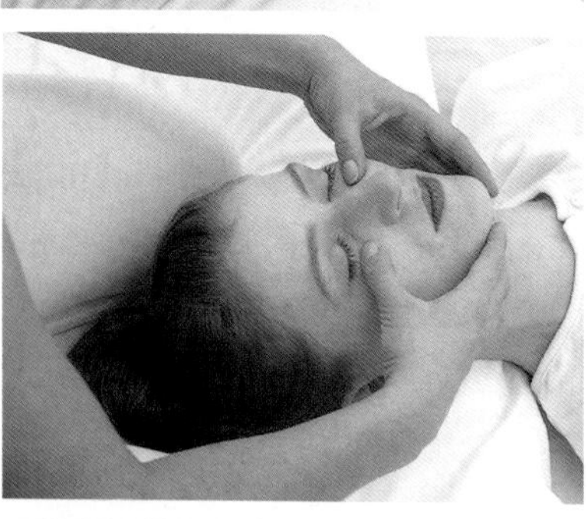

2 Working from inner to outer edge, apply a press/release motion under the ridge of both eyebrows using the tips of your index fingers. Then use your thumb pads to press/release over the top of the cheekbones, working out from the nose to the edge of the temples.

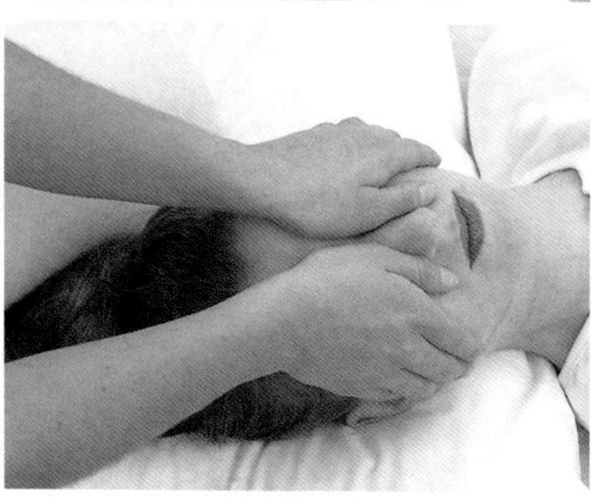

3 Briskly rub your hands together to create heat, then softly lay your slightly cupped palms over the eyes for a count of five to soothe and relax the eye muscles. Withdraw your hands slowly.

50 tension reliever

Tension headaches are a common consequence of stressful lifestyles. This massage can prevent muscle spasm and head pain, particularly if it is done at the onset of a headache.

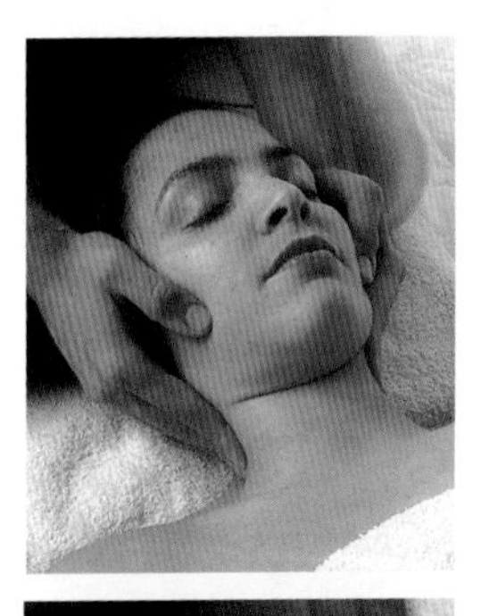

1 Kneel at your partner's head with the head in your lap or on a cushion. Begin by making circles with your fingertips on the muscles at each side of the neck. Continue around the sides of the head and behind the ears.

2 Use the backs of your hands to smooth tension away from the temples. Gently stroke the hands outwards across the forehead to soothe away worry lines.

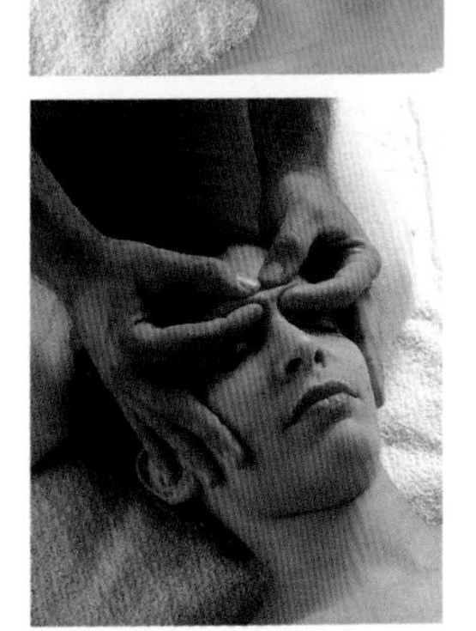

3 Pinch and gently squeeze along the line of the eyebrows, reducing pressure as you work outwards. These muscles may be very tender, so take care.

4 With your thumbs, use steady but firm pressure on the forehead, working outwards from between the eyebrows.

5 Work across the brow to the hair line. This also covers many acupressure points, and will release blocked energy.

index

acupressure points, 61
alcohol, 19
hangover cures, 41, 48
amethyst crystals, 24
antibiotics, 19
anxiety-calmer massage, 53
aromatherapy *see* oils, essential
asanas, yoga, 15
auras, 7

Bach flower remedies, 49
blue, colour, 26
breathing, relaxation, 28
bubble bath, lavender, 37

caffeine, 17, 23
candlelight, 29, 31
causes of headaches, 8, 41, 57
see also triggers, headache
celery juice, 19
chamomile, tea bags as eye packs, 39
chamomile oil, 23
massaging using, 50, 53
cluster headaches, 7–8
clutter, clearing, 32
colour healing, 26–7
computer users, 39
constipation, 19
crystal healing, 24
cucumber slices, 39

diets: elimination, 18
see also food
digestive problems, 19
drinks: before sleep, 21
caffeinated, 17, 21
fresh juices, 19
herbal teas, 44
wood betony infusion, 45
see also water
drugs: causing headaches, 19
to treat headaches, 8

endorphins, 55
eucalyptus, 42
evening primrose, 41
exercise, 13
eye-packs, 39
eyestrain, treating, 12, 39

face: face work-out, 14
massage, 53, 61
fatty acids, Omega-6, 41
feverfew, 40
flowers, scent, 9
food: as headache trigger, 8, 16
healthy eating, 16, 17, 19
intolerance, 18
foot reflexology, 55
frankincense, 31
fructose, 17
as hangover treatment, 48

gamma-linoleic acid (GLA), 41
geranium oil, 38
green, colour, 27

hangover cures, 41, 48
head: head and neck stretch, 14
massage, 50, 58, 61
self-massage, 57
see also face; neck; scalp
healing herb tub, 35
herbal sedatives, 46
herbal steam inhalation, 42

herbal teas, 44, 45
herbal tinctures, 47
homeopathy, 33, 42
for hangovers, 48
honey, 17
hops, 21
hormone replacement therapy (HRT), 7, 41
hydration, 12
hypoglycemia, brain, 48

incense, 31, 32
infusions: sedative, 46
wood betony, 45
insomnia, 21

laughter, 23
lavender, 21
bubble bath, 37
compresses, 43
inhalation, 36, 42
massage using, 50, 53
tinctures, 47
tisanes, 45
"leaky gut", 19
lemon balm tea, 44
lime blossom tea, 44

marjoram: massage using, 50, 53
oil, 21, 38
as sedative, 46
massage: anxiety calmer, 53

drawing out the pain, 60
face, 53
feet (reflexology), 55
head, 50, 58, 61
head self-massage, 57
for migraines, 58
neck and shoulders, 51, 52
scalp, 56
Shiatsu, 54
sleep, 59
temples and forehead, 50
using essential oils, 50, 53, 56, 58
meditation, 7, 29
menopause: headaches, 41
HRT in, 7
migraines, 7
evening primrose oil and, 41
herbal teas and, 44
massage using essential oils, 58
niacin (vitamin B3) and, 18
triggers, 8, 16, 18

neck: massage, 51, 61
stretches, 14
neroli oil, 38

oils, essential, 7
burning, 31, 38
inhalation, 36, 38

massage using, 50, 53, 56, 58
and sleep, 21
steam inhalation, 42

painkillers: drugs, 8
herbal, 42, 43
peppermint, 42
compresses, 43
massaging using, 50
pill, the, 41
PMS (premenstrual syndrome), headaches, 41, 44
posture, 54
premenstrual syndrome *see* PMS
protective bubble visualization, 30

reflexology, 55
Reiki treatment, 34
rose oil, 38
rosemary, 42, 45
inhalation, 36
massaging using, 50

sage, 38, 42
sandalwood, 31
scalp massage, 56
scents: candle, 31
flower, 9
sedatives: herbal, 46
massage as a, 59
serotonin, 40
sex, 20
Shiatsu massage, 54
shoulders, massage, 51, 52
sleep, 21
herbal sedatives, 46
sedative massage before, 59
"sleep cushions", 21
snacks, quick-fix, 17
spells, healing, 25
steam inhalation, herbal, 42
stretches, 14
sugar: causing headaches, 19

low blood sugar, 48
replacements, 17
swimming, 13

teas, herbal, 44
tension headaches, 6–7, 8, 50
colour healing, 26
compresses for, 43
crystal healing for, 24
massages for, 50, 51, 54, 61
stretches for, 14
thyme, 42
tincture, lavender, 47
tisanes, 44
toxicity, 19
treatment: conventional, 8
natural, 9
triggers, headache, 8, 16, 18
turquoise (colour), 27
types of headache, 6–8

valerian, 46
vegetable juices, 19
vervain, 46
visual disturbances, 7
visualization techniques, 29
protective bubble, 30
vitamins: B complex, 16, 19
C, 12, 48

water, 12, 19
and hangovers, 48
wood betony infusion, 45

This edition is published by Lorenz Books

Lorenz Books is an imprint of Anness Publishing Ltd
Hermes House, 88–89 Blackfriars Road, London SE1 8HA
tel. 020 7401 2077; fax 020 7633 9499
www.lorenzbooks.com
info@anness.com

Published in the USA by Lorenz Books, Anness Publishing Inc.
27 West 20th Street, New York, NY 10011; fax 212 807 6813

Published in Australia by Lorenz Books, Anness Publishing Pty Ltd
tel. (02) 8920 8622; fax (02) 8920 8633

This edition distributed in the UK by Aurum Press Ltd
tel. 020 7637 3225; fax 020 7580 2469

This edition distributed in the USA by National Book Network
tel. 301 459 3366; fax 301 459 1705; www.nbnbooks.com

This edition distributed in Canada by General Publishing
tel. 416 445 3333; fax 416 445 5991; www.genpub.com

This edition distributed in New Zealand by David Bateman Ltd
tel. (09) 415 7664; fax (09) 415 8892

A CIP catalogue record for this book is available from the British Library.

Publisher: Joanna Lorenz
Managing Editor: Helen Sudell
Editor: Simona Hill
Designer: Lisa Tai
Production Controller: Joanna King
Editorial Reader: Penelope Goodare
Photographers: Sue Atkinson, Michelle Garrett, Christine Hanscomb, Janine Hosegood, Alistair Hughs, Andrea Jones, Don Last, Liz McAulay and Debbie Patterson
1 3 5 7 9 10 8 6 4 2

GARDEN PONDS

DAVID SQUIRE

GARDEN PONDS

DAVID SQUIRE

Illustrated by Vana Haggerty

TIGER BOOKS INTERNATIONAL
LONDON

Designed and conceived by

THE BRIDGEWATER BOOK COMPANY LTD

Art Directed by PETER BRIDGEWATER

Designed by TERRY JEAVONS

Illustrated by VANA HAGGERTY FLS

Edited by MARGOT RICHARDSON

Managing Editor ANNA CLARKSON

CLB 3510

This edition published in 1995 by

TIGER BOOKS INTERNATIONAL PLC, London

Printed and bound in Singapore

ISBN 1-85501-418-1

CONTENTS

On the Grand Scale 6–7
Water gardening 8–9
Selecting and preparing the site 10–11
Installing a flexible liner 12–13
Installing a moulded pond 14–15
Concrete ponds 16–17
Raised ponds 18–19
Mini-ponds 20–21
Water features 22–23
Bog gardens and wildlife ponds 24–25
Constructing waterfalls 26–27
Pumps, fountains and lights 28–29
Planting 30–31
Fish for garden ponds 32–33
Other cold water pond fish 34
Other pond wildlife 35
Looking after ponds 36–37
Pigmy waterlilies 38
Small waterlilies 39
Medium waterlilies 40
Vigorous waterlilies 41
Oxygenating plants – Ceratophyllum-Tillaea 42–43
Floating plants – Azolla Stratiotes 44–45
Marginal plants – Acorus Hypercium 46–47
Marginal plants – Iris-Scirpus 48–49
Bog-garden plants – Astilbe Lythrum 50–51
Bog-garden plants – Onoclea Trollius 52–53
Repairs to ponds 54
Troubles with water 55
Troubles with fish 56–57
Troubles with plants 58–59
Garden-pond calendar 60–61
Glossary of water-garden terms 62–63
Index 64

ON THE GRAND SCALE

THE fascination of water is magnetic – few schoolboys can pass a puddle without walking through it! But water gardening has a nobler ancestry than this. Water has long been featured in Islamic gardens, particularly by the Arabs who overran Persia, now Iran, in the seventh century and took up the Persian tradition of water gardening. Later the Arabs conquered Syria and passed through North Africa to Spain, where there still remains a rich legacy of water gardens.

EUROPEAN TRENDS

The Italian Renaissance during the fourteenth century led to the construction of the gardens at Villa d'Este at Tivoli, which were begun in about 1560 and completed about fifteen years later. They include magnificent water features, including The Pathway of One Hundred Fountains.

French gardens in the sixteenth and seventeenth centuries featured some large canals, misguidedly explained later by historians just as moats for fortified chateaux, or as extended land drainage. However, the use of reflective pools at Versailles and water

VICTORIAN EXTRAVAGANZA

The combination of indoor water and plant gardening perhaps reached the peak of its development in Victorian times when Warrington plant cases and aquariums were featured in many homes. Fish and moisture-loving ferns were put in the same case, but as they were meant to be sealed units that recycled both oxygen and water, the practicality of this was doubtful: the air surface available to fish was dramatically reduced – and so too was the long-term health of the fish as they gasped desperately for air.

Nevertheless, the idea reveals the vitality and ingeniousness of Victorian gardeners.

DURING *the early 1800s, large estates in Europe often included distinctive water gardens. Here is a Dutch garden with a fountain.*

THIS *luxurious nineteenth-century German winter garden included a water feature.*

parterre at Chantilly suggest much more than water used just for a practical function.

The French style of water gardening was taken to Russia in the early eighteenth century, initially to St. Petersburg (later Leningrad but now called by its former name), where canals and fountains were introduced.

The English also took a liking to ornamental canals, notably at the palace of Hampton Court. However, large informal lakes were always more popular and invariably used on a grand scale.

MOVING RIVERS

Chatsworth House in England is noted for its water landscape which was remodelled several times, first in the French style, and later by the eminent English landscaper Lancelot 'Capability' Brown, who widened and altered the River Derwent. Sir Joseph Paxton, garden writer, plantsman, and designer in 1851 of the Crystal Palace, made further changes including a fountain with a jet of water 88m/288ft high.

FOUNTAINS

Self-acting indoor fountains became popular in Victorian window and indoor gardens. They were formed from two closed, inter-connecting reservoirs and the air pressure created by filling them forced water out in a jet. They were often based on designs by the first-century Greek writer and scientist, Hero of Alexandria. There is little to indicate that they were used in Ancient Rome, but his text Pneumatica *was printed in Latin in the sixteenth century, and in Italian a century later. Many Italian Renaissance water features employed his designs.*

WATER GARDENING

DURING the last fifty years, water gardening has become a popular hobby, with water features constructed in gardens of all shapes and sizes. Even ponds in tubs on small patios are a possibility.

Many different constructional materials are available, each with its advantages. Waterfall units have enabled rock and water gardens to be integrated, with pumps activating waterfalls and fountains. Few other aspects of gardening create as much attention and continued interest as a garden pond.

MATERIALS

Early pools were always concrete, and usually square or rectangular. Now informal shapes are also possible (see pages 16 and 17).

Flexible liners allow ponds of many shapes (pages 12 and 13).

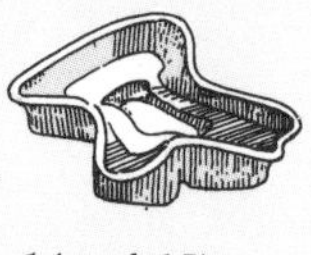

Moulded shells are available in various sizes, shapes and materials (pages 14 and 15).

MOST *ponds are set into the ground and can be viewed from all sides.*

FULLY *or partially raised ponds make construction easier. See page 18 for building methods.*

SOME *ponds are raised above ground level on one side, with the other side set into a bank. This helps to blend one level with another (see page 19).*

EVEN *the smallest patio can accommodate a pond in a large tub. Choose miniature plants (pages 20 and 21).*

CASCADING *water bubbling over a series of waterfalls enlivens the entire garden. When made from flexible liners or rigid waterfall units they are easily installed (pages 26 and 27).*

FOUNTAINS *can be purely functional, with water directly spouting out of a pond's surface. Alternatively, some statues issue forth water and create eye-catching features in gardens. Pumps and a water source are essential (pages 28 and 28).*

WILDLIFE *ponds create havens for aquatic creatures, from fish to frogs and newts. Bird life is also encouraged, as are insects and small mammals (pages 24 and 25).*

SELECTING AND PREPARING THE SITE

THE approximate position of a new garden pond is usually known: you have probably spent at least a year looking out of a window and visualizing the pleasures it will bring. But there are a few constructional as well as positional factors to consider.

The siting of a pond within an existing garden is often a conflict between the desired position and existing features. Selecting a suitable site is detailed on this and the opposite page.

POSITIONING THE POND

It is relatively easy to imagine a pond when installed in an existing garden, but also think of the future and the growth of nearby trees. Your present friendly and considerate neighbours might move and their successors not be so pleased with a wildlife pond immediately on their boundary.

Here are a few considerations regarding position:

• Avoid shaded positions. A pond needs at least seven hours of sunshine a day.

• Do not site it under trees; they not only cast shade but, if deciduous, shed their leaves in autumn. If these are not prevented from falling into a pond, they drop to the bottom, decay and create gases toxic to fish.

• When calculating the shade created by existing trees, try to think what their size might be in ten or more years' time.

• Do not position ponds near trees with poisonous berries or fruits, such as laburnum.

• Large trees nearby may have roots that will puncture liners, crack concrete or rigid shells.

• Some trees, such as plum and Blackthorn, harbour Waterlily Aphids and cause infestations repeatedly throughout summer.

THERE *are ponds of many shapes and for all gardens. Rectangular or square ponds, earlier made of concrete but now often formed from a rigid shell, are ideal in formal gardens and settings.*

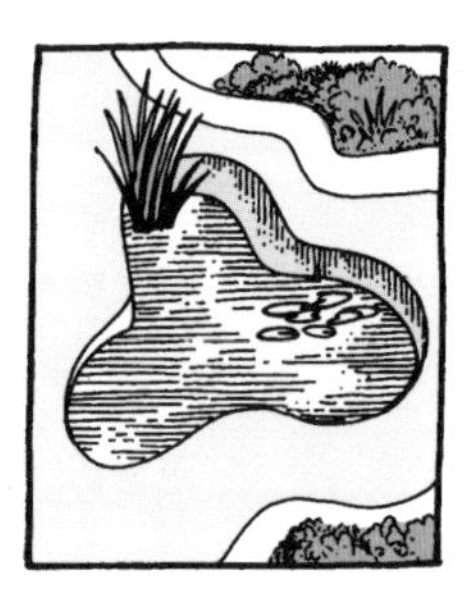

LOBE-SHAPED *ponds introduce an unusual form to patios, or other informal settings. Another possibility is to position a three-lobed pond at the junction of three broad, gravel paths.*

KIDNEY-SHAPED *ponds create an informal effect and are ideal for siting in a lawn, perhaps towards one corner where a focal point can be created. Position the lobed side towards the house.*

EARTH REMOVAL

Perhaps the most daunting task when constructing a pond is digging out the desired area and removing the soil. Landscape gardeners often use skips, but nevertheless barrowing large quantities of soil is a formidable task. Soil naturally bulks up when dug out: topsoil increases in volume by about one-quarter, subsoil by up to one-third. There is a temptation to convert excavated soil into a rock garden, but subsoil should always be removed entirely. Friable topsoil can be scattered thinly on borders, or later used in the construction of a rock garden after the area has been drained.

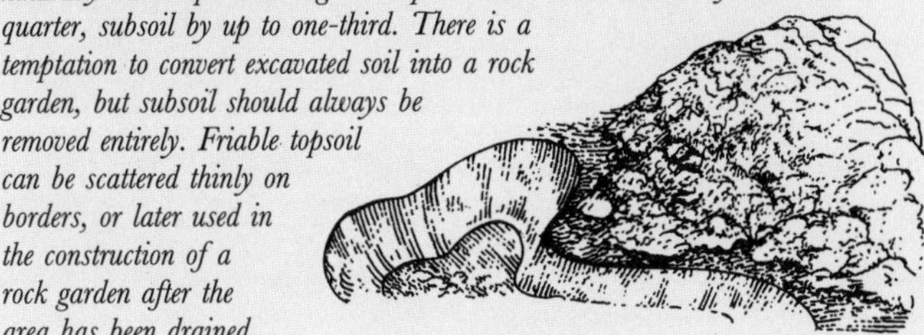

- Avoid positions exposed to cold winds; they singe the foliage of tender plants and may blow over tall, marginal ones. Also, in winter they cause temperatures to fall dramatically. Coniferous hedges 3–4.5m/10–15ft high and positioned 2.4–3m/8–10ft on the northerly side of the pond will reduce the wind's speed but will not cast shade.
- If the land slopes, do not dismiss its suitability for a pond: one side could be level with the surface, with the other raised.
- If waterfalls and fountains are desired, ensure water pipes and electricity cables can be installed without having to burrow under existing patios, walls or paths.
- Ensure the pond is not sited above land drains, power or telephone cables, or mains drainage pipes.
- If an informal pond with a bog garden is being constructed, it is best positioned at the lowest part of a garden. And, if possible, always put wildlife ponds towards the end of a garden.

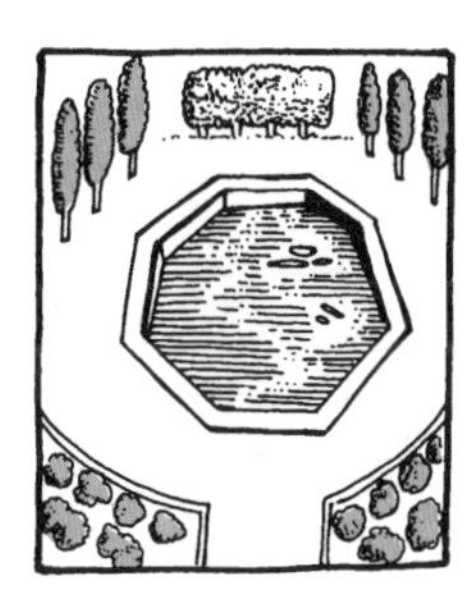

OCTAGONAL *ponds have a distinctly formal nature and are ideal in the centres of patios. Nevertheless, their regular outlines can be softened by using tall, spiky-leaved marginal plants.*

TRIANGULAR *ponds are unusual, formal and ideal in a corner of a patio. Use tall, iris-like marginal plants to soften the sharp corners, and softer, lax, sprawling plants between them.*

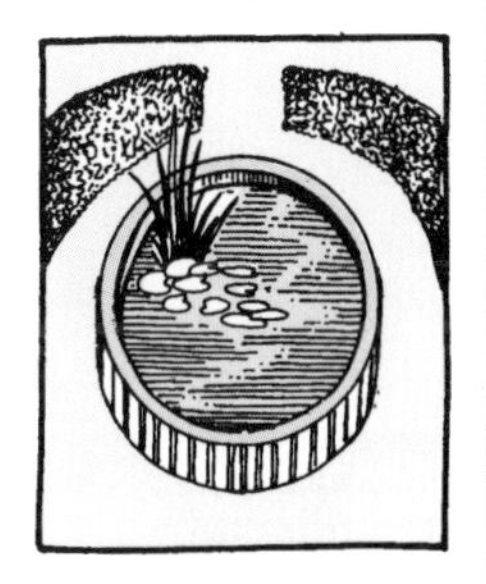

RAISED *ponds create unusual features and are ideal where the disposal of excavated soil is difficult. They can be round or irregular and used in both formal and informal settings.*

INSTALLING A FLEXIBLE LINER

ALSO known as a pool liner, a flexible liner enables ponds of all shapes and sizes to be constructed. Sheeting sizes up to 9m/30ft x 6m/20ft are readily available, while ones above this size can be ordered. Although these liners are flexible, it is more difficult to persuade them to assume the shape of a small corner, so this must be considered when designing the pond. Thin liners, however, are not as durable as thick ones.

MATERIALS
Basically, there are three types of materials for flexible liners. Always buy the best quality you can afford: after thoroughly preparing the site it is false economy to buy the cheapest liner, as it will soon need replacing.

All liners have either price or durability advantages:

• Polyethylene (more widely known as polythene) sheeting creates low-cost, relatively short-

ESTIMATING THE SIZE OF THE LINER

This is not difficult and best carried out after the area has been excavated. Measure the total width and length, then add twice the total depth to each measurement. As well, add 30cm/12in to both the length and width to allow for the flaps that will have paving stones put on them. Examples of these measurements are:

• *Pond A: 2.4/8ft long, 1.8m/6ft wide and 60cm/2ft deep. The sheet needed would be 4m/13ft x 3.3m/11ft.*

• *Pond B: 3.6m/12ft long, 1.8m/6ft wide and 90cm/3ft deep. The sheet needed would be 5m/19ft x 4m/13ft.*

If possible, buy a slightly larger sheet than necessary, so that there is no worry about it fitting and there is plenty of width for the flaps.

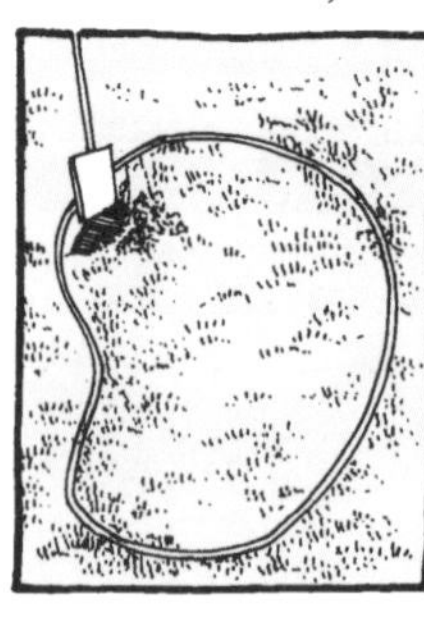

1. LAY *a hose-pipe or thick rope on the surface to form the size and shape of the pool and use a spade to cut the outline on the turf or soil. Then, mark an area about 25cm/10in larger all round the pool.*

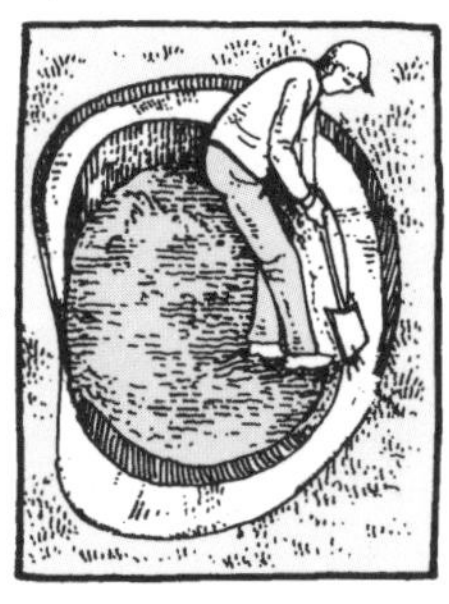

2. IN THE *outer strip, remove soil to 7.5cm/3in deep. Then, dig out the central area to about 75cm/2 1/2ft deep, at the same time forming a shelf about 23cm/9in wide at a similar depth.*

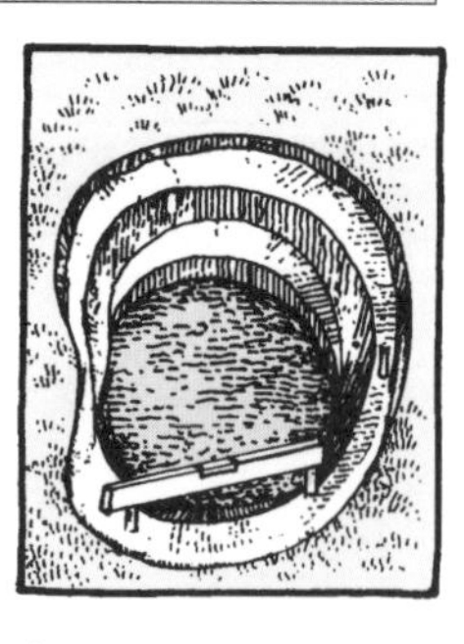

3. CUT *pegs 15cm/6in long and mark each one 5cm/2in from its top. Knock these in, 7.5cm/3in deep and 90cm/3ft apart, around the top and on the shelf. Use a spirit-level to check their tops are level.*

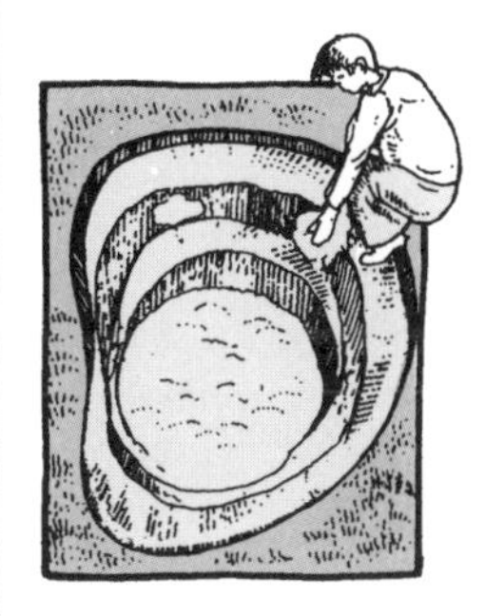

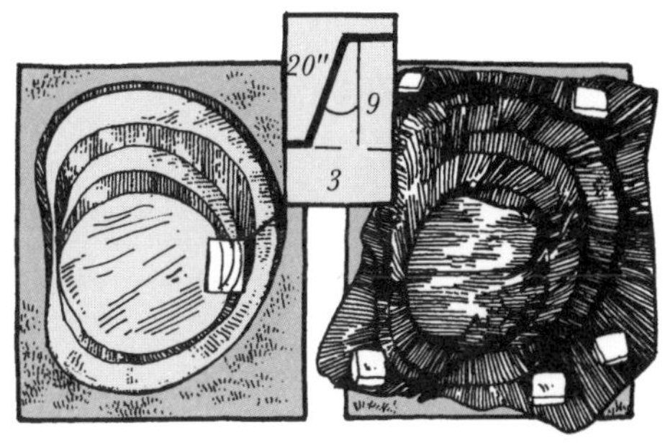

4. WHEN *the pegs are level, the pencil marks will indicate the levels. Make adjustments as necessary. Remove sharp stones and form a 3.6cm/ 1½in-thick layer of sand in the base. Fill holes in the sides.*

5. CONTINUE *to pad the sides with soft sand, forming a 20-degree angle. Put a layer of sand on the shelf. Slightly dampen the sand and smooth its surface. If not level, irregularities will show up later.*

6. PLACE *the flexible liner in the hole, taking care not to disturb the sand. As necessary, form pleats to reduce excess liner around the sides and in the base. When in position, place bricks along the edges.*

lived ponds. Ultra-violet rays soon degrade it, especially if there is always a gap between the top of the pond and the water's surface. For this reason, it is unsuitable for small ponds in a series of waterfalls. Better-quality types are guaranteed for ten or so years.

• PVC liners are strong and about half the price of butyl sheeting. Some PVC liners are reinforced with netting. Whatever the grade, look for a guarantee and buy the best quality you can afford.

• Butyl sheeting is the best of all lining materials and has the longest life expectancy. It is unaffected by ultra-violet rays and will last twenty years or more – even fifty. Look for a guarantee and indication of its life-expectancy. Because it is unaffected by sunlight, it can be used to form pools in waterfalls.

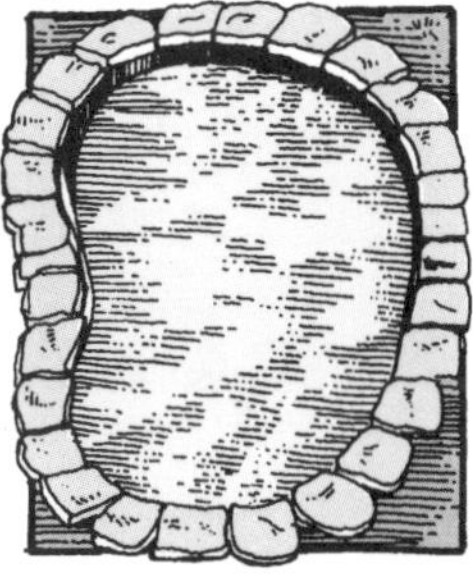

7. PLACE *a hose-pipe in the hole and gently fill with clean water. As the water level rises, remove the bricks and, as necessary, readjust the liner. Creases can often be removed by manipulating the liner.*

8. USE SHARP *scissors to cut off surplus liner to leave a 15cm/ 6in flap. This is turned over, flat on top of the sand. It is essential that this flap is quite wide as otherwise it may fall back into the pond.*

9. USE THREE *parts soft sand and one of cement as a bedding mixture for the edging slabs. Each stone's top must be flush with the surface and overhang the pool's edge by 5cm/ 2in. Cement between them.*

INSTALLING A MOULDED POND

SOMETIMES called rigid liners, pre-formed ponds and moulded shells, these 'ready-made' ponds enable water gardens to be established very rapidly. Sometimes, even the largest shells may be too small for a very large garden: however, several can be linked together by a series of waterfalls – but they do need the benefit of a sloping garden.

These shells can also be used fully or partially in the construction of raised ponds (page 18).

RANGE OF MATERIALS

Earlier, the life expectancy of a moulded pond was not great and many experts disapproved of them because of their lack of depth – sometimes less than 38cm/15in – and their garish, unnatural colours. Nowadays however, they are available in a wide range of colours and often have a life-span of twenty or more years. Also they are deeper.

During the 1950s, metal-shelled ponds were available, but these were only 25cm/10in deep and not practical as the water soon froze solid in winter. Today, there are three main materials:

- Plastic liners are the cheapest, only semi-rigid and usually created from vacuum-formed polythene in a range of colours, including green, brown, blue and grey. Careful installation is vital to ensure the surfaces are not stressed. With age, ridges and corners crack, while the surface deteriorates when exposed to strong sunlight. Therefore, it is essential that the edges are covered with paving slabs and the water-level is not allowed to drop so that the sides are exposed to direct light.
- Reinforced plastic liners are more rigid and resistant to damaging rays of

RANGE OF SHAPES

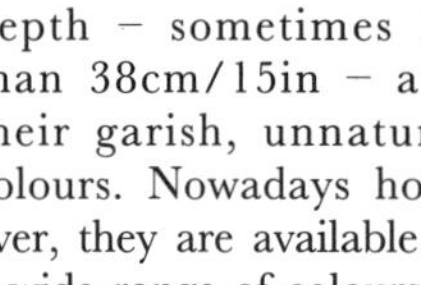

LEFT: *Rectangular shells are ideal for formal ponds, and available in several sizes.*

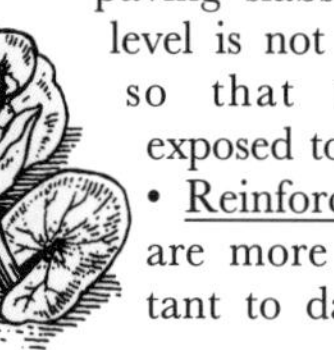

LEFT: *Round, formal shells with planting shelves are suitable for fitting into formal patios or paved surrounds. Position the shelf area on the far side.*

BELOW: *Squat, irregular-shaped ponds can often be fitted into corners.*

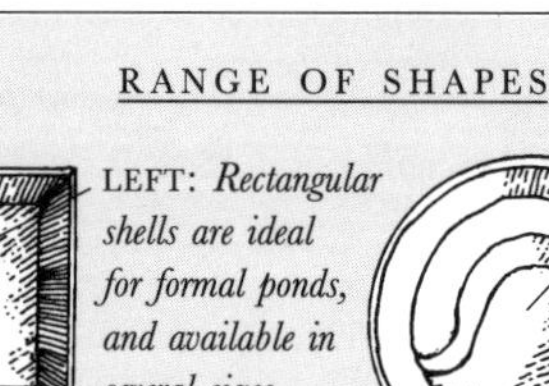

ABOVE: *Large, long, informal, irregularly shaped shells are ideal as a base pool for a water cascade.*

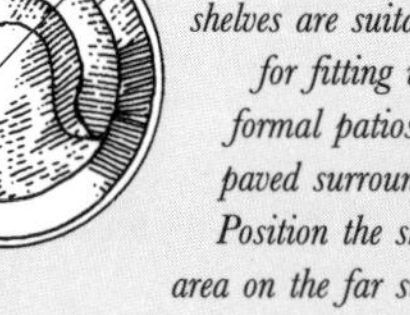

BELOW: *Half-round shells are especially attractive when installed against a wall.*

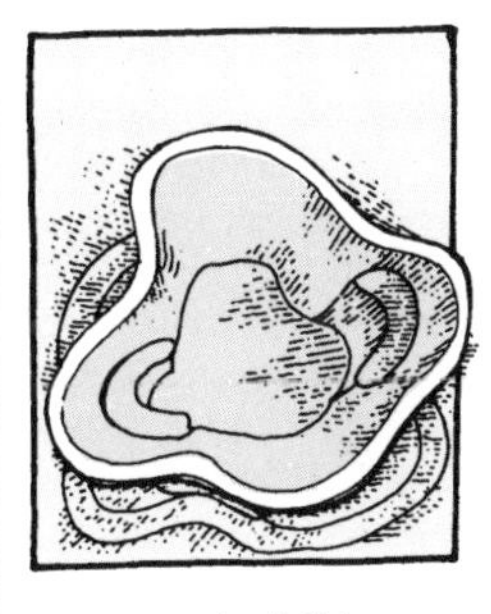

1. PLACE *the shell in position and mark on the ground the extremes of its outer edge, as well as those of the shelves. Also, mark an area 25cm/10in wide to accommodate a row of edging slabs.*

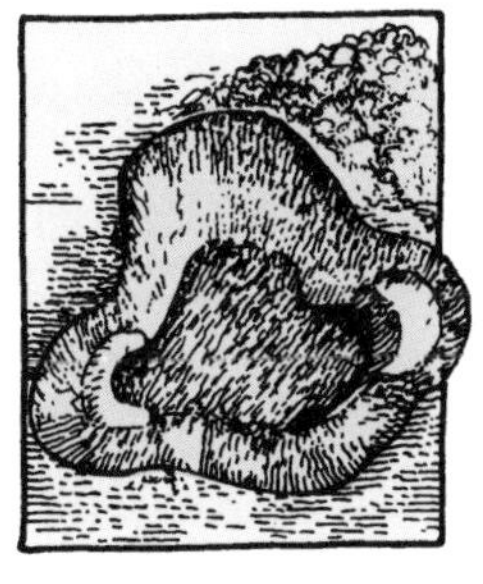

2. DIG OUT *the area to the total size of the shell, at the same time forming the shelves. The bottom should be about 5cm/2in deeper than the shell. Also, dig out the edging area, about 7.5cm/3in deep.*

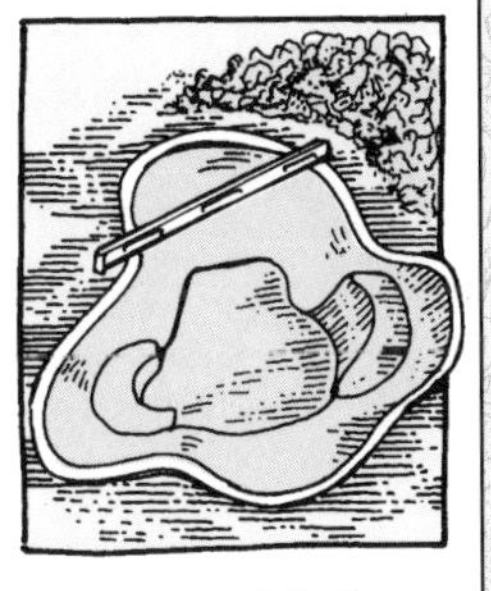

3. ENSURE *soil in the hole's base is firm and compact, then add a 5cm/2in-thick layer of soft sand. Put the shell in place and firm it on the sand. Use a builder's spirit-level to check that the rim is level.*

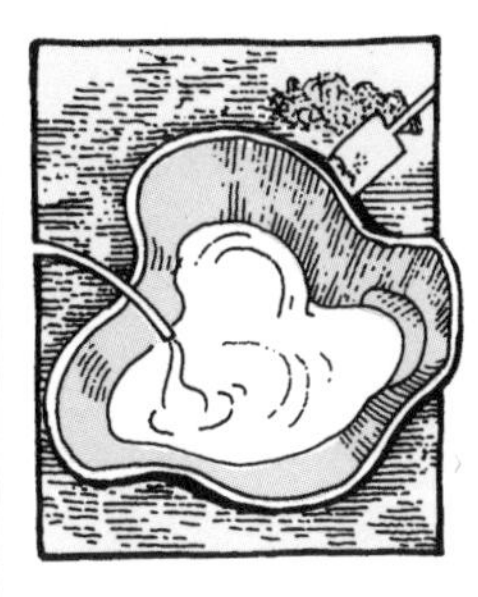

4. WHEN *level, pack soft sand or friable soil between the shell and ground. Firm it with the end of a stick. If gaps are left, they unduly stress the shell which may crack in the future. Fill the shell with clean water.*

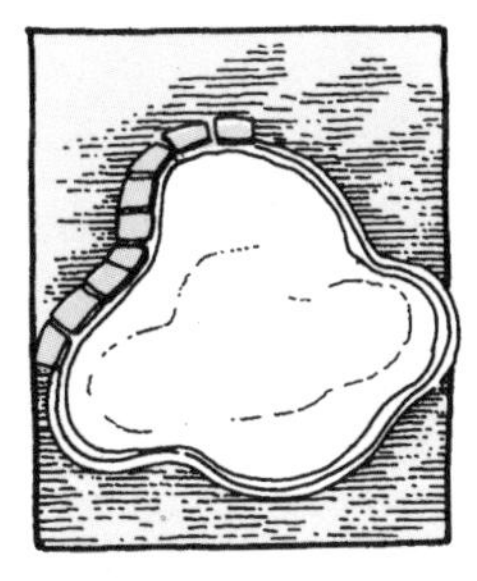

5. CHECK *that the shell fits on top of the outer area. If necessary, pack soft sand beneath it. Then, form and firm a layer of sand level with the shell's top but about 5cm/2in below the surrounding lawn.*

6. CEMENT *edging stones around the pond, so that they overhang by 5cm/2in. Use slabs 30cm/12in wide; narrow ones may topple into the pond. Then, add the plants (pages 30 and 31) and, later, the fish.*

ultra-violet. The best types are guaranteed for at least ten years, sometimes twenty, and provide a water depth of 45cm/1½ft or sometimes more.

• Glass-fibre shells are formed of glass fibres bonded with a resin, creating a material with a life expectancy of twenty or more years. These shells are not damaged by ultra-violet rays, nor do they rot or allow water to seep through. However, like all shells they need careful installation to prevent them being stressed when placed on an uneven base. They are available in green, blue, black and grey.

Because glass-fibre shells are not damaged by ultra-violet light, they are often used as pools in a series of waterfalls.

CONCRETE PONDS

GARDEN ponds have traditionally been made of concrete and were usually square or rectangular. This involved vast preparations with timbers and shuttering that were only used for a week or so. However, it is also possible to make a pond without the use of this shuttering, but it has to be a simple, uncomplicated oval or round shape, although a kidney outline is possible.

CHOOSING THE RIGHT DAY

Late spring and early summer or late summer and early autumn are the best times for amateurs to lay concrete, when there is no risk of frost and the weather is not excessively hot. It is essential that concrete dries slowly and evenly. After being mixed, concrete takes about two hours (less on a warm day) to begin setting, three or four days before it gains any appreciable strength and a week before it reaches half its ultimate strength. Many experts suggest it takes four or five weeks to reach full strength. During this time it is easily damaged, especially at the edges. If the weather gets too hot soon after concrete is laid, cover with planks and damp sacking, and keep it moist for about five days – slightly longer if very hot.

AFTER HARDENING

Tidy up the site, but do not tread on the edges. Paint the pond's surface with a proprietary sealant, allow to dry and fill with water. If a surface sealing material is not used, it is essential to fill and empty the pond several times to remove toxic chemicals from the concrete. Some experts also advise filling the pond a couple of times even when using a sealant. Kits to test the water to ensure it is not too alkaline are available (see page 55). Using a pump (see page 36) to empty the pond makes the task much easier. When the water is as required, introduce plants and fish.

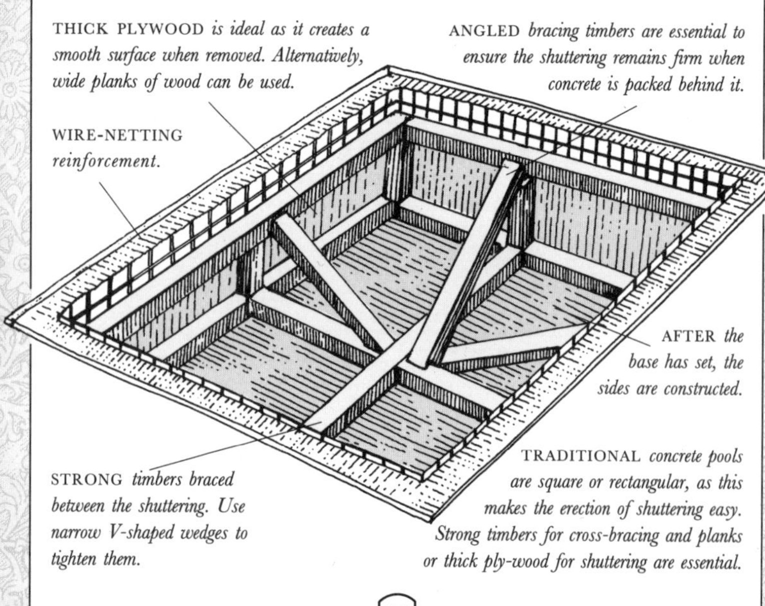

THICK PLYWOOD *is ideal as it creates a smooth surface when removed. Alternatively, wide planks of wood can be used.*

ANGLED *bracing timbers are essential to ensure the shuttering remains firm when concrete is packed behind it.*

WIRE-NETTING *reinforcement.*

AFTER *the base has set, the sides are constructed.*

STRONG *timbers braced between the shuttering. Use narrow V-shaped wedges to tighten them.*

TRADITIONAL *concrete pools are square or rectangular, as this makes the erection of shuttering easy. Strong timbers for cross-bracing and planks or thick ply-wood for shuttering are essential.*

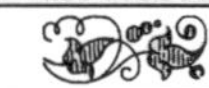

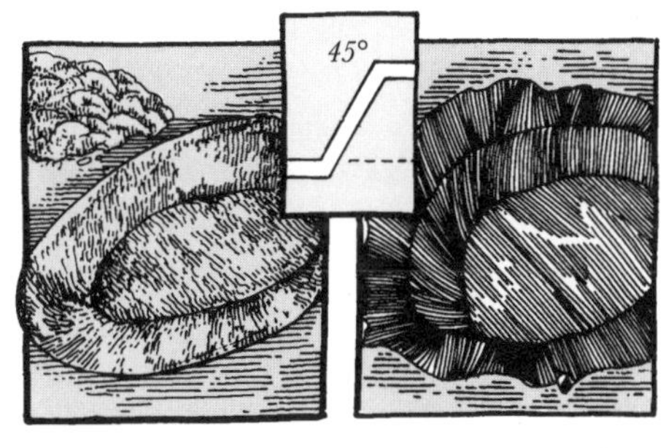

1. SELECT *a suitable site (see pages 10 and 11) and dig out a hole, making it 15cm/ 6in deeper than desired and 30cm/ 12in wider to allow for the thickness of concrete.*

2. CREATE *sides with a 45 degree angle. Shuttering is not used and if the sides are steep the concrete will slip down before it sets. Line the hole with thick polythene sheeting.*

3. SPREAD *a 10cm/ 4in-thick layer of concrete over the base and up the sides. Making the mixture slightly stiffer than normal ensures it does not slip off the sides. Spread it evenly.*

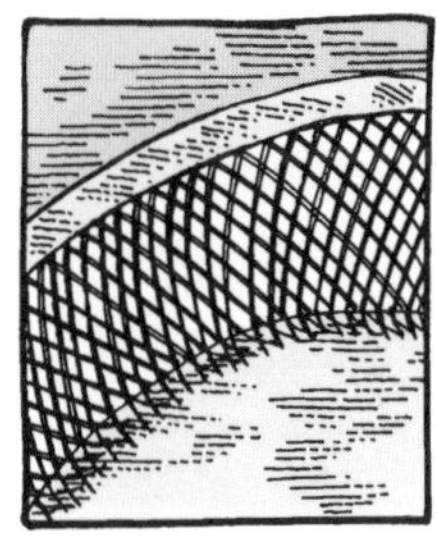

4. TO CREATE *added strength, press wire-netting reinforcement over the sides and 23cm/ 9in into the pond's centre. Cover the base with another piece of wire-netting, cut to shape.*

5. ENSURE *the wire-netting is bedded into the concrete. Then, lay a further layer of concrete, 5cm/ 2in thick, over it. Press it firmly into position using a trowel to smooth the surfaces.*

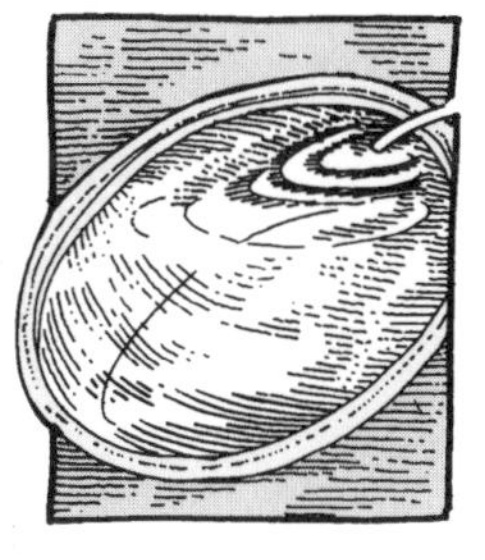

6. ALLOW *the concrete to dry thoroughly, but not too quickly as cracking may then occur. During hot weather, place planks across the pond and cover it with damp hessian to slow drying.*

MIXTURES AND MIXING TIPS

- *Use a mixture of 1 part cement powder, 1½ sharp sand, and 2½ 20mm aggregates. Alternatively, use 1 part cement powder and 3½ of combined aggregates.*
- *Always buy fresh bags of cement.*
- *Ensure the sand and aggregates are not contaminated with rubbish such as wood, soil or plants.*
- *Do not try to tackle the job on your own. With two or three people helping the job is a pleasure; on your own it can be a nightmare.*
- *An electrically-operated mixer is well worth hiring and will certainly repay its cost.*
- *Keep children and domestic pets indoors while working.*

RAISED PONDS

FEW water-garden features are as impressive as a raised pond with a fountain or cascade of water tumbling into a lower pool. The main problem with a raised pond is its exposure to low temperatures that damage brickwork. Also, there is a risk of ice pushing the sides outwards.

There is a greater risk of fish being killed, but a heater (see page 36) can reduce this.

It is often claimed that raised ponds are safer for toddlers than ground-level ones. However, as most children are inquisitive and would consider the sides of a pond as a challenge, the only way to make it safe is to form a strong wire-netting covering.

Ensure that an electricity supply is available during winter for water heaters, but first consider the cost of electricity during very cold weather.

MANY ADVANTAGES

Apart from the dangers from low temperatures and inquisitive, agile young children, raised ponds have many advantages:

- Wheelchair gardeners are able to admire the plants and fish without the risk of rolling in.
- Looking after the pool is often easier because it is not necessary to bend down continually.
- When the pond needs emptying, water is syphoned out cheaply and easily, rather than having to buy and power a pump.
- It also overcomes the major difficulty of disposing of subsoil that is necessary when constructing a ground-level pond. Topsoil can be spread over borders or eventually used to construct a rock garden, but thick, boot-loving clay is another matter and must invariably be removed, usually an expensive task.

RAISED PONDS *need not be formed at one level, but can be a combination of several. In this way, waterfalls and ornamental fountains may be used in the design and a more interesting water feature created.*

RAISED *ponds can be made from concrete and bricks. A wall two bricks thick (about 23cm/ 9in) is essential to withstand the outward pressure of water. Then, render the inside. Strong foundations for the wall and base are vital if parts are not to settle within a few years. Then, cover the inside with a couple of coats of a sealant; allow each coat to dry thoroughly.*

STRONG, *rigid shells are ideal and if made of glass fibre last twenty or more years. Place the shell on a 5cm/ 2in-thick layer of soft sand over a strong, well-compacted or concrete base.Construct a surrounding wall of house bricks or strong, ornamental blocks, ensuring the shell's rim fits over the top course. Cement a row of coping stones over the liner's edge.*

FLEXIBLE, *strong liners – preferably butyl sheeting – are essential, as they will have to withstand low temperatures. Strong foundations and walls two bricks thick are vital. Additionally, an insulating layer of polystyrene between the liner and wall is advisable. Ensure the top edge of the liner is secured in a layer of mortar, under the coping stones.*

PONDS ON SLOPES

Sloping gardens can make gardening difficult, but they also create opportunities to make your garden unique. A pond half-way down a slope, with a terrace of weather-worn flagstones on either side, always captures attention. It is vital to build strong foundations for the pond's wall which must be two bricks thick. Also, if the pond is made from a rigid shell, a base of compacted hardcore with a thick layer of soft sand on top ensures the weight of the shell and water is spread out.

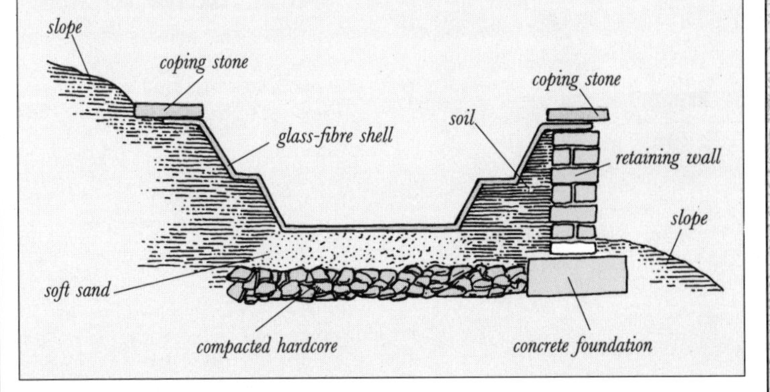

MINI-PONDS

IN SMALL gardens, where a large, raised or ground-level pond cannot be accommodated, a miniature pond in a large tub or stone sink is an ideal way to continue water gardening.

Unfortunately, a tub of water on a patio immediately attracts children and it therefore needs to be covered with wire-netting. Even though it is just a tub of water, children can still come to harm and care is always needed.

LARGE *tubs and deep, stone sinks are ideal containers in which to form miniature ponds. When planted with small waterlilies and other aquatic plants they create fascinating features. Fish can also be added to them.*

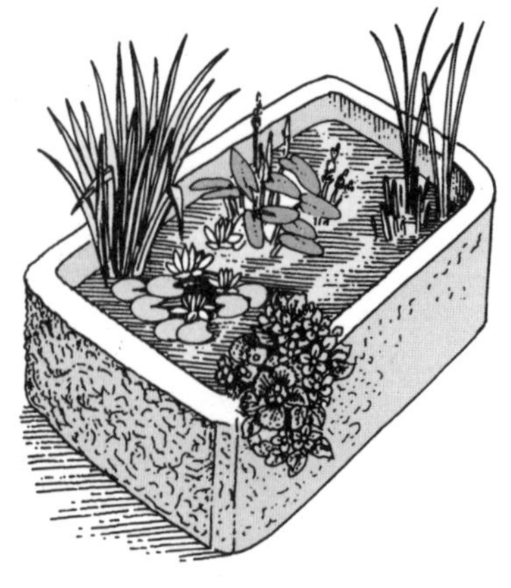

WINTER CARE

The greatest danger to miniature ponds in tubs or stone sinks is cold weather and extremely hot sunshine. Hot sunshine mainly affects stone sinks; water in them becomes much warmer than a similar volume in a wooden container. However, it is in winter that the main problem arises as the water may freeze and kill the fish and plants. The best solution is to move the container into a cool greenhouse in early winter. Alternatively, pack plastic bags filled and sealed with straw or hay around the sides, but this can look unsightly and is not very effective. Sometimes, miniature ponds are let into the patio's surface, with slabs around them. The water temperature drops less quickly than in tubs, but it is still necessary to remove the fish and plants in winter.

SAFETY-FIRST POOL

Another small water feature is a 'mini-pool'. This uses water but is not deep enough to cause danger to children.

These mini-pools are similar to larger features detailed on pages 22 and 23, but are smaller and just have water trickling over large, attractive pebbles.

It would be wishful thinking to believe that children would not be attracted by them, but as long as the only danger is wet knees or dungarees they become an acceptable risk to many parents. But do not consider blending this feature with a children's sand-pit as it soon becomes clogged and messy. Only strong jets of water will then clear out the sand.

1. BEFORE *positioning and filling a tub with water, check that it is sound and that the base is not rotten. If decayed, discard and seek a better container. If the tub is in good condition it can be immediately filled with water and thoroughly washed. At the same time, check that it is waterproof: occasionally, the sides shrink but when filled to the brim with water for a few weeks the wood expands and seals the gaps. During this period, place the tub in a cool corner, away from strong sunlight.*

2. IF THE *tub continues to leak it is necessary to line it with thick, black flexible polythene. Form pleats inside so that it looks neat. Fill with water and when full cut off the polythene level with the tub's top (below).*

3. WHEN *full of water, it can be planted. This is done in late spring or early summer, the latter being the best if the weather is still very cold. Use the same technique of planting as for plants in ponds (pages 30 and 31). Use plastic mesh baskets. Where the leaves of waterlilies would become submerged, initially place the basket on a few half-bricks. Remove the half-bricks progressively as the plant grows bigger, so that it is slowly lowered down into the water.*

PLANTS FOR MINI-PONDS

Waterlilies:

- Nymphaea *'Aurora': Colour first pinkish yellow, then orange and later red. It spreads about 45cm/ 1½ft.*
- N. *'Candida': Cup-shaped, white flowers with golden stamens.*
- N. *'Graziella': Free flowering, with orange flowers that reveal orange-red stamens. It has the bonus of having leaves splashed in purple.*
- N. pymaea alba: *Small, white flowers with yellow stamens. The flowers, usually only 2.5cm/ 1in across, are borne freely. It grows in shallow water and therefore needs protection against falling temperatures and freezing water in winter.*

Marginals:

- Carex stricta *'Bowles Golden': About 45cm/ 1½ft high, with narrow golden leaves. Plant it in water 5cm/ 2in deep.*
- Scirpus tabernaemontanii *'Zebrinus' (Zebra Rush/ Porcupine Quill Rush): Quill-like stems banded in green and white. Plant in 15cm/ 6in of water. Grows to 90cm/ 3ft.*

Floaters:

- Hydrocharis morsus-ranae *(Frog Bit): Bright green leaves; but be prepared to cut it back regularly.*
- Pistia stratiotes *(Water Lettuce): Floating leaves die down in winter.*

WATER FEATURES

SMALL water features are ideal on patios, where they can be selected to fit in corners, against walls or just as 'stand-alones' to introduce interest to central areas.

The style of this sort of water gardening is not to depend on fish and plants to create interest and vitality in a garden, but rather just cascading or splashing water falling on pebbles or trickling gently from a small fountain. It is a style of water gardening that is ideal when children are small.

WATER SPLASHING *on pebbles createsa soothing and relaxing feature. Neither fish nor plants are needed.*

The concept of fishless and plantless water gardens is rather Spanish and Arabian, and popular where water features are included in patios: properly, these are areas surrounded by the walls of a house. Windows overlook them, which continually have shelter and shade – essential in hot climates.

These water features need not be large and expensive. Indeed, the monotony of a large patio can be relieved by removing four or so paving slabs, installing a small, low-powered pump to power a jet of water 38cm/15in or so high and to fill the area with large attractive pebbles.

SITING THE FEATURE

Gardens need to be a mixture of features that immediately attract attention and those that one comes across perhaps after turning a corner. Clearly, one of these water features in the middle of a

WATERWHEELS

Water gardens can be enlivened by installing a miniature waterwheel. These also look good without a pond, perhaps merged into one side of an informal patio with water splashing on large pebbles and appearing to vanish underground. Alternatively, the wheel can be installed at the base of a series of cascades, or as a topmost feature where it can be used to introduce water to the top pond.

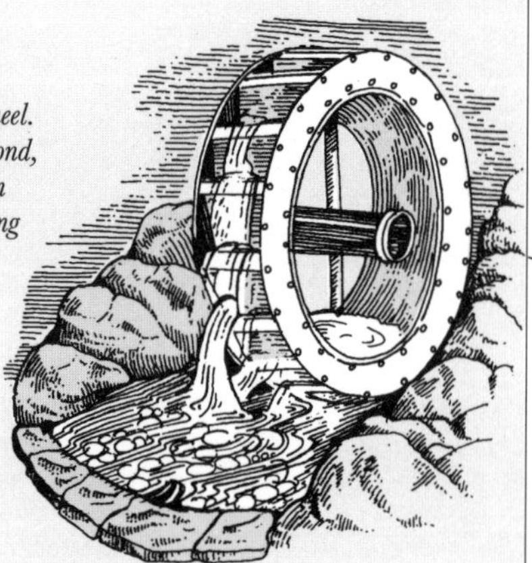

patio is less of a focal point than a centre of immediate interest. As such, it is useful for drawing people to an area from which the main garden can be seen, or for directing patio traffic to one side or another.

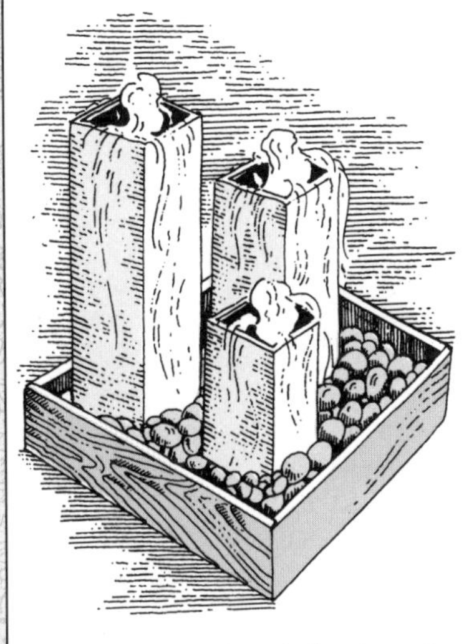

WATER FEATURES *can be formal as well as with rounded and relaxed edges. Here are box-like structures set in pebbles and ideal in a clinically formal setting. The water tumbles and cascades over the sides. This is more relaxing than tall jets of water.*

GARGOYLES AND LIONS

Spouts of water need not just be upright. Indeed, gargoyles or lions spouting water into a pond make unusual features and are certain to capture attention. To prevent these water spouts causing erosion, they should fall on water.

Some small statues, such as Peter Pan and cherub types, issue gentle spouts of water that are attractive and relaxing. The water usually appears from hand-held nozzles or Pan pipes. Others have flower-like receptacles that slowly spill water into a lower container.

FOUNTAINS

Fountains used in pool-less designs should be modestly sized, with spires of spouting water no more than half the height of the pedestal. For water features at ground level, a jet of water with a height no more than half the width of the feature is best. This does not then dominate the shape of the base.

In addition to traditional fountains with spires of water, some form a hemispherical film of water that does not excessively disturb the water itself. Others have geyser-like jets, forming columns of milky-white water. These fountains are superb where they splash on to large pebbles.

A further type is a jet of water which enters from a feature at the pond's side. Such a jet should fall on water rather than pebbles, as it is then less erosive.

Do not have tall spires of water, as they look unbalanced.

BOG GARDENS AND WILDLIFE PONDS

BOG GARDENS, when planted with moisture-loving plants, help informal ponds to blend with their surroundings. The range of plants suitable for planting in them is wide and includes moisture-loving ferns, herbaceous perennials and many primulas which, when planted in large drifts, create spectacular features.

Some boggy areas are natural, but they can also be created. Do not make the area too wide as it is then difficult to put in the plants and to look after them.

KEEP MOIST

From spring to autumn, regularly check that the soil remains moist, but not waterlogged. When constructing the boggy area, ensure that the surface of the compost is fractionally below the water surface in the pond. This often helps to keep it moist. Nevertheless, the compost must not become saturated as this quickly encourages the roots to rot and the plants to die. If soil drainage does become a little problematic, remove some soil and pierce the flexible liner several times with a garden fork. If at all possible, spread further grit directly over the liner.

The Skunk Cabbage (Lysichiton americanus).

BOG GARDEN

Soil that remains moist throughout the year is essential for the growth of bog garden plants. Many of these are illustrated and described on pages 50 to 53.

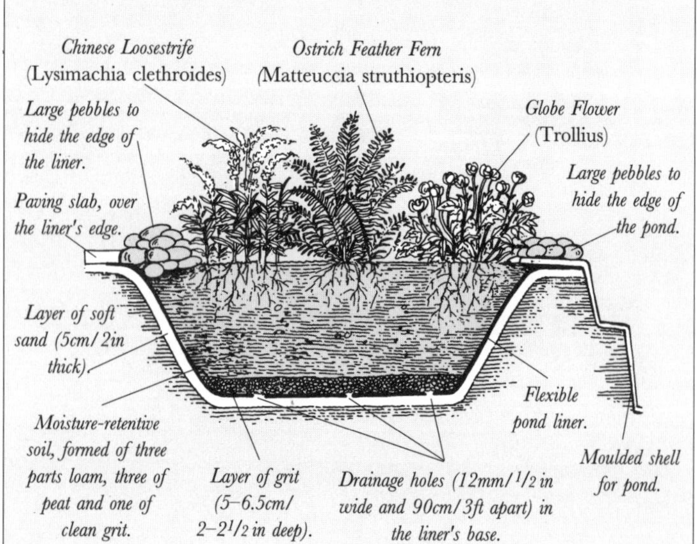

1. UNLESS *there is a natural boggy area alongside your pond, it is necessary to create one – but it should not become waterlogged. Dig out a hole 38–45cm/15–18in deep, giving each side a gentle slope.*

2. SMOOTH *the sides and form a 5cm/2in layer of soft sand over the base and sides. Place a flexible liner in it. The pool-side edge should be large enough to tuck under the side of the pond's rigid shell (or flexible liner).*

3. TRIM OFF *the liner's edges (other than at the pond's edge), but leave an overhang of at least 15–23cm/6–9in. Place a few pebbles on top to ensure the liner remains in place along the edges and sides.*

4. PIERCE *12mm/½in holes every 90cm/3ft in the liner's base, then form a 5–6.5cm/2–2½in-thick layer of clean grit in the base. Fill the hole with fertile, moisture-retentive compost.*

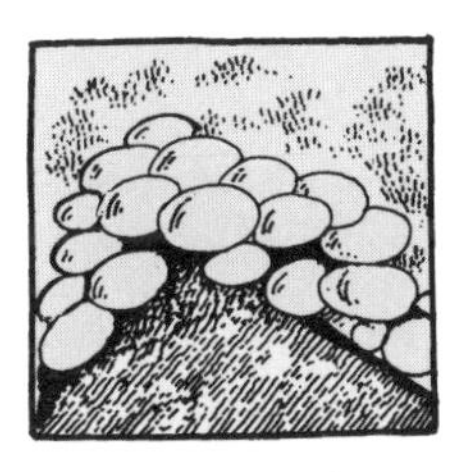

5. COVER *the edges of the liner with pebbles to make them more attractive and to prevent deterioration in the material from sunlight. Thin liners are especially prone to damage in this way.*

6. START *planting from the centre of the bog garden outwards. If necessary, stand on planks to prevent the soil becoming consolidated which eventually can make it become waterlogged.*

WILDLIFE POND AND NATURAL ENEMIES

Bog gardens are essential parts of wildlife ponds, enabling plants to be grown that provide havens for insects, birds, amphibians and small mammals. Use a flexible liner to make an informal pond, with a bog garden alongside it. Within a couple of years, frogs and newts will be established in the pond. There may also be a few mosquitoes in summer, but these will be controlled when fish eat their eggs. They are only a menace in stagnant water. Every inhabitant in your wildlife pond will have its own enemies, and that is part of nature. But garden chemicals are a different problem and, if used, they soon find their ways into wildlife food chains. Do not use chemicals to control insects or disease, nor herbicides to kill weeds in lawns alonside ponds.

CONSTRUCTING WATERFALLS

WATERFALLS add sparkle to water gardens, but they only usually look right in informal settings – fountains are better in clinical, formal ponds (see pages 28 and 29). Construction materials include concrete, pre-formed rigid shells, and flexible liners (butyl types, as they are resistant to ultra-violet rays).

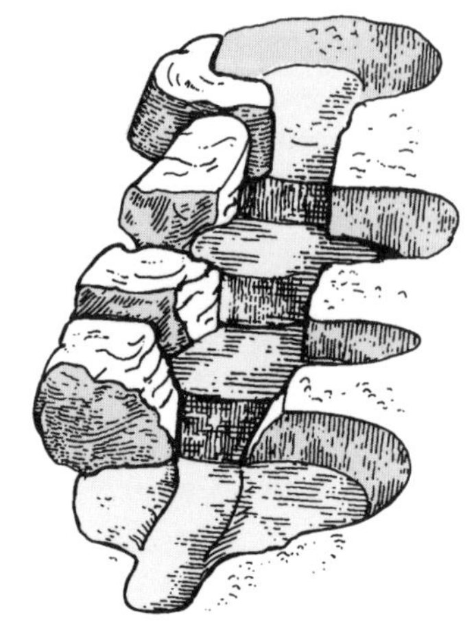

IN EARLIER *years, watercourses and waterfalls were always made of concrete. It is a versatile material and forms waterfalls of many shapes and sizes. However, mixing and moving concrete is heavy work and it may need reinforcement with wire-netting if the pool is large* *(see pages 16 and 17)**. Nevertheless, its ability to be moulded into shapes that personalise the waterfall to your garden often compensates for the added work. Paint all surfaces with a proprietary water-proofing sealant. Excavate each pool area deeper and wider than needed to allow for 10–15cm/4–6in-thick concrete. Do not economize on its thickness.*

SELECTING A PUMP

If the head – the difference between the water's surface in the main pond and that in the top waterfall – is less that 1.2m/4ft, the size of pump needed is much smaller than if the head of water is 1.8m/6ft or more. Also, if the width of each sill area is 15cm/6in or less this too reduces the volume of water that a pump needs to supply to the topmost pond. Keep all of the sills about the same width.

Do not select a small pump because it is cheap, as the resulting flow over the sills will be disappointing. Manufacturers of pumps will soon be able to advise about the size needed.

PRE–SHAPED SHELLS

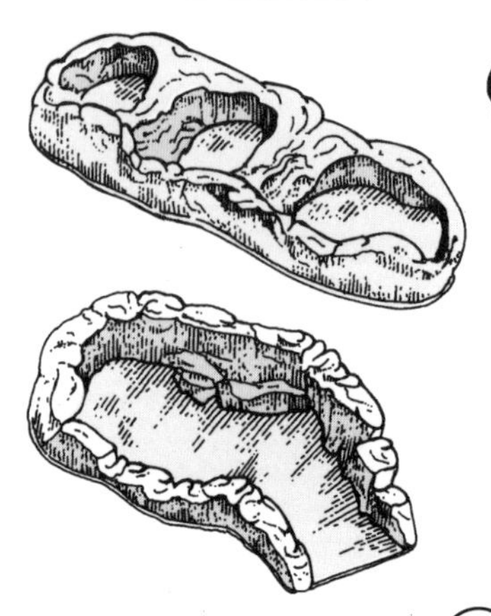

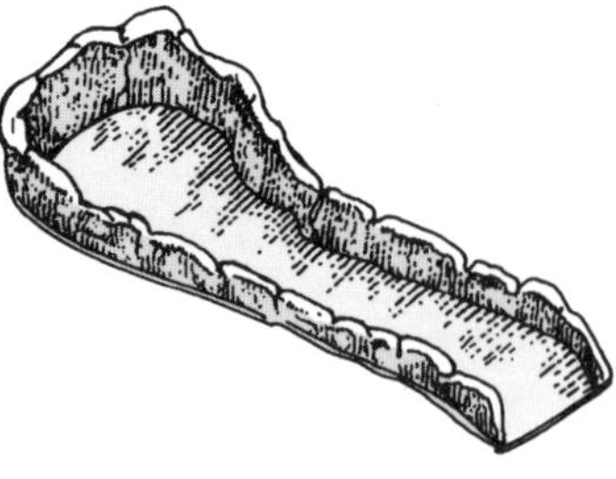

RIGID SHELLS, *formed of either individual cascades, several together or just a watercourse, are available in a wide range of shapes and sizes. Preferably, use glass-fibre types, as they are long lasting and resistant to ultra-violet rays.*

USING RIGID SHELLS

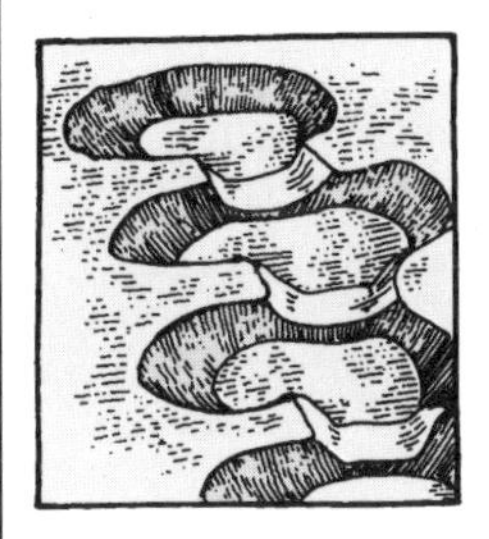

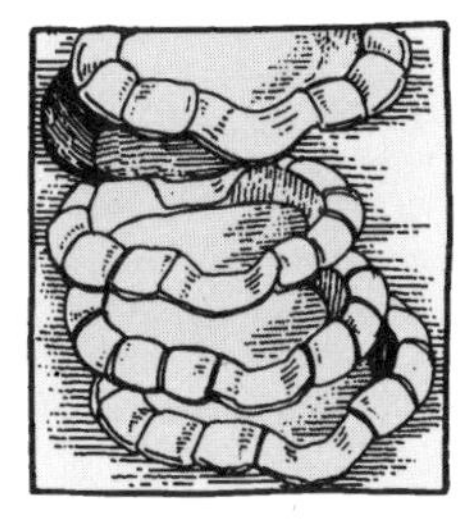

1. PUT *the shells into position, ensuring the mouth of each one overlaps the unit below. Mark an area slightly larger and dig out the soil. Use a spirit-level to check that the tops of the units will be level. Water must remain in them even when water is not being pumped. Install piping to carry water from the pump.*

2. INSTALL *the lower unit first, bedding it on an evenly thick, 5cm/2in layer of soft sand. Check that its top is level, then fill around it with sand or friable soil. Continue fitting the units, ensuring an overhang on each so that water tumbles into the lower one, at a position about one-third in from the side nearest the cascade.*

3. USE A *hose-pipe to fill the top pool to check if water splashes evenly, and that each unit retains water. If water cascades lopsidedly, adjust the levels. After checking them, repack and firm sand or friable soil around them. Install the pipe from the pump: camouflage its outlet with a large paving slab or rock.*

USING A FLEXIBLE LINER

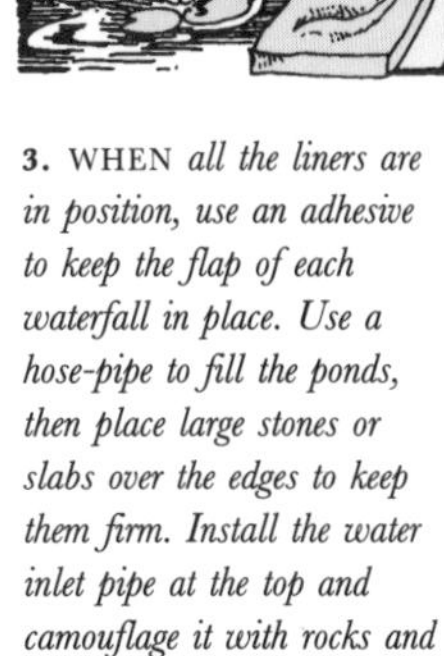

1. WHEN *making a flight of waterfalls from several pieces of flexible liner it is possible to make each cascade exactly the shape you desire. It does not have the size constraints imposed by rigid shells (see above). Starting from the base, dig out each pool. Make the watercourse curve slightly so that it appears to be a natural feature.*

2. SLOPE *the base of each pond backwards, so that it always retains water. Then spread 2.5cm/1in of soft sand over the surface. Starting from the lowest pool, position the liner to extend to the top of the waterfall on the higher side, and the base of the lower one. Spread the liner up the sides and overlap the top by about 20cm/8in.*

3. WHEN *all the liners are in position, use an adhesive to keep the flap of each waterfall in place. Use a hose-pipe to fill the ponds, then place large stones or slabs over the edges to keep them firm. Install the water inlet pipe at the top and camouflage it with rocks and plants. Then introduce plants to the pools.*

PUMPS, FOUNTAINS AND LIGHTS

WATER pumps and lights transform water gardens, but unless installed correctly water and electricity are a dangerous combination. Installing electricity is not cheap and needs the services of a competent electrician. Transformers which reduce the power to 12 or 24 volts make the installation safer, but where large volumes of water for waterfalls are involved a mains-powered pump is essential.

The installation must include a residual current device (RCD) that will trip out the power supply should a fault occur. Also, heavy-duty cables and plastic conduit are essential to connect the pump to the power supply. Always make the installation capable of providing power for additional equipment, such as pool heaters.

WATER PUMPS

There are two types of pump used in water gardens: the most popular one is the submersible, the other the surface type.

- *Submersible pumps are designed to be totally submerged and to work silently. They are easily installed: after placing a pump in a pond, run the cable to the pool's side and use a waterproof connector to link it to a mains electricity supply or a transformer. There is a wide range of submersible pumps and the type required depends on the volume of water to be moved.*
- *Surface pumps are needed where large, or several, fountains are to operate and large volumes of water are needed.*

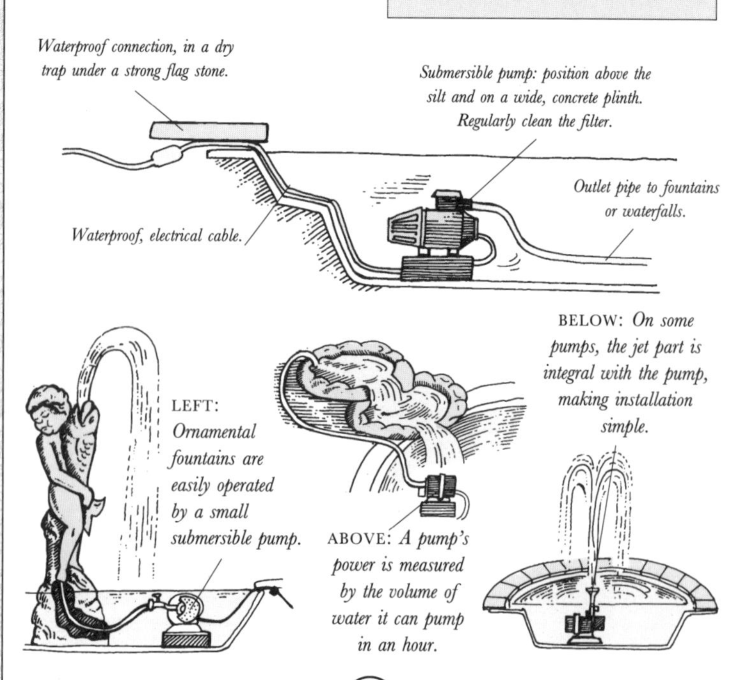

LEFT: *Ornamental fountains are easily operated by a small submersible pump.*

ABOVE: *A pump's power is measured by the volume of water it can pump in an hour.*

BELOW: *On some pumps, the jet part is integral with the pump, making installation simple.*

FOUNTAINS

Few garden features are as fascinating as a fountain. There are many shapes and sizes of jet and a few are shown here. As well, there are statues that spurt water, while pool-side ornaments such as gargoyles disgorge spouts immediately into pools. Where fountains are sited directly in ponds there are a few important considerations:

• The height of the spray should not rise to a distance of more than half the pond's total width.

• The spray should not fall on waterlilies or marginal plants, as they are quickly damaged by this, especially when flowering.

• Floating water plants are soon disturbed by spray and pushed around the pond.

• In windy areas, install fountains that produce large water droplets. Some fountains function like a geyser and produce a tumbling mass of water.

• Keep water-filters clean as fine-nozzled fountains soon become clogged.

• Consult a specialist before buying a fountain and pump. Explain the fountain's size, if it is a secondary feature (the other, perhaps, a waterfall) and the length and size of the piping.

• For safety, fountains with sprays of water up to 1.2m/4ft high can be powered by low-voltage submersible pumps. However, for water sprays 2.1m/7ft high a mains-powered submersible is needed, while for fountains over this height a surface pump is essential. Do not economize on the pump's size. It is always better to have a pump too large.

LIGHTS

Patio and pond lighting transforms gardens and creates a more pleasurable area for evening entertainment in summer.

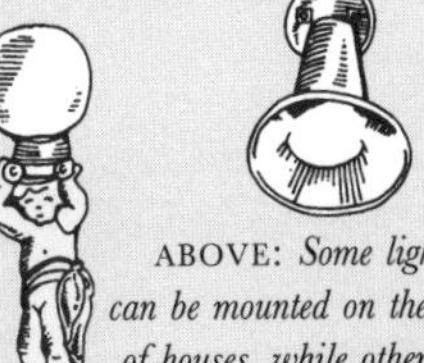

ABOVE: *Some lights can be mounted on the sides of houses, while others are more suited to the tops of low walls, often those surrounding a patio. Alternatively, secure them to the patio's surface.*

ABOVE: *Lights permanently secured to the tops of low walls can be directed to flood large areas with light. Some are fitted with spikes and can be temporarily secured in the ground, perhaps to highlight waterfalls. Some lights (right) can be submersed in ponds.*

PLANTING

WATERLILIES are best planted during late spring or early summer. Choose early summer in cold climates. However, they are often available from water-plant nurseries throughout summer, although it is certainly easier to plant them early in the year. Other aquatics are planted between late spring (or early summer in cold areas) and late summer, when actively growing.

In earlier times it was usual to put plants in soil in a pond's base: this is not now recommended. Always plant into a plastic mesh basket or similar container as this allows more control over the plant, such as adjusting the depth to which it is placed in the water. It also makes lifting, repotting and feeding easier.

As soon as plants arrive from suppliers, unwrap them and submerge the roots in a bucket of clean water. Plant them as soon as possible. If there is a delay, leave them in water and place the bucket in a cool, shaded, well-illuminated corner.

RANGE OF PLANTS

In addition to waterlilies, there is a wide range of other plants that can be planted in ponds. Some of these are solely decorative, others are essential to the lives of fish and other creatures in ponds (see pages 34 to 37).

- *Waterlilies are the most popular of water plants. Always choose varieties that suit the pond's depth (pages 38 to 41).*
- *Marginals are positioned on shelves within the pond (pages 46 to 49).*
- *Oxygenators, as their name suggests, are submerged and help to aerate and keep the water clean (pages 42 and 43).*
- *Floaters live freely in the water (pages 44 and 45).*
- *Bog plants live in moist soil around ponds (pages 50 and 51).*

1. PLASTIC MESH *planting baskets are widely available in several sizes, from 10cm/4in to 30cm/12in wide. Most are square, others round, some kidney-shaped and ideal for shelves in round ponds.*

2. TO PREVENT *compost falling through holes in a planting basket, line the inside with coarse hessian. Ensure it completely covers the inside of the basket. Louvred types of basket do not need lining.*

3. FILL *the basket two-thirds full with heavy loam enriched with a sprinkling of bonemeal to assist in the development of roots. Ensure the loam is free from decaying debris, such as old roots.*

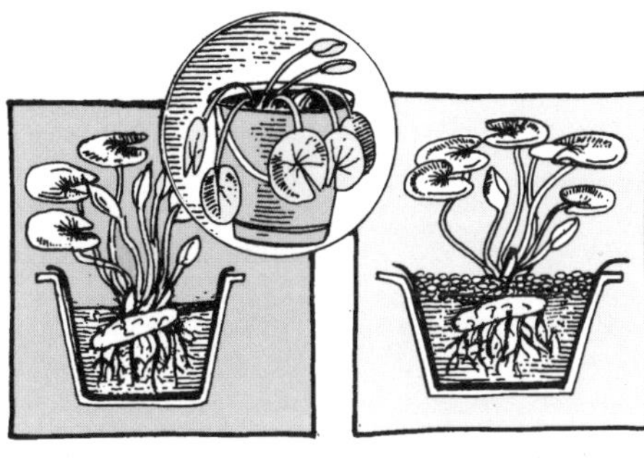

4. UNWRAP *a new plant and place in a bucket of water. Do not allow its roots to become dry. Form a hole in the loam and spread out the roots. Cover with loam and firm it over them.*

5. BURY *the plant's crown slightly deeper than before (mark on stem), with the surface of the loam 3.6–5cm/ 1½–2in below the basket's top. Then add 2.5cm/ 1in of grit.*

6. WATER *the compost, then place in a pond. The leaves of waterlilies must float on the surface. Place the container on bricks. As plants grow, progressively remove the bricks (see below).*

PLANTING DEPTHS

Planting aquatic plants in plastic mesh or other proprietary containers creates the opportunity to position them at the right depth throughout their first year. The containers for waterlilies and marginal plants can be slowly lowered throughout the first season. Always select waterlilies that suit the depth of water as, if too vigorous – even when the plant's container is on a pond's base – the leaves protrude out of the water, dominating the pond and smothering less demanding water plants. Ensure the baskets will not fall over.

PLACE *marginal plants on blocks to raise their leaves above the surface.*

NEWLY PLANTED *waterlilies need raising to prevent leaves being submerged.*

OXYGENATING *plants can be immediately placed on the pond's base.*

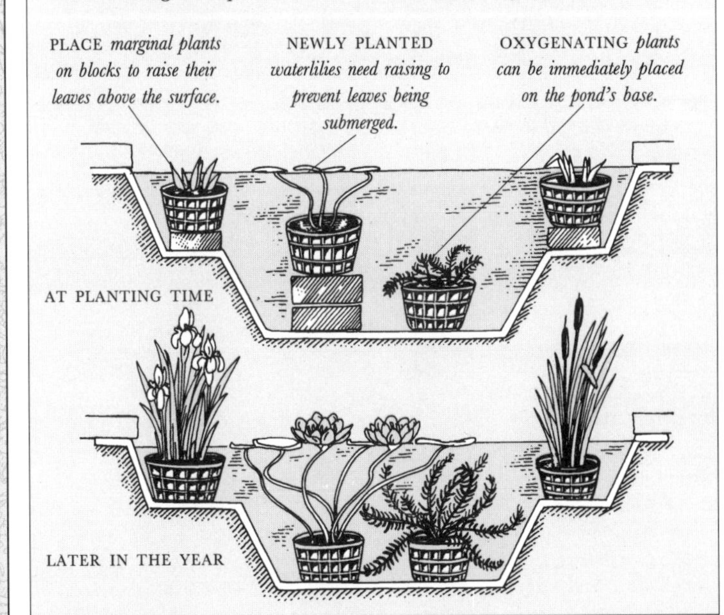

FISH FOR GARDEN PONDS

UNLIKE any other garden feature, water gardening involves plants and living creatures. Most people – and especially children – are captivated by fish, while dragonflies and damselflies are enchanting as they hover over a pond in summer. Frogs and newts also have their charm, but are more esteemed by enthusiasts of wildlife ponds than owners of a small, ornamental pool on a patio.

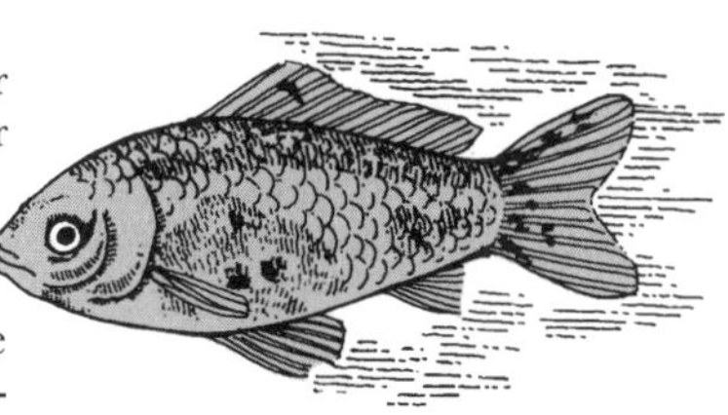

THE LONDON SHUBUNKIN *has a similar shape to a goldfish. The blue form is seen most often, but there are others.*

RANGE OF FISH

Goldfish are the most popular fish for ponds, widely available and colourful. Shubunkins, which were developed from goldfish, are also widely seen. Most are hardy, such as the Common Goldfish, London Shubunkin, Bristol Shubunkin, Comet and Sarasa Comet, and can be left outside throughout winter. In cold areas, the Bristol Shubunkin must be taken indoors in winter.

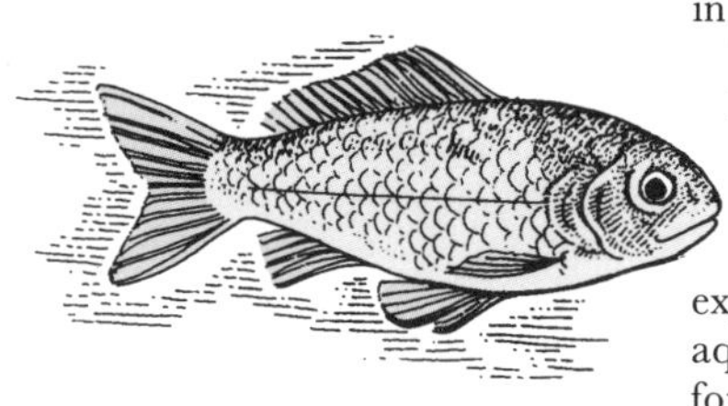

THE COMMON GOLDFISH *is a popular fish for garden ponds. It is inexpensive to buy, reliable and never fails to create interest.*

The fancy forms, such as Fantain and Moor, are even less hardy and must be taken indoors at the onset of cold weather in autumn; some experts only recommend them for aquariums indoors. Other fancy forms, such as Veiltail, Oranda, Ranchu, Celestial, Lionhead and Pompon, are only kept in aquariums indoors.

BASIC FISH ANATOMY

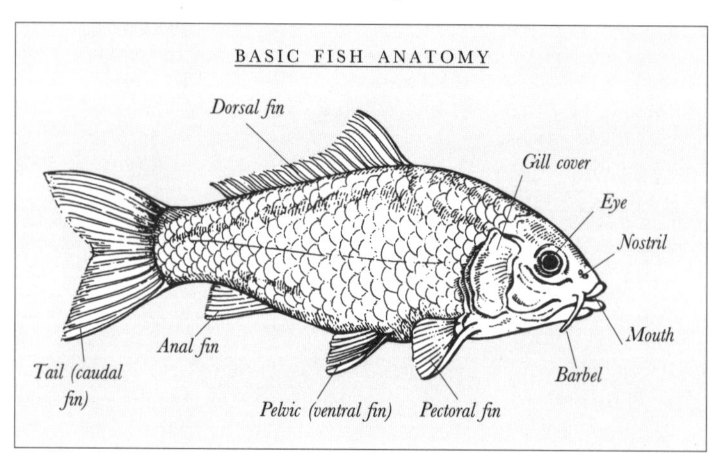

BUYING AND STOCKING

Introduce fish to ponds between late spring and late summer, when the pool's temperature is at least 10°C/50°F. Leave about five weeks between planting and adding fish. Large fish are usually expensive, while small ones are sometimes difficult to establish. Therefore, look for fish 7.5–13cm/3–5in long. Always buy from a reputable source. As fish are companionable, buy several.

They are usually sold in water in a polythene bag, plus a burst of oxygen. Fish can exist in this way for twenty-four hours if kept cool, but put them into the pond soon.

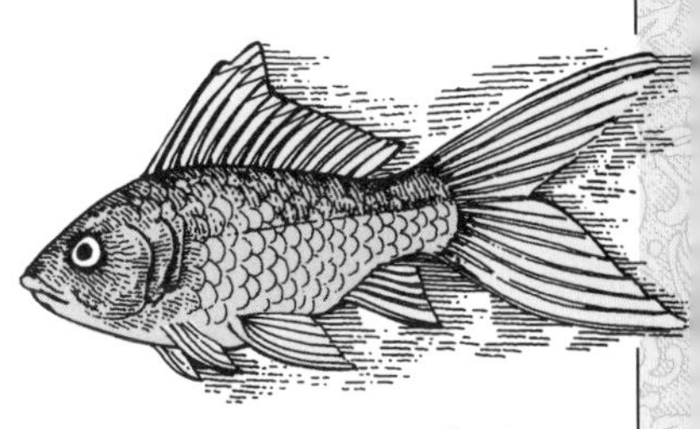

THE COMET GOLDFISH *has large fins that enable rapid movement for short distances.*

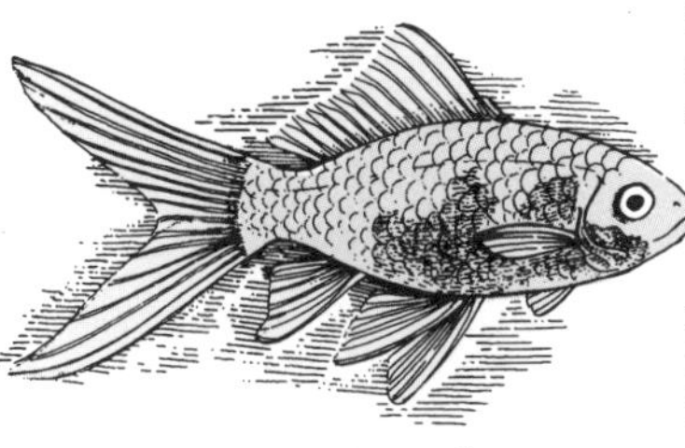

THE SARASA COMET *is very ornamental, with a white and red body and flowing tail.*

Place a clean plastic bucket in the pond – covered by water – and gently lower the unopened polythene bag into it. After three or four hours the temperatures in the bag and pool will be about equal. Carefully open the bag and allow the fish to swim out. Do *not* just tip them into the water.

Do not introduce too many fish: an approximation of the number is 2.5cm/1in of fish (including tail) to each 30 x 30cm/12 x 12in of pond surface. *Koi* carp need more space, about one fish to every 1.5 x 1.5m/5 x 5ft of surface area.

KOI – FANCY CARP

These are increasingly popular fish for ponds, but only if large. Koi *is the Japanese name for carp, with the ornamental forms correctly known as* Nishikigoi *but invariable just called* koi. *Occasionally they are called Fancy Carp.*

Koi *were recorded in Japan more than 1600 years ago and have been bred for a wide range of colours, from white, through gold and blue to black. Some are speckled in several colours, including silver, and may reach 75cm/2½ft or more in length. They therefore need a large pond, at least 1.2m/4ft deep.*

Koi *stir up silt on the pond's base, clouding the water. This makes it necessary to cover the base with shingle or gravel and to install a filter. Keeping* koi *has become a cult and before putting them into a pool it is wise to study a specialist book.*

OTHER COLD-WATER POND FISH

THE WIDE range of fish suitable for ponds includes, as well as goldfish and shubunkins, a choice of other cold-water fish. Most are best suited to life in large ponds and especially to those intended to attract wildlife.

There is often a temptation to put *any* fish into a garden pond, but don't: first check their suitability and certainly avoid introducing predatory types such as Pike, Perch and Catfish.

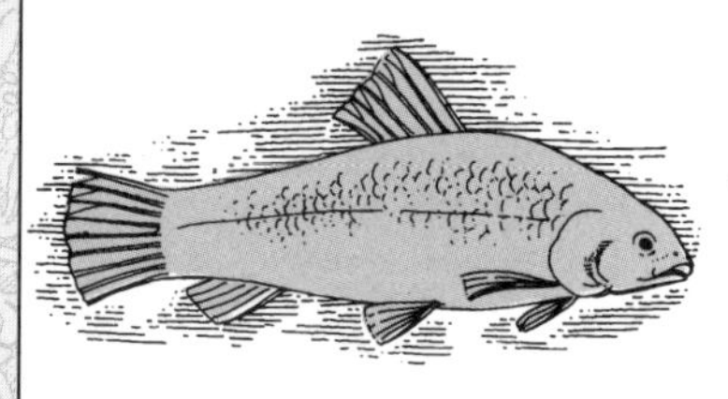

GOLDEN ORFE *has an attractive salmon or pale golden body. They are timid and best bought in pairs. Only put them in well-aerated ponds with a surface area of 3.75sq m/ 40sq ft or more.*

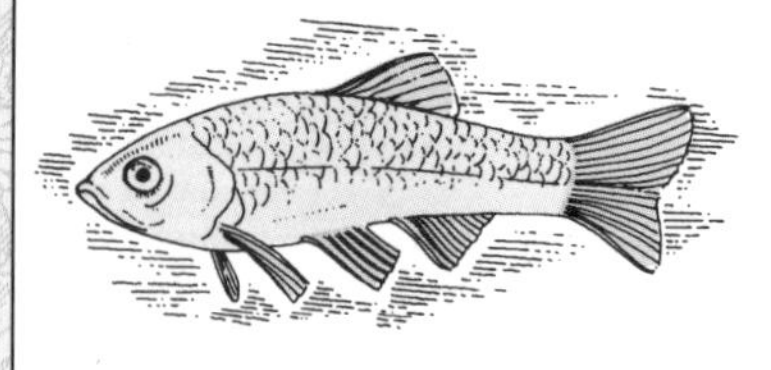

TENCH *live towards the bottom of a pond, clouding the water as they stir around in the silt. They are often known as 'doctor' fish, but there is little evidence to suggest they have curative powers. The Golden Tench (above) has a pale orange body, while the Green Tench (below) is much duller.*

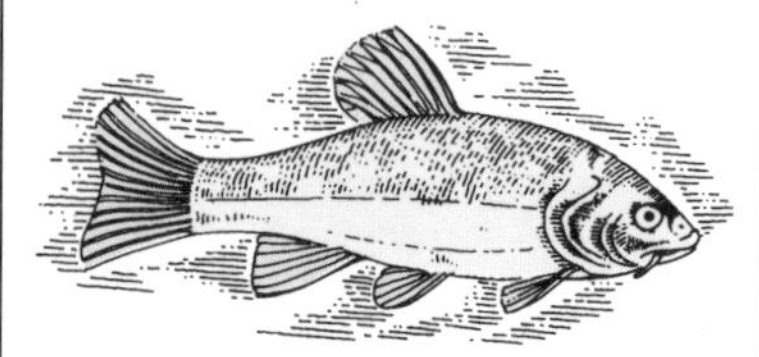

GOLDEN RUDD *eventually grows 15–20cm/ 6–8in long in a garden pond, although it has been known to reach 45cm/ 18in in lakes and rivers. It has an attractive silvery shade overlaid with a golden hue, but is best identified by its reddish fins.*

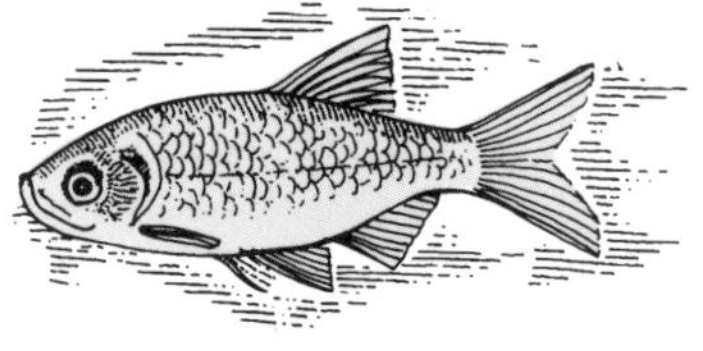

MINNOWS *are popular and widely found in ponds, streams and lakes. They grow about 7.5cm/ 3in long, are normally silvery but during the breeding season in spring the male changes to green, strongly flushed with red. They live in shoals.*

OTHER FISH

- *Blue Orfe: Increasingly popular, with pale blue scales and a rather slower nature than the golden type.*
- *Gudgeon: A scavenger and prone to stirring up silt.*
- *Rosy Minnow: A North American fish with a salmon-orange body. It does not grow very large and is well able to survive the coldest winter.*
- *Three-spined Stickleback: It has an aggressive nature, grows 6–8cm/ 2½–3½ in long and is best put in wildlife ponds. Male fish have a metallic blue hue and a red belly when spawning.*

OTHER POND WILDLIFE

WILDLIFE ponds are more likely than formal ponds – which are usually found on patios – to encourage animals such as common frogs, toads, newts, small mammals, dragonflies, pond skaters and damselflies.

Birds are also encouraged and if these are song birds there is no problem. Wildlife ponds, however, entice herons and they soon steal fish. If birds are a continual problem, the only solution is to cover the pond in netting.

THE COMMON FROG *is a frequent visitor to garden ponds.*

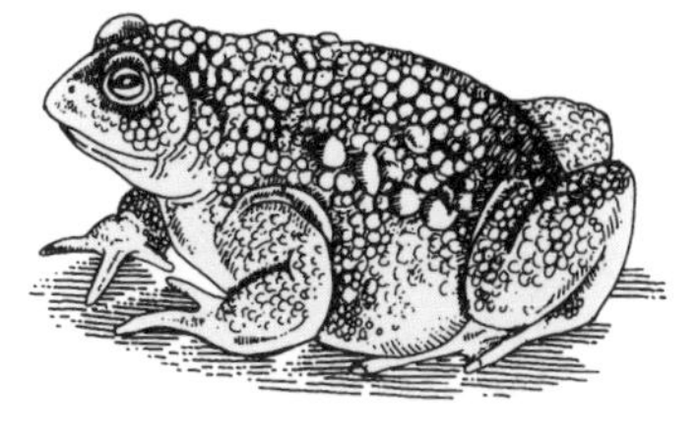

THE COMMON TOAD *has a voracious appetite for slugs and snails.*

THE COMMON *or Smooth Newt visits ponds to breed during spring and summer.*

OTHER POND LIFE

- *Damselflies: Pretty and dainty, they flit over the pond in summer. They lay eggs in the water, some of which become food for fish. They are encouraged by plants that surround wildlife ponds.*
- *Ram's Horn Snail: Unlike the plant-damaging Great Pond Snail with its corkscrew-shaped shell, the Ram's Horn Snail has a Catherine-wheel outline. It feeds on debris and helps to keep the pond clean.*
- *Small mammals: Water Voles and Water Shrews are encouraged by food supplies near wildlife ponds, as well as by water. Grass snakes also frequent the same area.*

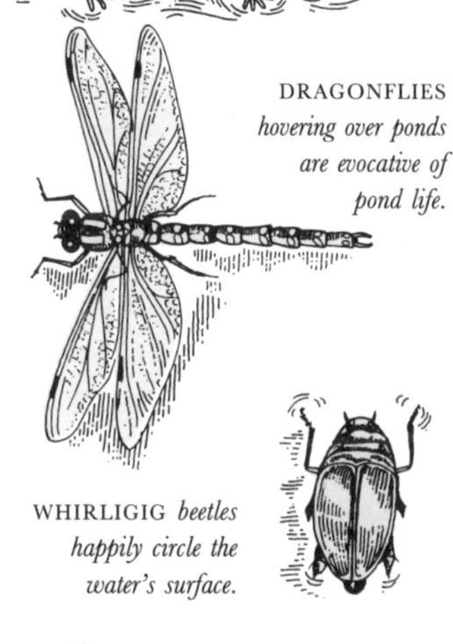

DRAGONFLIES *hovering over ponds are evocative of pond life.*

WHIRLIGIG *beetles happily circle the water's surface.*

POND SKATERS *skid to and fro across the surface.*

LOOKING AFTER PONDS

LOOKING after fish and plants is not difficult and here are a few techniques and pieces of equipment that can make water gardening even more of a pleasure. Water heaters and filters are expensive initially, but some of the ideas here are inexpensive and just common sense.

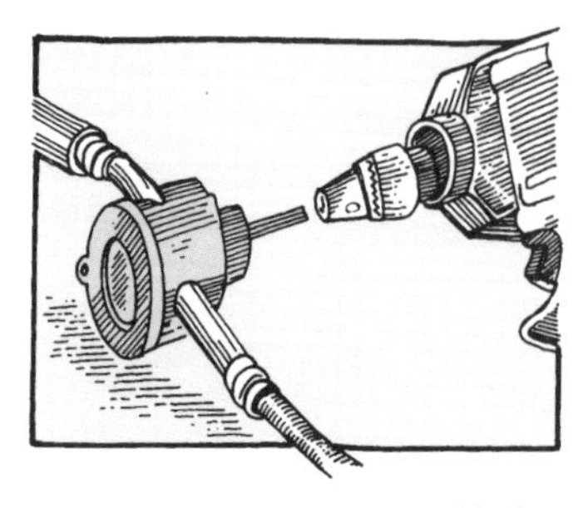

EMPTYING PONDS *at ground level is laborious when buckets are used. However, inexpensive pumps are available and some can be powered by electric drills. They pump about 750 litres/200 gallons an hour, but cannot be used continuously to operate fountains. Raised ponds can be emptied by syphoning the water through a length of hose pipe.*

FEEDING RINGS *prevent fish food floating all over the pond. Uneaten food can then be removed easily.*

FISH FEEDERS *that float and discharge food when rocked by fish help to prevent food wastage and water contamination.*

POND HEATERS *are essential in winter to create an open area of water amid ice. This allows an exchange of gases between the water and air – essential for the health of fish – and eases the pressure of ice on the pond's sides. Some heaters are operated by mains electricity, others through a transformer, these are much safer to instal and use. Most float and therefore need to be in position at the first sign of ice. To ensure that sufficient ice-free space is created, a heater is needed for every 2.3–4.6sq m/25–50sq ft of the pond's surface, but this is determined by the heater's actual power.*

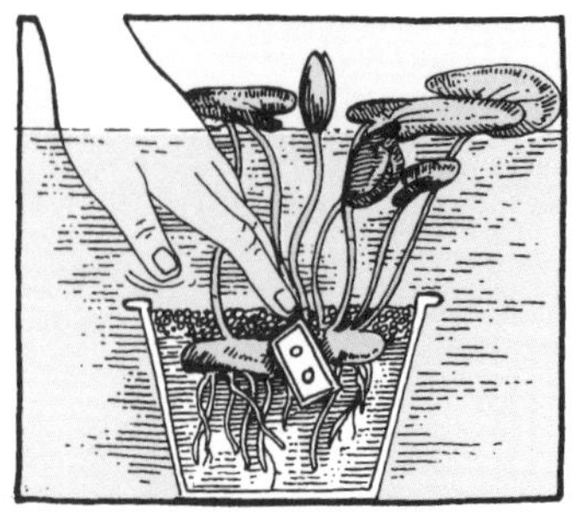

FEEDING WATER *plants is made easy when fertilizer is applied in sachets that can be inserted into compost in spring and mid-summer.*

NEVER *create ice-free areas by breaking ice. Instead, install a water heater (below left) or repeatedly place a metal kettle of boiling water on the ice. It is usually necessary to refill the kettle several times with more boiling water. Ensure it does not fall in – tie it to a rope!*

INITIAL *filling and topping up of ponds must be carried out with care. If empty, place a bucket on the pond's base and allow water to gently trickle over the rim. For topping up, tie a piece of canvas over the pipe's end. This prevents the water being unduly disturbed.*

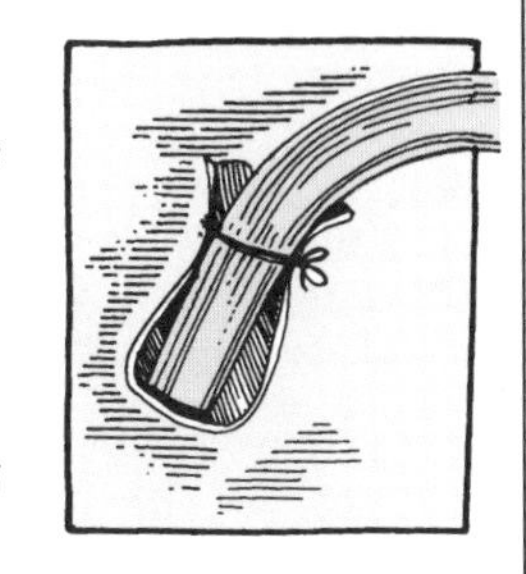

NETTING *attached to a wooden or metal framework is ideal for preventing children falling in, birds fishing and leaves blowing into ponds in autumn. Wire netting impaled on a garden fork is superb for removing leaves.*

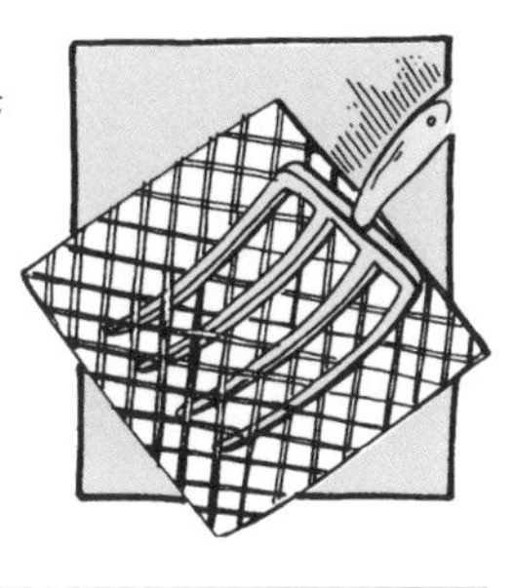

FILTERS

Most ponds do not have filters, but if you keep Koi *Carp they are essential. There are two main types, mechanical and biological. The mechanical type uses a pump to draw water through filters that remove dirt floating in the water. It is usually installed in the pond and water passing through it is piped to fountains and waterfalls. Biological types need a pump that must run continuously. The unit operates outside the pond.*

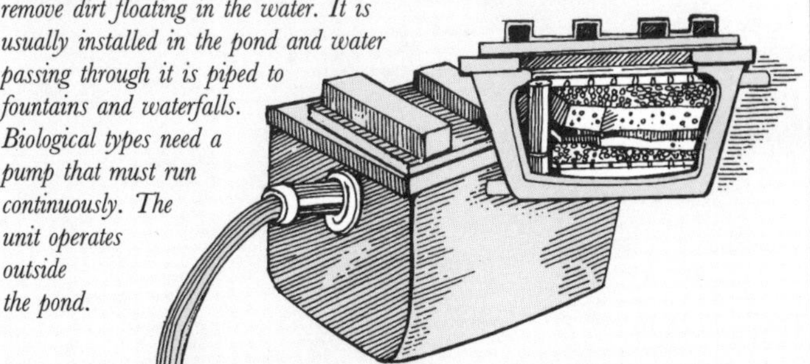

PYGMY WATERLILIES

PYGMY, dwarf or miniature waterlilies have a diminutive nature and are ideal for planting in small ponds as well as in water gardens in tubs and stone sinks. They are excellent for planting in ponds where the water depth does not exceed 23cm/9in above the surface of the planting container. This means that if the container is 10cm/4in deep, the total depth of the pond should be no more than 32cm/13in. Even though classified as pygmies, there are many variations in their vigour. However, each plant covers 1/10–1/4 sq m/1–3 sq ft of the water's surface. The flowers are diminutive, 5–10cm/2–4in wide, but are no less attractive than their larger cousins. They are, however, more easily damaged in severe winters than waterlilies with a thicker insulation of water above their roots.

Whatever the sizes of the plants, when planting always ensure the leaves are not drowned (see pages 30 and 31 for planting instructions).

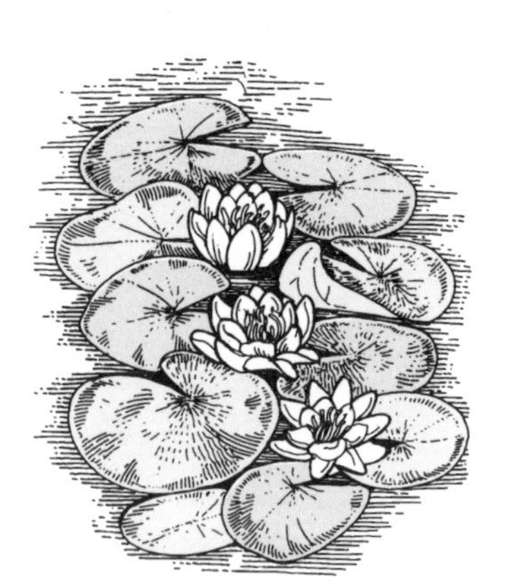

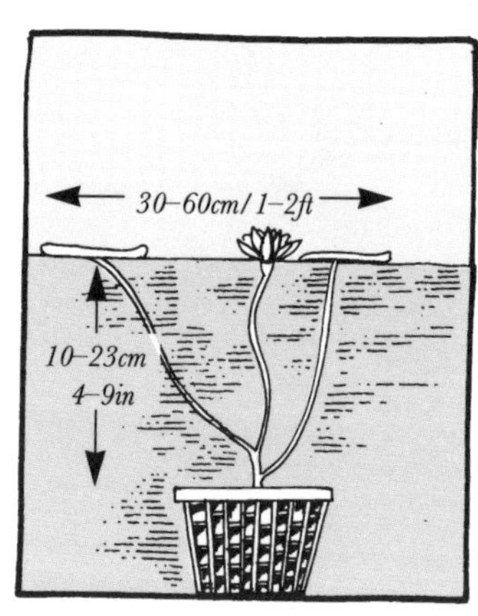

PYGMY WATERLILIES

- Nymphaea *'Aurora': Changeable colours, first pinkish yellow, then orange and later red.*
- N. *'Graziella': Free flowering, with orange flowers.*
- N. laydekeri *'Lilacea': Scented; pale pink and darkening with age.*
- N. odorata minor: *Scented; white and star-shaped, with yellow stamens.*
- N. odorata *'W. B. Shaw': Fragrant, with pink flowers.*
- N. *'Paul Hariot': Changeable colours, first yellow, later orange and then red. The leaves are attractively marked in brown.*
- N. pygmaea alba: *A real miniature, with flowers only 2.5cm/1in wide. White with yellow stamens.*
- N. pygmaea *'Helvola': The smallest yellow-flowered waterlily. Star-shaped, pale yellow flowers with golden stamens.*

SMALL WATERLILIES

THESE are more vigorous than the pygmy types, although some authorities name some pygmy waterlilies as small ones; and conversely, a few small ones are classified as pygmies. For example, some experts classify the deep pink flowered *N. laydekeri* 'Purpurata' as a pygmy type, others as a small one. Clearly, plants are not as uniform as sausages and depending on their circumstances some may be more or less vigorous.

Small waterlilies are ideal for ponds with 15–45cm/6–18in of water above the container. Their spread is more dramatic than pygmy types, each plant occupying 1/4–1 1/10 sq m/3–12 sq ft of the water's surface.

As the vigour of waterlilies increases, so too does the range of plants from which to choose. Colours include white, pink, red, yellow and orange. The widths of their flowers are 10–15cm/4–6in and therefore they are more dominant than the pygmy types.

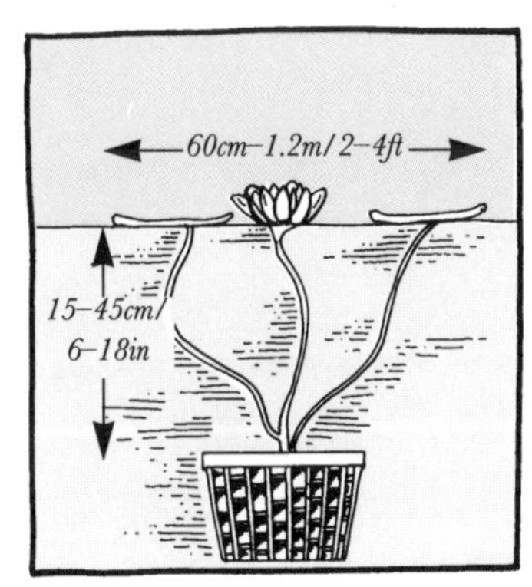

SMALL WATERLILIES

- Nymphaea *'Albatross': Large white flowers with golden centres. Attractive leaves, purple when young.*
- N. *'Firecrest': Fragrant, bright pink flowers streaked orange and red. The stamens are tipped in red.*
- N. *'Froebeli': Wine red flowers with orange stamens. Olive green leaves. Widely grown and very reliable flowering.*
- N. *'James Brydon': Popular and prolific, with orange-red flowers and orange stamens. Purplish leaves. It grows well in light shade.*
- N. laydekeri *'Fulgens': Bright red flowers, darkening with age, and orange-red stamens. It bears many flowers, but only a few leaves, which have purple undersides. The leaves have long stems.*
- N. *'Sioux': Buff-yellow flowers that turn to peach then copper-orange. The green leaves are peppered deep brownish-red.*
- N. *'William Falconer': Dark red cup-shaped flowers with yellow centres. Also has attractive, dark green leaves.*

MEDIUM WATERLILIES

MEDIUM waterlilies need a position where their crowns are covered with 30–60cm/12–24in of water. Their spread is variable, often 1.2–1.5m/4–5ft and a single plant could cover 2 sq m/20 sq ft or more. The flowers are larger than the small types and up to 21cm/7in across.

Some waterlilies are classified in either medium or vigorous categories. For example, 'Mrs Richmond' is sometimes placed with vigorous types, other times with the medium ones. Clearly, although some vigorous types can be grown in relatively shallow water, they invariably spread further than the medium-vigorous ones, and some need deeper water to prevent their leaves for ever rising above the surface.

These are popular varieties and suit most ponds. However, always check the water's depth before introducing them. Ponds soon become a jungle of leaves if too vigorous types are selected.

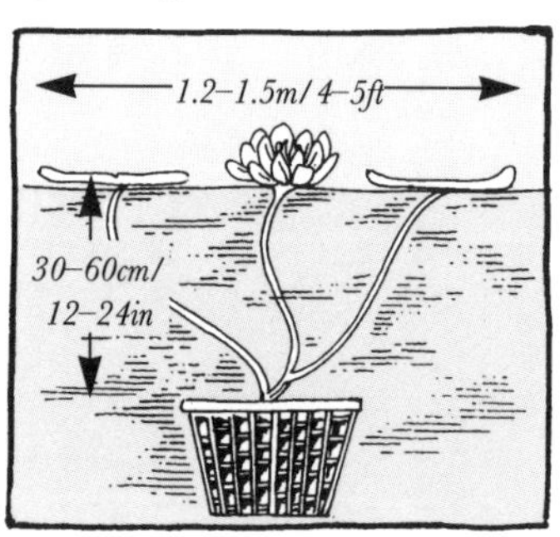

MEDIUM WATERLILIES

- Nymphaea *'Amabilis': Fragrant, rose-pink flowers with a flat, star-shaped nature. The pink darkens with age.*
- N. *'Gloriosa': Scented, rose-red flowers that are borne prolifically throughout summer.*
- N. *'Gonnere': Also known as 'Snowball', it has semi-double white flowers with yellow stamens.*
- N. marliacea *'Albida': Fragrant, white flowers with yellow stamens.*
- N. marliacea *'Rosea': Fragrant, with pale pink flowers and golden stamens. The young leaves are an attractive purplish-green.*
- N. *'Mme. Wilfron Gonnere': Cup-shaped, rich pink flowers flushed with white. Attractive deep rose centres.*
- N. *'Moorei': Pale yellow flowers with yellow stamens.*
- N. *'Rene Gerard': Large, star-shaped, red flowers with golden stamens. A very free flowering variety.*
- N. *'Rose Arey': Large, star-shaped rose-pink flowers with yellow stamens.*
- N. *'Sunrise': Scented, bright yellow flowers with golden stamens. The underside of each leaf is an attractive red shade.*

VIGOROUS WATERLILIES

THESE are the amazons of the waterlily fraternity and some, such as *N. alba, N.* 'Charles de Meurville', *N.* 'Gladstoniana' and *N. tuberosa* 'Richardsonii', are more suited to lakes than ponds in gardens. They may lack the dainty nature of the pygmy types, but in a large setting form dominant features.

They need 45–90cm/1½–3ft of water over their roots. The size of their flowers is up to 25cm/10in wide and each plant will easily spread 1.5–2.4m/5–8ft or more. Therefore, it could be disastrous to think big in a small pond! Some of the most vigorous lilies can easily cover 4.5 sq m/50 sq ft. And as such lilies obliterate the surface it is impossible to grow more choice plants if the pond is small. With the selection of all lilies, the maxim must be to choose varieties that suit the pond's size.

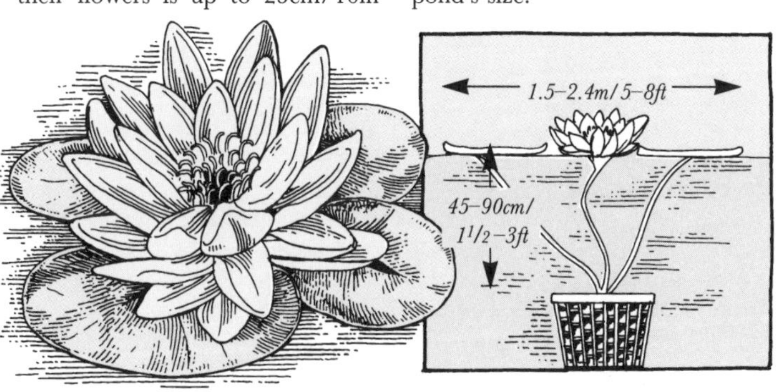

VIGOROUS WATERLILIES

- Nymphaea alba: *Vigorous and only suited to very large ponds. Cup-shaped, white flowers with yellow stamens. It is often seen in lakes, and when seen* en masse *is very dramatic.*
- N. *'Attraction': Free flowering with red flowers, flecked in white and displaying golden stamens.*
- N. *'Charles de Meurville': Vigorous plant with large, rich red flowers which can grow to 25cm/10in across. Only suitable for a very large pond.*
- N. *'Colonel A. J. Welch': Yellow flowers and lightly marbled leaves. It needs a very large pond with deep water.*
- N. *'Colossea': Scented, blush pink flowers that fade to white. Golden stamens. Only for large ponds.*
- N. *'Gladstoniana': Enormous white flowers with golden stamens. Only suitable for deep, large ponds.*
- N. marliacea *'Chromatella': Bowl-shaped, yellow flowers. Very free flowering and reliable.*
- N. *'Mrs Richmond': Globular pink flowers that turn red with age. Golden stamens. Vigorous and needs a deep pond.*
- N. tuberosa *'Richardsonii': Very vigorous and perhaps best suited to lakes and deep water, about 90cm/3ft or more. Cup-shaped white flowers with yellow stamens.*

OXYGENATING PLANTS

Ceratophyllum–Tillaea

ALSO known as submerged aquatics and water weeds, these oxygenating plants are essential for the well-being of fish and to keep water clean. They also provide shelter for spawning fish and their fry. Additionally, they release oxygen into the water and absorb mineral salts, that if not removed, would encourage the presence of algae. Ponds that do not have oxygenating plants become green and dirty, with fish that have to come to the surface repeatedly to gasp in air. This is especially likely to happen during warm weather in summer.

NATURE CONSERVATION

There is always the temptation to pull up plants from a countryside pond or stream and to put them into your pond. However, never do this as natural plant life is then rapidly depleted. Also, by taking wild plants it is possible to introduce diseases and pests into ponds. Instead, buy cultivated plants from a reputable water-plant nursery.

Most oxygenating plants grow completely below the surface, but a few, such as the Water Violet (*Hottonia palustris*), display attractive flowers. Some have a vigorous nature and unless regularly pruned will soon engulf a pond. All these plants need to be planted in a meshed container placed on the pond's base. This enables easy removal should it become necessary, as well as aiding propagation by division of plants and taking of cuttings.

CERATOPHYLLUM DEMERSUM *(Hornwort) has bristle-like, dark green leaves arranged in whorls. In autumn, stems fall to the pond's base and the plant survives in a dormant form. It is hardy and grows well in cold water. It is increased in summer by division or cuttings.*

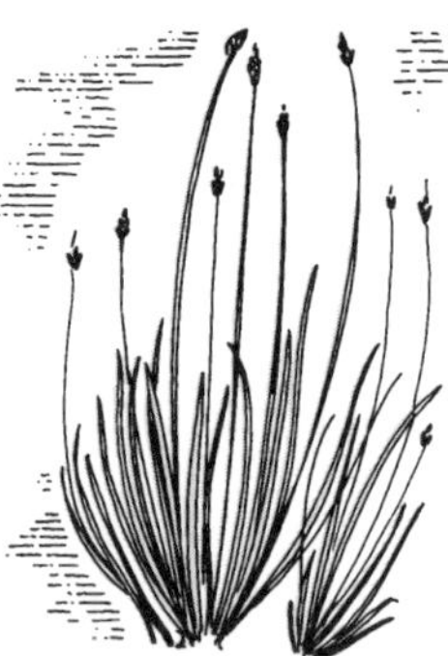

ELEOCHARIS ACICULARIS *(Hairgrass) is a dainty grass with green, hair-like stems and a tufted habit, 13–15cm/5–6in high. It forms runners and is easily increased by breaking off parts in late spring or summer and replanting them. It is not suitable for very cold regions.*

ELODEA CANADENSIS *(Canadian Pondweed) is a popular oxygenator, but very invasive and therefore needs regular trimming in summer. Stems are packed with dark green leaves, with the bouquet of thyme. When severed they can be used in spring or summer as cuttings, and soon develop roots.*

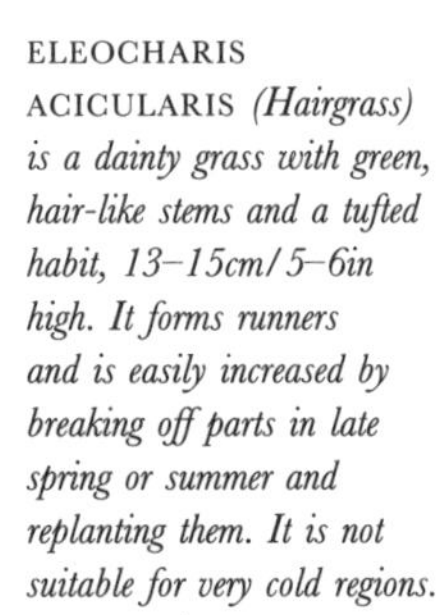

FONTINALIS ANIPYRETICA *(Willow Moss) is slow-growing, with evergreen, dark green leaves that spread out to form mossy patches. It is an ideal oxygenator to have when fish are spawning. Increase it by dividing large clumps in late spring or summer. It is superb in full sun or light shade. A related species,* F. gracilis, *is not so common but is smaller, with thread-like leaves.*

HOTTONIA PALUSTRIS *(Water Violet) is a British native plant and one of the few oxygenating plants that develops flowers. These are lilac coloured and borne in whorls on stiff, upright stems, about 20cm/8in above the water's surface in early summer. In autumn, stems die down and the plant overwinters as dormant buds. The beautiful leaves are feathery, finely divided and bright green.*

LAGAROSIPHON MAJOR *(Goldfish Weed), also known as* Elodea crispa, *is a superb oxygenating plant from South Africa that survives in all but the coldest climates. The small, narrow, curling leaves tightly clasp the stems and although it can be invasive in large ponds – where plants cannot be easily reached for pruning – is definitely worth growing. Take cuttings every three years, in spring or summer.*

MYRIOPHYLLUM SPICATUM *(Water Milfoil) develops long, trailing stems clothed in finely cut, feathery whorls of light green leaves. It is ideal for a small pond and is increased easily by weighting small pieces into compost during summer. It develops red-petalled flowers, slightly above the surface.*

POTOMOGETON CRISPUS *(Curled Pondweed) has large, curly-edged, reddish, lance-like leaves attached to wiry stems. It is invasive and needs regular pruning. In early summer it develops small, pinkish-white flowers slightly above the water's surface. Take cuttings in summer.*

TILLAEA RECURVA, *now properly known as* Crassula recurva *and widely sold as* Crassula helmsii, *is native to Australia and known as the Swamp Stonecrop. It is vigorous, with evergreen, fleshy leaves. Divide congested plants in spring, or take cuttings during late spring or early summer.*

FLOATING PLANTS

Azolla–Stratiotes

THESE are plants with no visible means of support: they float freely with leaves and stems on or just below the surface. Their roots trail beneath them. Some have flowers, most do not.

Some of these plants are native to warm countries and therefore, in order to survive in temperate climates, they must be put in a bucket of water and soil during winter and placed in a frost-proof greenhouse or light shed. In spring, after the risk of frost has passed, they can be reintroduced to a pond. Some other floaters just drop to the pond's base in autumn and survive as buds that re-grow in spring.

These plants can be bought from early to late summer and need only be carefully dropped into the water.

VITAL ROLE

In addition to being attractive and playing a minor role in aerating water, in large ponds they help to fill blank areas between waterlilies and marginal plants along a pond's edges. The main problem with vigorous floaters in large and deep ponds is that it is difficult to reach them to prune them if they become invasive. Fortunately, however, it is plants from warm countries that are most invasive and these are invariably reduced or killed by frosts.

Vigorous forms of floaters such as the Greater Duckweed (*Lemna polyrhiza* but now properly known as *Spirodela polyrhiza*) and Thick Duckweed (*L. gibba*) should not be introduced to ponds. For this reason, always buy plants from reputable nurseries.

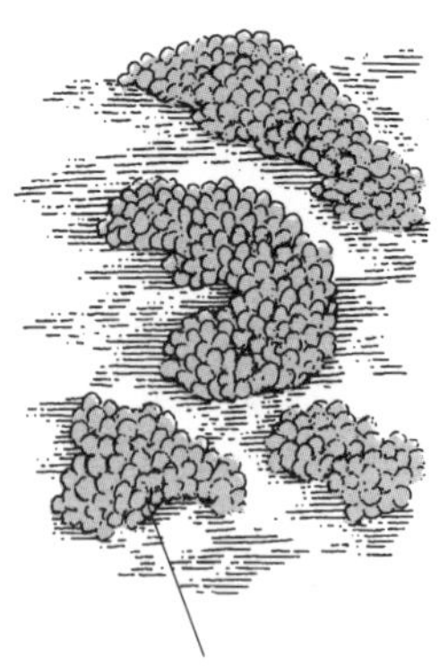

AZOLLA CAROLINIANA *(Fairy Moss/Mosquito Plant) is a popular floater, forming dense mats of pale green fronds that turn red in autumn. Winter frosts and ice cut back plants; in very cold areas, over winter plants in containers filled with water and soil. In spring, replace in a pond.*

EICHHORNIA CRASSIPES *(Water Hyacinth) is notorious in warm countries for blocking waterways. Frost, however, soon kills it and therefore plants are overwintered in buckets of water and soil in greenhouses. In warm areas, blue, lavender and yellow flowers appear in summer.*

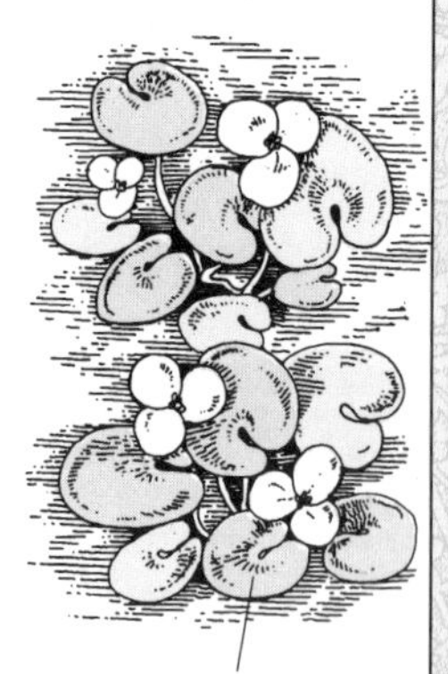

HYDROCHARIS MORSUS-RANAE *(Frog Bit) is ideal for small ponds and develops bright green, kidney-shaped leaves and small, white three-petalled flowers during mid-summer. It is hardy and survives outdoors in most winters. Congested plants can be divided in summer.*

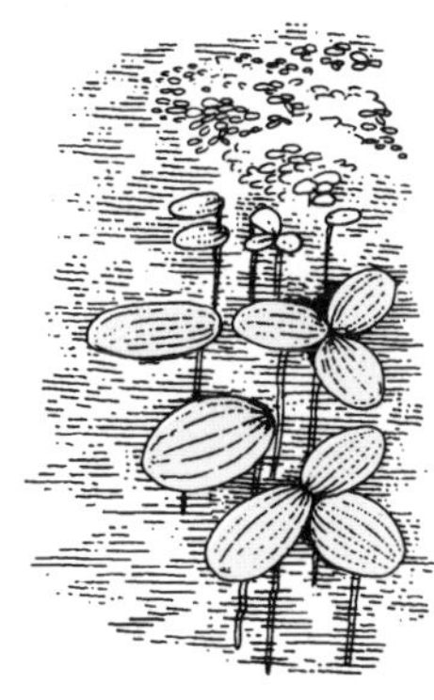

LEMNA MINOR *(Common Duckweed/ Lesser Duckweed) is an invasive floater that soon covers the surface with small, bright green leaves. Small roots dangle below the leaves. It grows rapidly and therefore needs repeated thinning. It is hardy and survives most winters.* Lemna trisulca *is the least invasive species.*

PLAGUE PROPORTIONS

The Water Hyacinth (Eichhornia crassipes) *is native to tropical America but has spread widely throughout the tropics, growing and reproducing so quickly that it is a serious pest on many rivers and waterways. The bladder-like bases of the leaf stalks act as floats, keeping the plant high out of the water and making its eradication very difficult.*

Although initially introduced as an ornamental plant, in some countries it is now banned. Waterways in Florida, Panama, Java, India, Australia and Argentina have frequently been choked. However, in India these plants have been used in making paper and pressed boards. Additionally, although of little nutritive value, leaves and stems have been fed to cattle, and it is a source of cellulose. The leaf stalks are rich in potash and in Asia have been used as manure; in Malaya they are fed to pigs, and in the West Indies to donkeys.

TRAPA NUTANS *(Water Chestnut/ Trapa Nut) is a tender, annual floater, soon killed by frost. The black, chestnut-like fruits fall to the pond's base in autumn, overwinter and produce new plants in spring. However, fruits are only produced in warm countries.*

PISTIA STRATIOTES *(Water Lettuce/ Shell Flower) needs a warm climate and is soon killed by frost. Place in a bucket of water and soil and put in a greenhouse during winter. It develops floating rosettes formed of felted, lettuce-like leaves. It is ideal for the rarity enthusiast.*

STRATIOTES ALOIDES *(Water Soldier/ Water Aloe) is hardy, with sword-like leaves borne in clusters that resemble the tops of pineapples. White flowers appear in mid-summer, then the plant sinks to the pond's base. In late spring, plants rise naturally until they are just below the surface.*

MARGINAL PLANTS

Acorus – Hypericum

THESE are adaptable plants, mainly living in shallow water at the edges of ponds. Some will also grow in boggy areas around ponds. The depth of water needed varies, and therefore this is indicated for each of the plants featured on these and the following pages.

These plants have their roots submerged, but the major parts of their stems and leaves are above the surface.

UNIFYING THE POND

Growing water-loving plants at the edges of a pond helps to unify it with the surrounding garden, although this can also be achieved by planting bog plants in moist soil around a pond (see pages 50 to 53). Marginal plants also introduce informality to the edges of formal ponds.

To ensure marginal plants are in the correct depth of water (the distance between the top of their soil ball to the surface of the water) plant them in individual containers. Then, either place them on shelves within the pond or on bricks or upturned clay pots. Ensure that this added base will not tear lining materials. Never plant water-plants directly into soil in the pond's base, and do not put several different plants in the same container.

In dry, hot summers, ensure that the water level in the pond does not fall and expose the roots of these plants.

ACORUS CALAMUS *'Variegatus' (Variegated Sweet Flag) develops erect, green leaves with cream stripes along their edges. It grows 60–90cm/2–3ft high and forms a dominant feature. Bruised leaves have a tangerine-like bouquet. Plant in 7.5–15cm/3–5in of water. Divide congested plants in spring or summer.*

ALISMA PLANTAGO-AQUATICA *(Water Plantain/Mad Dog Weed) has flower heads that often reach 75cm/2½ft high but has leaves only 15–30cm/6–12in tall. The small, pink flowers produce seeds that create self-sown plants. Plant in 5–15cm/2–6in of water. Divide clumps in late summer after the flowers fade.*

BUTOMUS UMBELLATUS *(Flowering Rush/Water Gladiolus) is a superb plant, with inverted, umbrella-like heads of rose-pink flowers on stems up to 1.2m/4ft high during mid-summer. The leaves are rush-like. Plant in 7.5–13cm/3–5in of water. To prevent plants becoming congested, divide in spring every two or three years.*

CALLA PALUSTRIS *(Bog Arum/ Water Dragon) develops dark green, heart-shaped leaves and white flowers like Arum Lilies during summer. However, they are smaller than those of the Arum Lily. Plant it in 5–10cm/ 2–4in of water. Lift and divide congested plants in spring.*

CALTHA PALUSTRIS *'Plena' (Double Marsh Marigold/ Double Kingcup) has round to heart-shaped, green leaves and large, double, yellow, buttercup-like flowers during late spring and early summer. It grows in water up to 5cm/ 2in deep. Divide clumps in late summer.*

CAREX STRICTA *'Bowles Golden' grows about 45cm/ 1½ft high and develops narrow, golden leaves. Plant it in water up to 5cm/ 2in deep, and divide congested clumps in spring.* Carex riparia *'Variegata' has leaves variegated green and white, and is 45–60cm/ 1½–2ft high.*

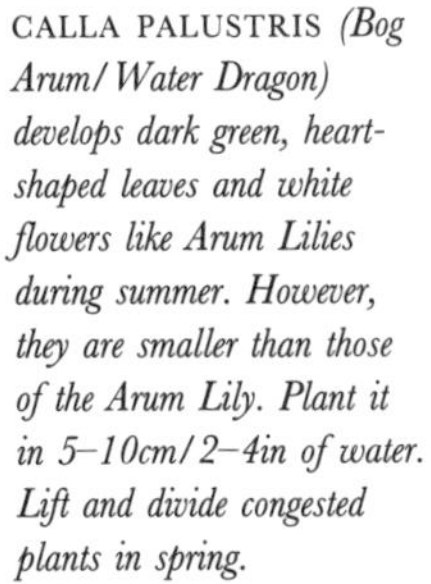

GLYCERIA SPECTABILIS *'Variegatus' (Manna Grass/ Variegated Water Grass) is vigorous, up to 90cm/ 3ft high and with green leaves variegated white and yellow. Plant it in water up to 15cm/ 6in deep. Its invasive roots will penetrate plastic sheeting. Divide congested plants in spring every two or three years.*

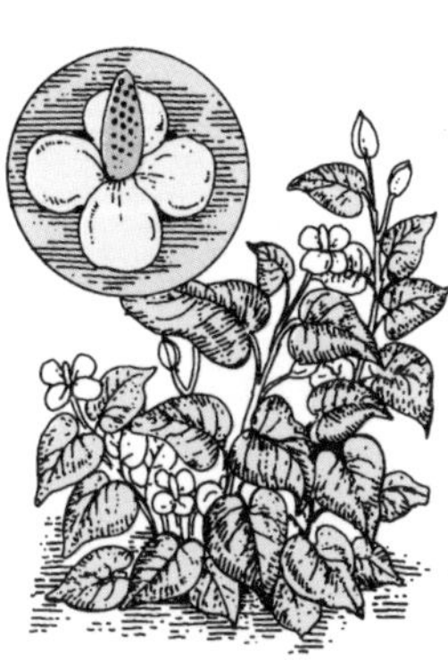

HOUTTUYNIA CORDATA *'Plena' smothers the ground in bluish-green, heart-shaped leaves. Double white flowers appear in early summer. There is also a variegated form, with reddish-green leaves splashed cream and yellow. Plant in water 5–10cm/ 2–4in deep and lift and divide congested clumps in spring.*

HYPERICUM ELODES *(Marsh Hypericum/ Marsh St. John's Wort) grows 23–30cm/ 9–12in high, spreads up to 45cm/ 1½ft and has rounded, woolly leaves. During mid to late summer it develops yellow, bowl-shaped flowers. Plant it in water up to 5cm/ 2in deep, and divide congested clumps in spring.*

MARGINAL PLANTS

Iris – Scirpus

IRIS LAEVIGATA *(Japanese Water Iris) grows 45–60cm/ 1½–2ft high and displays royal blue flowers during early summer. There are several superb varieties. As well,* I. laevigata *'Variegata' has cream and green leaves. Plant these in water up to 15cm/ 6in deep and lift and divide congested clumps as soon as their flowers fade.*

IRIS PSEUDACORUS *(Yellow Flag Iris) grows 90cm–½ m/ 3–4ft high and in early summer develops yellow flowers. Plant them in water up to 30cm/ 12in deep. The form 'Variegata' has green leaves striped in yellow. Lift and divide congested clumps as soon as their flowers fade. Plant the variegated form in water 15cm/ 6in deep.*

MENYANTHES TRIFOLIATA *(Bog Bean/ Buck Bean/ Marsh Trefoil) grows about 25cm/ 10in high, with three-lobed, broad bean-like green leaves. White, starry flowers appear on stiff, upright stems during early summer. It can be invasive, so always plant it in a container. It is ideal for water up to 7.5cm/ 3in deep. Lift and divide in spring.*

SACRED LOTUS

This superb water plant grows wild from Southern Asia to Australia. Earlier known as Nelumbium speciosum, *later* Nelumbo speciosum *and now* Nelumbo nucifera, *it has been widely grown in the Orient for its edible rhizomes and seeds, while the flowers are sacred to Buddhists. The leaves, which rise above water and shed all the moisture, encouraged an early Hindu proverb:*
'The good and virtuous man is not enslaved by passion nor polluted by vice; for though he may be immersed in the waters of temptation, yet like a lotus leaf he will arise uninjured by them.'

OTHER MARGINAL PLANTS

• Cotula coronopifolia *(Golden Buttons/Brass Buttons) is an annual with small, button-like yellow flowers throughout summer. Plant it in water up to 10cm/4in deep.*
• Lobelia cardinalis *(Cardinal Flower) is a short-lived perennial, about 60cm/2ft high and with scarlet flowers during mid and late summer. Plant it in shallow water.*
• Mentha aquatica *(Water Mint) has a creeping nature, with small, lavender-coloured flowers. Plant it in water up to 7.5cm/3in deep.*
• Mimulus ringens *(Lavender Water Musk/Allegheny Monkey Flower) grows about 75cm/2½ft high and reveals lavender-blue flowers during mid and late summer. Plant it in water 5–10cm/2–4in deep. There are other related water-loving species.*
• Myosotis scorpioides *(Water Forget-me-not and also known as* M. palustris*) grows 23cm/9in high and develops pale blue, yellow-eyed flowers from late spring to mid-summer. It grows in water up to 7.5cm/3in deep.*

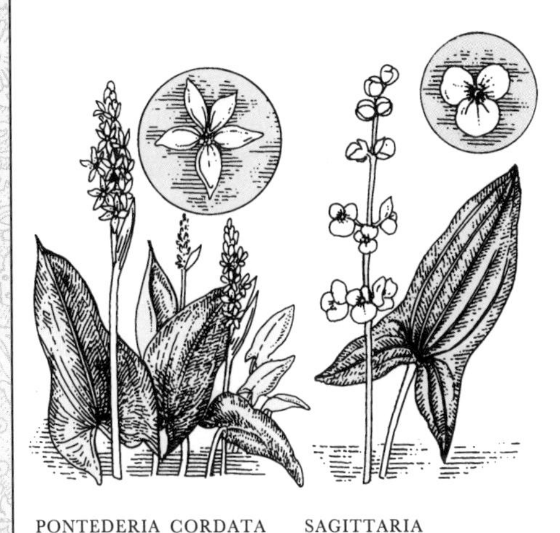

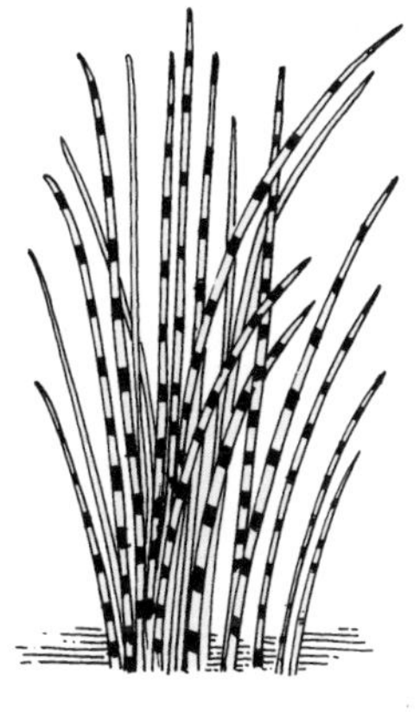

PONTEDERIA CORDATA *(Pickerel Weed) has heart-shaped, glossy green leaves up to about 60cm/2ft high and produces spikes of purple-blue flowers on upright stems during mid and late summer. Each flower has a yellow eye. This marginal plant can be planted in water up to 23cm/9in deep. In late spring or early summer, lift and divide congested plants. It is ideal for formal as well as informal ponds, where it softens sharp edges. Its leaves will arch over edging slabs.*

SAGITTARIA SAGITTIFOLIA *(Arrowhead/Swamp Potato/Swan Potato) has light green, arrow-shaped leaves. From mid to the early part of late summer it develops white flowers, clustered on upright, 60cm/2ft-high stems. Plant it in water up to 23cm/9in deep, and in early and mid-summer lift and divide congested plants. The Japanese Arrowhead* (S. japonica leucopetala) *is another water-loving and less invasive species.*

SCIRPUS TABERNAEMONTANII *'Zebrinus' (Zebra Rush/Porcupine Quill Rush) grows up to 90cm/3ft high and forms quill-like stems banded in green and white. With age, they tend to become completely green. Plant it in water up to 15cm/6in deep and divide congested plants in late spring or early summer, every year. If plants are left to become congested, the stems are not so attractive. Cut down old stems in autumn.*

BOG-GARDEN PLANTS

Astilbe–Lythrum

LIKE marginal plants (pages 46 to 49), bog plants can be used to create interest around a pond. However, unlike marginal plants, bog plants are planted outside the pool, in soil that is continually moist, but not waterlogged and airless.

VARIOUS NAMES

Bog plants are also known as waterside plants, moisture-loving plants and poolside plants, and need soil that is always moist, not just when it rains. Although many plants survive continually moist soil, not all are truly bog plants. Some are amenable to life in wet soil, but do equally well in damp soil in borders. Therefore, they are not true bog plants.

Boggy areas can be made by burying a flexible liner and filling it with moisture-retentive soil (see pages 24 and 25).

These plants have their roots permanently in moist soil, but their stems, leaves and flowers are totally above the surface. There are many suitable plants, some low and ideal for fusing the pond with the edge of a bog garden. Others are tall and dominant and best positioned at the extreme edge of the area. Do not attempt to plant the area with plants that are too large, as the pond will then cease to be the dominant feature. The Prickly Rhubarb *(Gunnera manicata)*, for example, needs careful positioning if it is not to appear oppressively domi-

ASTILBE *x* ARENDSII *(Perennial Spiraea) grows 60–90cm/2–3ft high and develops lax spires of red, pink or white flowers amid deep green, fern-like leaves from early summer to late mid-summer. It thrives in moist soil and full sun or light shade. Divide large and congested plants in mid to late spring.*

EUPATORIUM CANNABINUM *(Hemp Agrimony) grows 60cm–1.2m/2–4ft high and develops slightly rounded heads of small, reddish-purple flowers from mid to late summer. The form 'Plenum', with double, purple-pink flowers, is more widely grown. Divide congested clumps in mid-spring.*

GUNNERA MANICATA *(Prickly Rhubarb/Giant Rhubarb) grows up to 3m/10ft high and is famed for its large, dark green, rhubarb-like leaves. It is only suitable for the largest garden, perhaps as a focal point towards the back of an informal pond. Remove crowns from the mother plant and replant in late spring.*

HEMEROCALLIS *(Day Lily) is superb at the side of a pond, in full sun or light shade. The garden hybrids are best, in colours including yellow, orange, red and pink from early summer to late mid-summer. Plant it in large groups. Cut stems to slightly above the ground after the flowers fade, and divide plants in autumn or spring.*

HOSTA FORTUNEI *'Albopicta' (Plantain Lily) grows about 45cm/ 1½ft high, with pale green leaves broadly variegated in light yellow. There are many other low-growing hostas with variegated leaves. Grow in full sun or light shade. Lift and divide congested plants in early spring.*

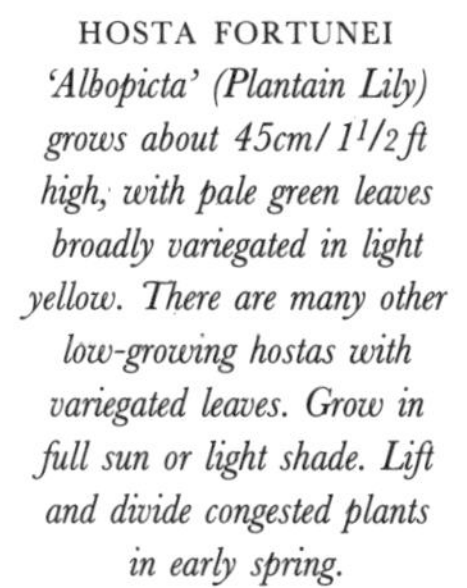

IRIS SIBIRICA *(Siberian Iris) grows 60–90cm/ 2–3ft high and develops blue flowers during early summer. There are several varieties. The mid-green, slender, sword-like leaves die down in autumn. It thrives in moist soil. Lift and divide congested clumps every four or five years in autumn, or during spring when growth has resumed.*

LYSIMACHIA CLETHROIDES *(Chinese Loosestrife/ Gooseneck Loosestrife) grows 75–90cm/ 2½–3ft high and reveals arching spires of star-shaped, small, white flowers from mid to late summer. It is ideal in moist soil and full sun or light shade. Divide large, congested clumps in autumn or early spring.*

LYSICHITON AMERICANUS *(Skunk Cabbage) develops pointed, oval, grass-green leaves up to 90cm/ 3ft long and spectacular arum-like, yellow spathes on stems 23–45cm/ 9–18in high during spring and early summer. It thrives in moist soil. Raise new plants from seeds.* L. camtschatcensis *has pure white spathes.*

LYTHRUM SALICARIA *(Purple Loosestrife/ Spiked Loosestrife) grows 1.2–1.5m/ 4–5ft high and develops long spires of small, reddish-purple flowers throughout summer. Varieties include 'The Beacon' (rose-crimson), 'Robert' (rose-red) and 'Lady Sackville' (rose-pink). Divide congested plants in autumn.*

BOG-GARDEN PLANTS

Onoclea–Trollius

MOISTURE-LOVING FERNS

Many ferns grow happily in almost barren soil on the sides of mountains, while others cling to life in natural stone walls. However, there are several that grow well in moist soils.

The Sensitive Fern (Onoclea sensibilis) *and the Royal Fern* (Osmunda regalis) *are featured below, but the Ostrich Feather Fern* (Matteuccia struthiopteris) *is also ideal for waterside planting or in bog gardens. It is also known as the Shuttlecock Fern, and this aptly describes the elegant, arching fronds. It spreads by means of underground rhizomes and after a few years forms a large, imposing clump.*

The Hart's-tongue Fern (Phyllitis scolopendrium) *can also be grown in moist soil and is especially suited to shaded positions. There are several forms, some with attractively crinkled edges to the leaves.*

Onoclea sensibilis

Osmunda regalis

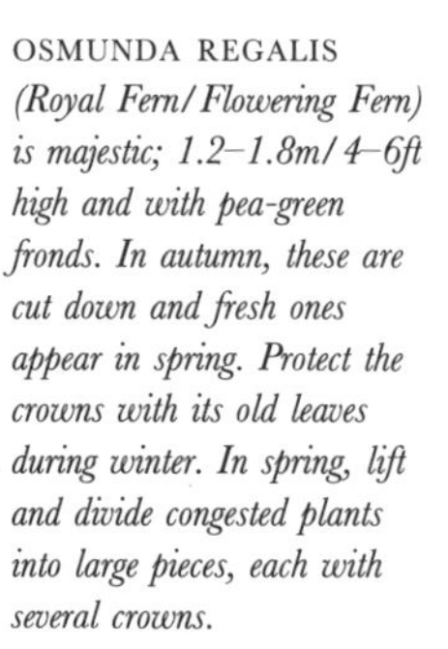

ONOCLEA SENSIBILIS *(Sensitive Fern) is a hardy, vigorously spreading fern that grows up to 60cm/2ft high and develops pale green fronds. At the first frost in autumn, the fronds turn brown, but new ones appear again in spring, first coloured pink, then green. Lift and divide large and congested plants in spring.*

OSMUNDA REGALIS *(Royal Fern/Flowering Fern) is majestic; 1.2–1.8m/4–6ft high and with pea-green fronds. In autumn, these are cut down and fresh ones appear in spring. Protect the crowns with its old leaves during winter. In spring, lift and divide congested plants into large pieces, each with several crowns.*

PELTIPHYLLUM PELTATUM *(Umbrella Plant) grows 90cm–1.2m/3–4ft high; in spring it develops tall, upright stems that bear umbrella-like heads of pink flowers. Then, the very dominant, large lobed leaves appear, initially green but slowly changing to bronze in autumn. Divide the roots in spring.*

FURTHER BOG-GARDEN PLANTS

- Aruncus dïoicus *(Goat's Beard), also known as* A. sylvestris, *grows about 1.5m/5ft high and develops lax plumes of creamy white flowers in early and into mid-summer.*
- Filipendula ulmaria *(Meadowsweet/Queen of the Meadow) grows about 75cm/2½ft high, with fragrant, creamy white flowers from early to the latter part of mid-summer. However, it is best grown in the form 'Aurea'; it is smaller, with golden green leaves.*
- Schizostylis coccinea *(Crimson Flag) grows about 75cm/2½ft high and develops star-shaped, bright scarlet flowers during late summer and autumn – and often later. The variety 'Major' has deep red flowers and those of 'Mrs Hegarty' are clear pink.*
- Zantedeschia aethiopica *(Aurum Lily/Trumpet Lily/Calla Lily) is well known for its large, white, somewhat trumpet-like flowers. It is best seen in bold clumps.*

POLYGONUM BISTORTA *(Knotweed/Snakeweed/Bistort) is invasive and grows up to 90cm/3ft high. The form 'Superbum' has long spires of pink flowers, mainly in early summer although a further flush sometimes appears later. Divide congested plants in autumn or spring.*

PRIMULA DENTICULATA *(Drumstick Primula) grows 30cm/12in high; although a perennial, it is usually grown as an annual or biennial and creates deep purple flowers in spring and early summer. It looks good when planted in large drifts. There is also a white-flowered form. Raise fresh plants from seeds.*

RODGERSIA PINNATA *is a hardy herbaceous perennial, 90cm–1.2m/3–4ft high with large, deep green, sometimes bronzed, leaves throughout summer. Choose a place in full sun and slightly sheltered from wind. And ensure that the soil does not become dry during summer. Divide congested plants in spring.*

TROLLIUS *x* HYBRIDUS *(Globe Flower), an herbaceous plant, grows 45–60cm/1½–2ft high and during early summer reveals masses of large, buttercup-like flowers amid deeply divided mid-green leaves. Varieties include flowers in pale yellow, fiery-orange and yellow. Divide plants in autumn or spring.*

REPAIRS TO PONDS

LEAKING ponds are a disaster for fish and plants, as well as creating an unplanned boggy area around the pool. Throughout summer there is usually a normal water loss from ponds: when waterfall pumps start up they cause a sudden but temporary drop in the water's surface. Also, marginal plants and waterlilies transpire masses of water on a hot day, in addition to normal evaporation from the pond's surface.

If the loss of water occurs suddenly through a crack or tear, ensure the fish and other pond life are put in large containers, and the plants in buckets of water. If the damage to the pond is serious, consider putting a liner inside it. This is often the surest way to stop water leaking, although repairs are possible (see below).

REPAIRS TO RIGID SHELLS

Good quality materials are much less likely to become damaged than low-cost economy ones.

- *Vacuum-formed polythene types are especially likely to be stressed and twisted. Also, ultra-violet rays degrade thin materials.*
- *Shells are stressed if not positioned on a firm bed of soft sand. Ensure the edges are well supported with compacted soil or soft sand.*
- *Dropping a shell causes cracks. Repair kits are readily available.*
- *Do not allow children to walk inside the shells when they are empty, and before being formed into a pond, as they will crack.*

CRACKS IN CONCRETE: *These occur through old age, subsidence, inquisitive tree roots, frost or incorrect construction. If the surface is being worn away, rub with a wire brush to remove loose material, then use several coats of a sealant. For cracks, use a chisel to ensure the gap is wider an inch or so below the water surface than level with it. This ensures that waterproof mastic remains in the crack. For large and extensive cracks, it is necessary to use mortar.*

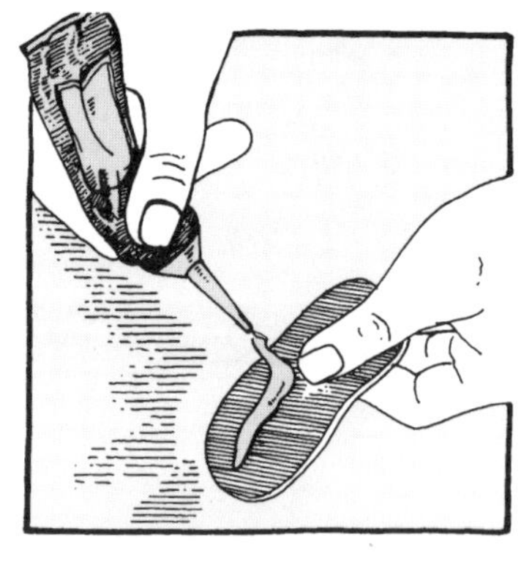

TEARS IN FLEXIBLE LINERS: *Inferior quality liners have a limited life-span: some last only a few years, whereas others have a life of twenty or more seasons. Damage occurs because the liner was laid on stony ground, through roots of invasive trees or marginal plants, or by standing in the pool while wearing spiky shoes. Proprietary repair kits are available – follow the instructions carefully, and do not economize on the size of the patch.*

PROBLEMS WITH WATER

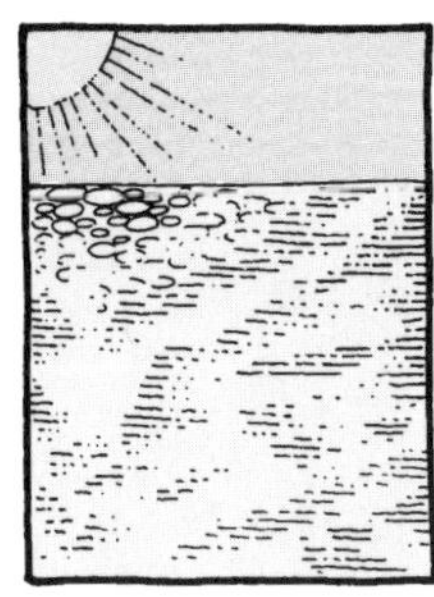

GREEN *water is usually caused by excessive amounts of algae; fish love this soup, but it is unsightly. Excessive sunlight, insufficient surface space and a lack of submerged plants create the problem. Chemical controls are possible.*

THE *water's surface sometimes becomes polluted. Place a sheet of newspaper on it, then lift off. Pollution from dead leaves, drowned and decayed animals and garden chemicals means that draining, cleaning and refilling are essential.*

DIRTY *water is often caused by scavenging fish stirring up sediment or causing compost to seep from containers. Therefore, replant the containers, using hessian liners. Put gravel on top. Remove existing sediment with a pool vacuum.*

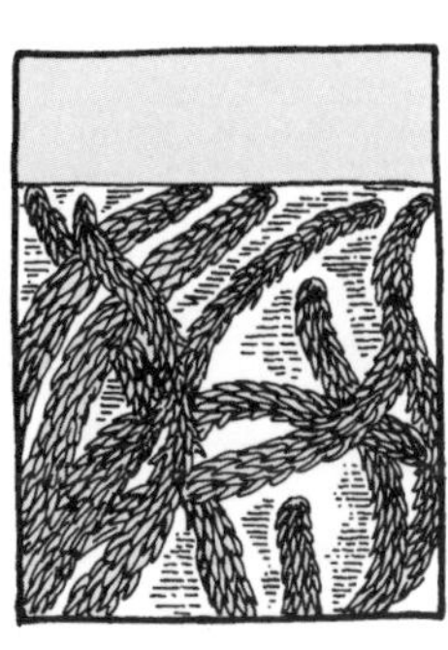

SOME *water weeds grow rapidly during hot summers and cause congestion, impeding fish and suffocating other plants. Whatever the type, cut them back carefully and slowly so that fish and their fry have time to escape the area. The weeds can be placed on a compost heap. Some marginal plants are also invasive, and need to be thinned out during summer. But take care not to spoil their shape.*

EXCESSIVELY *acid or alkaline pond water occasionally affects fish. Use a proprietary water pH tester: a pH value between 6.5 (slightly acid) and 8.5 (slightly alkaline) is all right. If the water is too acid add limestone; if too chalky paint alkaline pool surfaces with a sealant. Alternatively, use pH buffering agents to correct the balance.*

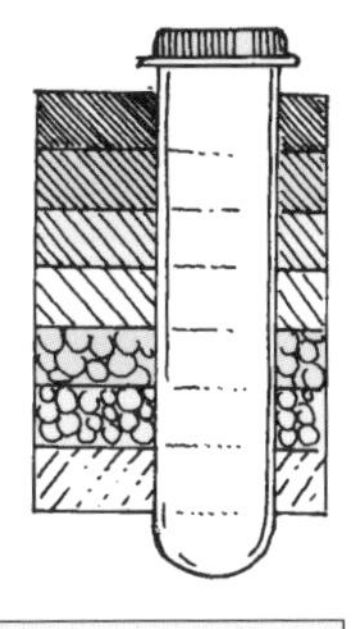

CLEANING OUT PONDS

To clean a pond, choose a warm day in early summer. First clear a path to the pool, if necessary removing a few marginal plants. Place them in buckets or fabricated ponds in a shaded position. Also, remove fish and other pond life to shaded holding pools. Pump out the water (see page 36), continually checking for fish. Scrub the pond's sides with clean water, remove the water and refill with tap water (see page 37). Replace plants and slowly reintroduce the fish (see page 37).

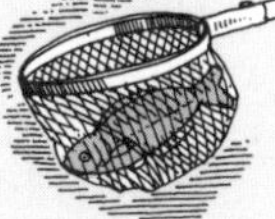

PROBLEMS WITH FISH

AFTER a glimpse of the problems with fish, illustrated here, it may be thought that fish are continually ill. In reality, most fish are never infected or attacked. Most problems are a result of inadequate feeding, dirty conditions and damage through bad handling. Buying fish from a reputable stockist prevents the introduction of many diseases, and they should be introduced carefully into a pond so that they are not unduly stressed (see pages 32 and 33). Ensure that chemical sprays do not contaminate the pond, and that seeds from poisonous plants can not fall in. Additionally, in winter, do not knock the ice: install a water heater so that part of the surface remains free of ice (see pages 36 and 37).

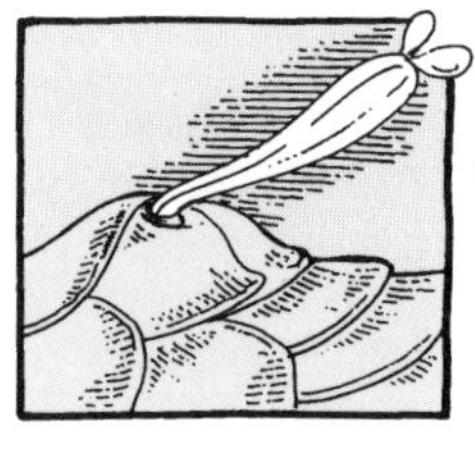

ANCHOR WORMS *cause distress to fish, creating small lumps into which the head of the 12mm/1/2in-long worm is anchored. The worm's body hangs loose, causing the fish to swim in circles. Proprietary treatments are widely available.*

BIRDS *are a constant pest, especially if gulls or kingfishers are present. Herons even wade into ponds to take fish. Either construct a wire-netting cover or form a 45cm/1 1/2ft-high wire fence to prevent them walking directly into the pond.*

DROPSY *is not common, but very serious. Scales project from a bloated body. Also, the fish's eyes protrude. The kindest action is to kill the fish humanely. Remove it and hold it firmly, then sharply knock the back of its head several times.*

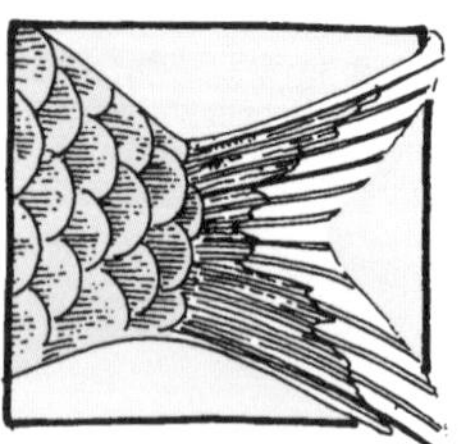

FIN ROT *is a bacterial disease that causes tissue between the two outer parts of the tail to decay. If neglected, fish die. Fish with long tails are likely to become infected. There are proprietary cures – but treat the fish early to prevent extensive damage.*

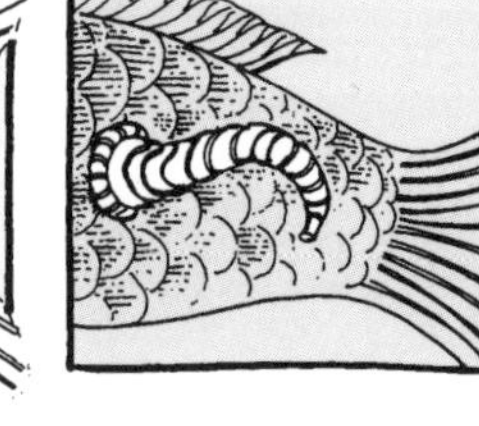

FISH LEECH *are about 2.5cm/1in long and attach themselves to the sides of fish, causing an injury that encourages the presence of diseases. Proprietary cures are available. Fortunately, it is not a common problem in garden ponds.*

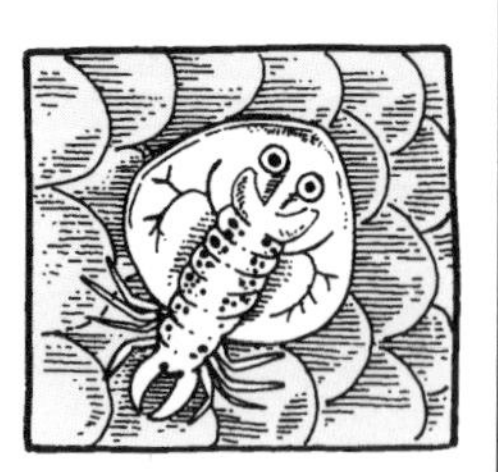

FISH LOUSE, *about 6mm/1/4in wide, has a sucker-like disc to attach itself to fish. The first sign is when a fish swims in circles and rubs itself against the pond. Initially, dab with paraffin to remove, then use an antiseptic.*

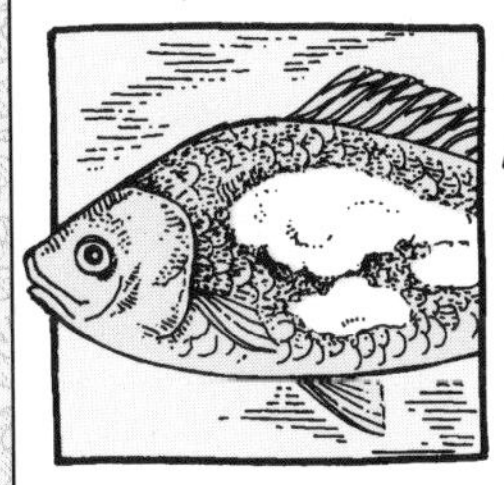

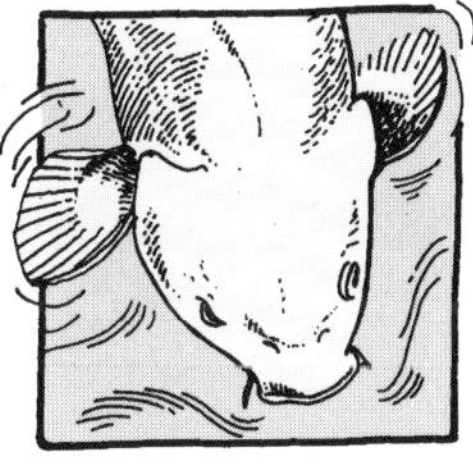

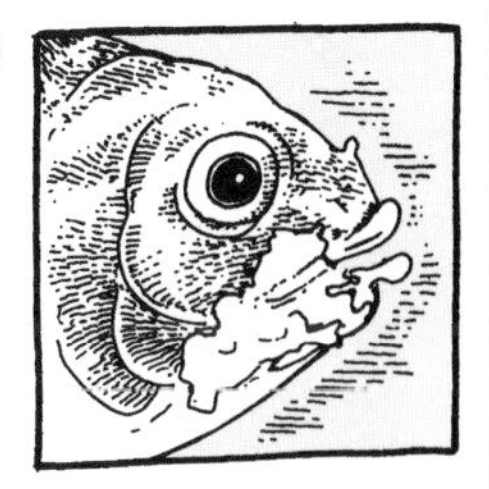

FUNGUS *is a common problem; a growth looking like cotton wool develops on fish. Weak, underfed or stressed fish are especially likely to be infected. Avoid rapid temperature changes and immediately use a proprietary fungicide.*

GILL FLUKE *is a microscopic worm that becomes fixed to gills and is first noticed when a fish moves violently and constantly. Proprietary remedies are available and it is invariably necessary to treat all of the fish in the same pond.*

MOUTH FUNGUS *is seen as white growths on the jaws. It soon spreads and causes decay. Use a proprietary treatment immediately the fungus is seen. Pollution of the water contributes to the problem, and undernourished fish are vulnerable.*

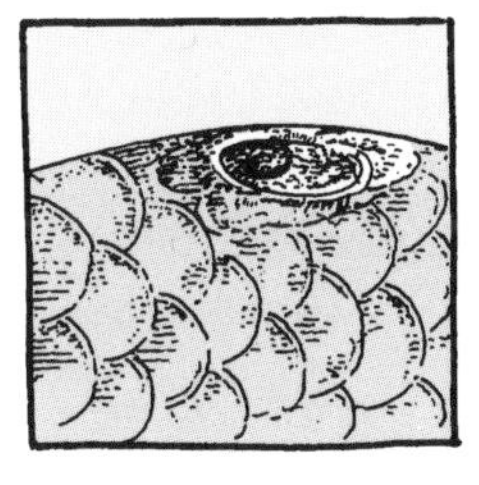

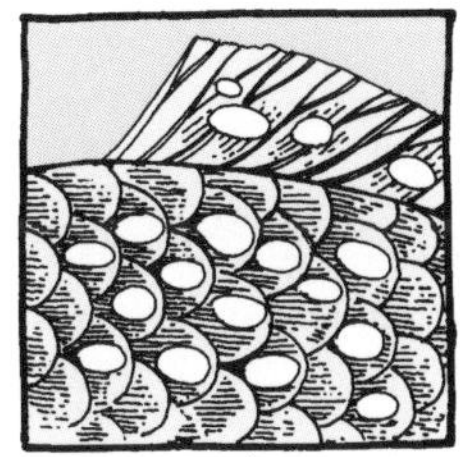

PROTOZOAN SKIN PARASITE *is microscopic. It is only discernable when fish rub themselves against the sides and have a slimey, bluish-grey covering on their bodies. Proprietary treatments are available, but seriously infected fish should be humanely killed.*

ULCERS *are caused by bacteria which gain entry through damaged scales, spreading and causing fish distress. Humanely kill badly infected fish, but if noticed early, treat the others with a proprietary remedy. For large fish, consult a veterinary surgeon as soon as possible.*

WHITE SPOT *is first noticed when fish zig-zag around the pond in a frenzy. They become peppered with white spots. It is essential to use a proprietary cure immediately. But it is kinder to humanely kill badly infected fish than to allow them to suffer.*

WATER PESTS

Many agile insects are predators in ponds, eating eggs and fry and even wrestling with small fish. There is little that can be done to control them, especially in large and wildlife ponds. However, if these desperadoes of the pond world are seen, remove them. The water boatman, great diving beetle and dragonfly larvae are the main problems, but so is the water scorpion, which sucks dry the body of its victim.

Water Boatman

Dragonfly Larva

Great Diving Beetle

PROBLEMS WITH PLANTS

PREVENTING and controlling pests and diseases of water plants by the use of chemicals is not possible, as fish and other pond life would soon be killed. Therefore, vigilance is necessary to remove affected leaves and to use clean water to spray them, forcing insects into the water where fish can eat them. When washing leaves, do not use strong jets that would damaged unaffected leaves. And take care not to spray flowers. Alternatively, weight down leaves so that insects and their larvae are washed off.

WATERLILY LEAVES

Various insects and diseases (in addition to Waterlily Crown Rot on the opposite page) decimate waterlily leaves. Here are a few of them:

BROWN CHINA MARK MOTHS *lay eggs on leaves of waterlilies and other aquatic plants. They hatch and the cream-coloured larvae, about 2.5cm/1in long, chew oval to round areas out of leaves to make protective cases for themselves. Remove eaten leaves, together with the leaf cases. If left, they soon start to rot.*

LEAF-MINING MIDGES *eat narrow, serpentine lines all over lily pads, eventually skeletonizing them. Pull up and destroy leaves seriously affected. The fish in well-stocked ponds will soon eat the larvae.*

WATERLILY APHIDS *quickly spoil the appearance of waterlily pads. They suck sap from stems and leaves that are above the surface, as well as from flowers. Additionally, the aphids excrete honeydew on which a black mould grows. As soon as aphids are seen, wash them off plants by using a spray of clean water. The fish will then eat these common pests.*

WATERLILY BEETLES *are important pests: both the adult beetle and larvae chew holes in pads. Spraying with chemicals is not advizable; submerging them for a day washes them off. Also, spraying with clean water helps in their eradication. In autumn, cut down plants to reduce the beetle's chance of surviving during winter.*

WATERLILY LEAF SPOT *is caused by a fungus that first forms red patches near to the edges of pads, which later turn black. Damp, warm weather often initiates an infection. Eventually, these areas decay and form holes. Remove leaves as soon as the problem is seen.*

BLOODWORMS, *about 2.5cm/1in long, are the larvae of midges. They create unsightly raised areas on a pond's base, especially if it is liberally coated in silt. Occasionally, they chew roots of water plants such as lilies, but are never an intolerable nuisance as the larvae are invariably eaten by fish. The midge is a non-biting type and of no danger to humans.*

CADDIS FLIES *can be a problem, but if fish are present the larvae are soon eaten. The caddis flies visit ponds during evenings, laying eggs in or near the water. They hatch and the larvae, when making homes, first use pieces of dead plants, shells and sand. Sometimes, they tear away roots, leaves and flower buds of water plants. Usually, the problem disappears if the pond is well stocked with fish, although it is possible to remove them by hand should the infestation be severe and the pond depleted of fish.*

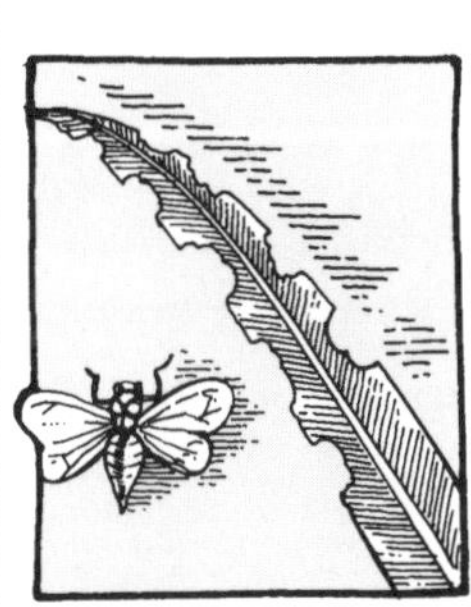

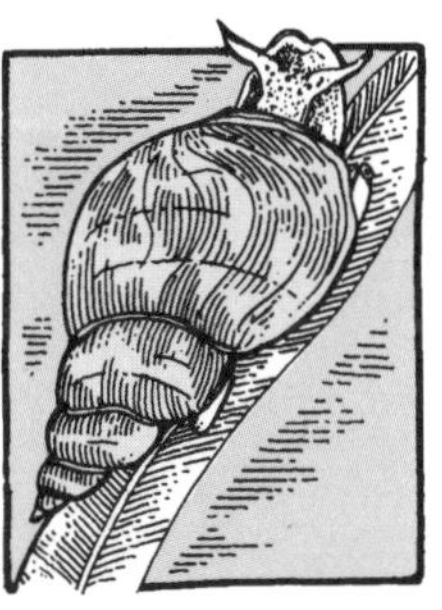

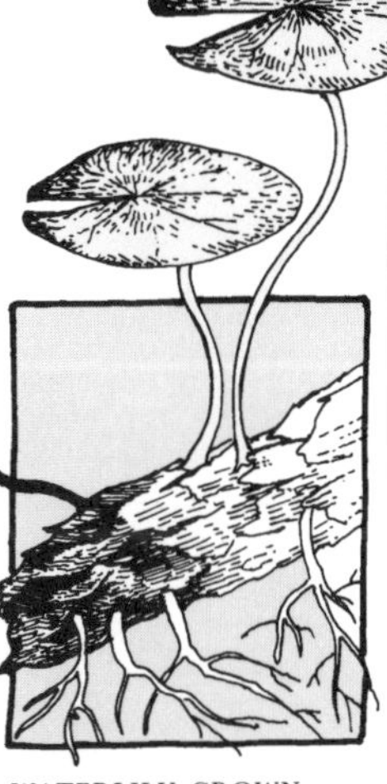

IRIS SAWFLY LARVAE, *about 18mm/3/4in long and dull, bluish-grey, chew the leaves of certain irises, especially the Japanese Water Iris* (I. laevigata) *and Yellow Flag Iris* (I. pseudacorus). *They tear leaves, leaving saw-like edges. Cut off seriously damaged leaves and pick off and destroy the grubs. Early control is essential to prevent further infestations.*

SNAILS *such as the Great Pond Snail, with a somewhat corkscrew shell, can wreak havoc with waterlilies and marginal plants. Usually their diet is dead leaves and stems, but if present in vast numbers, by necessity they resort to live plants. Pick them off, mainly from the undersides of leaves. Do not use chemical sprays or slug pellets in ponds, as they will cause pollution.*

WATERLILY CROWN ROT *is a fungus that causes crowns of waterlilies to rot, turn black and to develop appalling smells. The base decays, causing leaves to yellow and become detached from the crown. Once present, infected plants are best removed. Check all plants and watch for symptoms. Buy plants from reputable water-plant nurseries and inspect them before planting.*

GARDEN-POND CALENDAR

SPRING

In spring, water plants resume growth. Herbaceous plants in bog gardens around ponds send up shoots and flowers; the Double Marsh Marigold (*Caltha palustris* 'Plena') is usually the first one to flower, with others following in early summer.

Fish will also become more active and as this happens they need to be fed. Do not use more food than they need. At this stage it is a tonic to them and helps to prevent the onset of diseases.

Dead leaves and stems should have been removed in autumn, but if this was neglected, do it now. Also, tidy up bog gardens.

If a pond has become neglected and the water assumes the nature of a primeval soup, the only solution is to remove all plants and fish and to clean it out. This is really a task for early summer, but in warm areas it is possible in late spring.

Crowded waterlilies and other aquatics can be lifted and divided now; in cold areas this is best left until early summer.

Inspect water pumps and engage a qualified electrician to check all wiring and fuses. Ensure that a Residual Current Device is included in the circuit.

IN THE GARDEN

- Construct a concrete pond in late spring, after all risk of frost has passed yet before the onset of hot weather (16–17).
- Ensure the soil in bog gardens does not become dry (24–25).
- Plant new waterlilies in late spring (30–31).
- Remove water heaters in early spring, as soon as all risk of frost has passed (36–37).

SUMMER

Basking in the beauty of a well-stocked pond is the aim of all garden-pond enthusiasts, but vigilance is essential to ensure vigorous pond weeds do not choke other plants. Also, algae attempts to dominate ponds during midsummer and this needs to be removed periodically.

During summer, be alert to birds who may take fish from ponds early in the morning, or at other times when you are not present. *All* birds are protected by the law, some more than others, and they must be deterred from stealing fish, rather than being killed or wounded.

IN THE GARDEN

- Use feeding rings to prevent food spreading over the pool's surface (36–37).
- Construct a concrete pond in early summer, before the weather becomes too warm and possibly damages freshly laid concrete (16–17).
- Ensure that the soil in bog gardens does not become dry in warm weather (24–25).
- Plant waterlilies in early summer in cold areas (30–31).
- Construct a concrete pond in late summer, when the weather is cool (16–17).
- Feed water plants in early summer, using sachets of food (36–37).
- Regularly top up ponds with water during summer (36–37).
- When filling an empty pond, place a bucket on its base and put in the hosepipe, so that water gently cascades into the pond (36–37).
- Repair tears in flexible liners as soon as they are seen (54–55).

AUTUMN

This is the time for sorting out ponds and removing dead plants. Do not leave this task until winter sets in, as few people then want to put their hands and arms into near-freezing water.

It is essential to prevent leaves falling in ponds, as well as being blown into them from lawns.

Leave hardy floating plants as they are: some later just drop to the pond's base, where they over-winter as dormant buds. In spring they rise to the surface and produce fresh shoots.

Remove non-hardy aquatic plants to the comfort of buckets of water in a cold but frost-proof greenhouse.

If nearby trees create problems with fallen leaves this autumn, consider trimming them back. Also, trees that cast too much shade must be pruned.

If a neighbour's tree is too large, with branches over your garden, ask if they can be cut back close to the tree's trunk. This is better than cutting them off at the boundary, which invariably leaves unsightly stumps.

IN THE GARDEN

- Construct a concrete pond in autumn, while the weather is not hot but before there is risk of frost.
- Ensure that the soil in bog gardens does not become dry (24–25).
- In late autumn, either take miniature water gardens in tubs or stone sinks into a greenhouse, or put straw in plastic bags around them for protection (20–21).
- In autumn, fit a wire-netting screen over a pond to prevent leaves falling into the water, decaying, and contaminating the pond (36–37).

WINTER

If all dead stems, leaves and flowers were removed from the pond in autumn, there is little to do in winter. Check that all plants have been removed from small ponds in a series of waterfalls, as the water may freeze and the plants be damaged or killed.

Listen to weather forecasts and turn on water heaters. These usually float on the pond's surface, with the heating element slightly below. Therefore, they must be operating before the formation of frost. If freezing takes place while you are away from home, first de-ice an area by placing a metal kettle filled with boiling water on the ice (see page 33). Once an ice-free area is created, install the heater and leave it operating continuously until the temperature rises above freezing.

The sides of concrete ponds are especially vulnerable to the pressure created by freezing water. Floating a wide, thick plank across the widest part of a pond helps to relieve the pressure. Ensure the wood is clean and not contaminated with chemicals, such as wood preservatives.

Armchair water gardening is possible in winter, by glancing through catalogues and books and noting plants and fish to buy in spring. Also, it is a time to plan extensions to a pond, perhaps a bog garden or wildlife pond. And if a wildlife pond is possible, also consider introducing native plants to an area around the bog garden. Many of them are available in mixtures of seeds, or on their own. These are very widely detailed in specialist seed catalogues.

IN THE GARDEN

- Install water heaters in early winter, before ice first forms on the surface (36–37).

GLOSSARY OF WATER-GARDEN TERMS

AQUATIC: *A plant that lives totally or partly submerged in water.*

BOG GARDEN: *An area, natural or constructed, formed of moist soil. Many herbaceous plants and ferns enjoy these conditions.*

BOG-GARDEN PLANTS: *Also known as waterside plants, moisture-loving plants and poolside plants. They grow in soil that remains moist throughout the year.*

BUILDER'S SPIRIT-LEVEL: *A strong spirit-level, usually about 90cm/36in long.*

BULKING UP: *When soil is dug out of a hole, its volume then exceeds the hole's volume. Clay soil bulks up more than sand.*

FI HYBRID: *A plant raised from crossing two distinct and unrelated parents. Such plants have additional vigour.*

FLEXIBLE LINER: *Also known as a pool liner, it is waterproof material formed of polythene, PVC, or butyl (synthetic rubber).*

FLOATERS: *Also known as floating plants, these float freely with their leaves and stems on or just below the water's surface. As well as creating further interest in a pond, they provide useful shelter for fish.*

FLORE-PLENO: *Flowers with more than the normal number of petals. Some are semi-double, others double.*

FRIABLE: *Soil that is crumbly and easy to cultivate.*

GLASS FIBRE: *A composite material formed of glass fibres bonded with a resin. The result is a strong material not deteriorated by ultra-violet rays.*

GOLDFISH: *The most common fish in ponds and the first choice for novice water-gardening enthusiasts. Shubunkins were developed from goldfish (please see page 32). Although goldfish and shubunkins are very common, they are beautiful fish with varied markings.*

HEAD: *The distance between the surface of water in the lowest pond and that of the water in the topmost one. This influences the type of water pump needed.*

HERBACEOUS PERENNIAL: *A plant that each autumn dies down to ground level and in the following spring develops fresh shoots and new leaves, and then flowers.*

KOI: *A type of ornamental carp, sometimes 75cm/2½ft or more long. They are specialist fish and need expert treatment as well as special equipment.*

MAINS TRANSFORMER: *A piece of electrical equipment that alters voltage. For example, the mains electricity supply is usually 240/220/110 volts. If this is reduced to 12 or 24 volts, the installation or equipment becomes safer. Some water pumps and lighting equipment for outdoor use are designed to operate at these reduced voltages.*

MARGINAL PLANTS: *These live in shallow water at the edges of ponds. Some will also thrive in boggy soil around a pond.*

MINI-POND: *A miniature pond, usually in a wooden tub or stone sink, but also at ground level in a small patio.*

MINI-POOL: *A small water feature, often incorporating a figure but not including a pond. Usually, water flows over large, ornamental pebbles.*

MOISTURE-LOVING PLANTS: *Another term for bog-garden plants.*

MOULDED POND: *Another term for a rigid pond liner.*

NYMPHAEA: *The genus in which most waterlilies are classified. However, the Brandy Bottle or Yellow Waterlily, native to wide areas of the Northern Hemisphere, is called* Nuphar lutea.

OXYGENATING PLANTS: *Also known as submerged aquatics or water weeds, these usually grow completely below the water's surface. They help to oxygenate water, keeping it clean and creating shelter for young fish.*

PLANTING DEPTH: *The distance from the top of a plant's container to the water surface.*

PLANTING LINER: *Usually of coarse hessian and placed inside a plastic mesh basket to prevent soil falling out.*

PLASTIC MESH BASKET: *A square, round or kidney-shaped plastic container for planting aquatic plants such as waterlilies, through which water can pass freely. They are available in several sizes.*

POOL LINER: *Another term for a flexible pond liner.*

POOLSIDE PLANTS: *Another term for bog-garden or moisture-loving plants.*

PRE-FORMED POND: *Another term for a rigid pond liner.*

PUDDLED CLAY PONDS: *An early way to create a pond. After digging a hole, the surfaces were dusted with soot to deter worms and then coated in wet clay. For success, they had always to be kept full; if the surface level dropped, the clay dried and cracks encouraged leakages.*

PUMP: *see Water Pump.*

RAISED POND: *A pond that is wholly or partially raised above the ground.*

RCD: *Stands for Residual Current Device, an* essential *part of* all *outdoor electrical systems. Should a fault occur, the device immediately cuts off the power. They are often integral parts of fuse boxes, but can also be fitted as an additional device. They are also sometimes known as circuit-breakers.*

REINFORCED POLYTHENE: *A type of flexible liner. It is used to form ponds and is a superior material to ordinary polythene. It is also more resistant to the damaging effects of ultra-violet rays.*

RIGID LINER: *Also known as a pre-formed pond, moulded pond, pre-cast pond or just a shell. It is pre-formed and used to make a pond or a series of waterfalls. Its shape is clearly defined.*

SHELL: *Another term for a rigid liner.*

SHUBUNKIN: *A type of fish developed from goldfish. There are several forms. Some can be left outside throughout the year while others are better off indoors.*

STAMEN (STAMENS): *The male part of a flower, formed of the anthers which bear the pollen. Sometimes they are coloured and enhance a flower's beauty.*

SUBMERGED AQUATICS: *Also known as oxygenating plants.*

SUBMERSIBLE PUMP: *A type of water pump designed to operate under water. It can be used to pump water to fountains and waterfalls.*

SURFACE PUMP: *A type of water pump designed to operate in a dry chamber. It is more powerful than a submersible type and is used where a large volume of water needs to be moved or where the head is high.*

TRANSFORMER: *see Mains Transformer.*

WATER HEATER: *Used to keep an area of pond water free from ice during the winter.*

WATER PUMP: *There are two main types of pump used to move water in garden fountains, waterfalls and water spouts: submersible and surface.*

WATERSIDE PLANTS: *Another term for bog-garden or poolside plants.*

WATER WEEDS: *Also known as oxygenating plants or submerged aquatics.*

WILDLIFE POND: *An informal pond – usually positioned towards the far end of a garden – which encourages the presence of wildlife such as frogs, birds, insects and small mammals. These ponds also encourage the presence of birds.*

INDEX

Acorus calamus 46
Alisma 46
Arrowhead 49
Aruncus dioicus 53
Astilbe 50
Aurum Lily 53
Azolla caroliniana 44

Birds 35, 56
Bog Arum 47
Bog Bean 48
Bog gardens 11, 62
  making 24–5
  plants 30, 46, 50–3
Butomus umbellatus 46
Butyl sheeting 13, 19

Calendar 60–1
Calla palustris 47
Caltha palustris 47
Canadian Pondweed 42
Cardinal Flower 49
Carex stricta 21, 47
Ceratophyllum 42
Children 17, 18, 19, 20
Chinese Loosestrife 24, 51
Concrete ponds 19, 54
  constructing 16–17
Cotula coronopifola 49
Crassula 43
Curled Pondweed 43

Day Lily 51
Dragonflies 35, 57
Duckweed 44, 45

Earth removal 11
Eichhornia crassipes 44–45
Electricity 18, 28–9, 36, 62
Eleocharis 42
Elodea canadensis 42
Emptying ponds 18, 36–7, 55
Eupatorium 50

Fairy Moss 44
Fancy Carp 33
Feeding 36
Ferns 6, 24, 52
Filipendula 53
Filling ponds 21, 26, 36–7
Filters 29, 37
Fish 6
  care 18, 20, 36, 42
  problems 10, 56–7
  types 32-3, 34
Flags 46, 48, 53
Flexible liners 62
  bog gardens 24, 50
  installing 12–13
  raised ponds 19
  repairs 54
  waterfalls 27
Floating plants 21, 30, 44-5, 62
Flowering Rush 46
Fontinalis 43
Fountains
  types 9, 23, 28–9
  Victorian 7
Frog Bit 21, 44
Frogs 35
Frost 18, 20, 36

Gargoyles 23, 29
Glass-fibre shells 15, 19, 62
Globe Flower 24, 53
Glossary 62–3
Glyceria 47
Goat's Beard 53
Golden Buttons 49
Golden Orfe 34
Golden Rudd 34
Goldfish 32–3, 62
Goldfish Weed 43
Gunnera manicata 50

Hairgrass 42
Hart's-tongue Fern 52
Head 26, 62
Heaters 18, 28, 36, 63
Hemerocallis 51
Hemp Agrimony 50
Hornwort 42
Hosta fortunei 51
Hottonia 42, 43
Houttuynia cordata 47
Hydrocharis morsus-ranae 21, 44
Hypericum elodes 47

Ice 18, 20, 36
Iris 48, 51

Knotweed 53
Koi 33, 37, 62

Lagarosiphon major 43
Lavender Water Musk 49
Lemna 44, 45
Lighting 28–9
Lilies 51, 53
Liners 12–15, 19, 63
Lions 23
Lobelia cardinalis 49
Loosestrife 24, 51
Lotus 48
Lysichiton 24, 51
Lysimachia 24, 51
Lythrum salicaria 51

Manna Grass 47
Marginal plants 21, 31, 46–9, 62
Marsh Marigold 47
Marsh St John's Wort 47
Matteuccia 24, 52
Meadowsweet 53
Mentha aquatica 49
Menyanthes 48
Mimulus ringens 49
Mini-ponds 20–1, 62
Mini-pool 20, 62
Minnows 34
Moisture-loving plants 30, 46, 50–3, 62
Moulded ponds/shells 14–15, 26, 62
Myosotis 49
Myriophyllum 43

Nelumbo nucifera 48
Netting 18, 37
Newts 35
Nymphaea 62
  medium 40
  pygmy 21, 38
  small 39
  vigourous 41

Onoclea sensibilis 52
Osmunda regalis 52
Ostrich Feather Fern 24, 52
Oxygenating plants 31, 42–3, 63

Patios 8–9, 20, 22–3
Peltiphyllum 52
Perrenial Spiraea 50
Pests 25, 56–7, 58
Phyllitis 52
Pickerel Weed 49
Pistia 21, 45
Plantain Lily 51
Planting 30–1
Plastic liners 14, 63
Plastic mesh baskets 21, 30–1, 63
Polyethylene liners 12, 63
Polygonum bistortum 53
Pond Skaters 35
Ponds
  cleaning 55
  emptying 18, 36, 55
  filling 21, 26, 36–7
  looking after 36–7
  position 10
  repairs 54
  types 8, 10–11, 14
Pondweed 42, 43
Pontederia cordata 49
Potomogeton crispus 43
Pre-formed ponds 14–15, 26, 63
Prickly Rhubarb 50
Primula 24, 53
Pumps 28–9, 63
  emptying ponds 36
  waterfalls 26
Purple Loosestrife 51
PVC liners 13

Raised ponds 8, 11, 18–19, 63
Repairs 54
Residual Current Device 28, 63
Rigid liners 14–15, 19, 26, 54, 63
Rodgersia pinnata 53
Royal Fern 52
Rushes 21, 46–7, 49
Safety 18, 20, 28–9
Sagittaria 49
Schizostylis 53
Scirpus 21, 49
Sensitive Fern 52
Shubunkins 32, 63
Shuttlecock Fern 52
Siberian Iris 51
Sinks 20–1, 38
Sites
  preparation 10
  sloping 11, 19
Skunk cabbage 24, 51
Sloping sites 11, 19
Spirodela polyrhiza 44
Statues 9, 23, 29
Stratiotes aloides 45
Submerged aquatics 31, 42–3, 63
Swamp Stonecrop 43

Tench 34
Tillaea recurva 43
Toads 35
Topsoil 11, 18
Trapa nutans 45
Trees 10
Trollius 24, 53
Tubs 9, 20–1, 38

Umbrella Plant 52

Warrington cases 6
Water
  pests 57
  problems 55
Water Chestnut 45
Water features 22–3
Water Forget-me-not 49
Water gardening 6–9
Water Gladiolus 46
Water Hyacinth 44, 45
Water Lettuce 21, 45
Water Milfoil 43
Water Mint 49
Water Plantain 46
Water Soldier 45
Water Violet 42, 43
Water weeds 31, 42–3
Waterfalls 8–9, 18, 26–7
Waterlilies
  diseases 58–9
  medium 40
  planting 30–1
  pygmy 21, 38
  small 39
  vigourous 41
Waterwheels 22
Whirligig Beetles 35
Wildlife ponds 11, 24–5, 63
  visitors 9, 35
Willow Moss 43
Winter care 18, 20, 36

Zantedeschia 53
Zebra Rush 21, 49

The New
CAMBRIDGE
English Course Readers

LIFE ON EARTH

and other pieces

CHRISTINE LINDOP
DOMINIC FISHER

The other titles in this series are:
Fortune for Free and other pieces
Hidden Pictures and other pieces
Family Business and other pieces

Published by the Press Syndicate of the University of Cambridge
The Pitt Building, Trumpington Street, Cambridge CB2 1RP
40 West 20th Street, New York, NY 10011–4211, USA
10 Stamford Road, Oakleigh, Melbourne 3166, Australia

First published 1994

Printed in Great Britain
at the University Press, Cambridge

ISBN 0 521 42538 7

GO